The Audubon Society Field Guide to North American Seashore Creatures

A Chanticleer Press Edition

The Audubon Society Field Guide to North American Seashore Creatures

Norman A. Meinkoth, Professor Emeritus of Zoology, Swarthmore College.

Alfred A. Knopf, New York

This is a Borzoi Book
Published by Alfred A. Knopf, Inc.

 Published in the United States by Alfred A. Knopf, Inc., New York, and simultaneously in Canada by Random House of Canada Limited, Toronto. Distributed by Random House, Inc., New York.

Prepared and produced by
Chanticleer Press, Inc., New York.

Color reproductions by Nievergelt Repro AG, Zurich, Switzerland.
Printed and bound by Kingsport Press, Kingsport, Tennessee.
Type set in Garamond by
Dix Typesetting Co. Inc., Syracuse, New York.

First Printing

Library of Congress Catalog Number:
81-80828
ISBN: 0-394-51993-0

CONTENTS

13 Introduction

Part I Color Plates
35 How to Use This Guide
37 Key to the Color Plates
38 Thumb Tab Guide
57 *Color Plates*
Sponges and Corals 1–33
Plantlike Animals 34–90
Sea Squirts 91–111
Encrusting Animals 112–129
Flowerlike Animals 130–165
Sea Anemones 166–198
Flatworms, Nudibranchs, and Sea Hares 199–234
Wormlike Animals 235–273
Barnacles 274–288
Oysters, Mussels, and Clams 289–345
Scallops and Cockles 346–369
Chitons 370–381
Limpets and Abalones 382–393
Whelks, Conchs, and Augers 394–438
Olives and Doves 439–456
Periwinkles and other Snails 457–477
Octopods and Squids 478–486
Jellylike Animals 487–516
Urchins and Sand Dollars 517–534
Sea Stars 535–579
Lobsters and Shrimps 580–627
Crabs 628–690

Part II Text

321 Sponges
335 Cnidarians
395 Comb Jellies
399 Flatworms
405 Nemertean Worms
411 Segmented Worms
445 Aschelminthean Worms
447 Peanut Worms
451 Echiurid Worms
453 Mollusks
583 Arthropods
663 Echinoderms
707 Bryozoans
717 Entoprocts
719 Phoronids
721 Brachiopods
725 Acorn Worms
727 Chordates

Part III Appendices

749 Phylum Illustrations
761 Glossary
775 Picture Credits
779 Index

THE AUDUBON SOCIETY

The National Audubon Society is among the oldest and largest private conservation organizations in the world. With over 425,000 members and more than 450 local chapters across the country, the Society works in behalf of our natural heritage through environmental education and conservation action. It protects wildlife in more than seventy sanctuaries from coast to coast. It also operates outdoor education centers and ecology workshops and publishes the prizewinning AUDUBON magazine, AMERICAN BIRDS magazine, newsletters, films, and other educational materials. For further information regarding membership in the Society, write to the National Audubon Society, 950 Third Avenue, New York, New York 10022.

THE AUDUBON SOCIETY FIELD GUIDE SERIES

Also available in this unique all-color, all-photographic format:

Birds (two volumes: *Eastern Region* and *Western Region*)

Butterflies

Insects and Spiders

Mammals

Mushrooms

Reptiles and Amphibians

Rocks and Minerals

Seashells

Trees (two volumes: *Eastern Region* and *Western Region*)

Wildflowers (two volumes: *Eastern Region* and *Western Region*)

THE AUTHOR

Norman A. Meinkoth is Professor Emeritus of Zoology at Swarthmore College where he was Chairman of the Department of Biology and taught for more than thirty years. He has worked with marine invertebrates from Woods Hole, Massachusetts, to Bangkok, Thailand. Dr. Meinkoth is the author of numerous papers and articles on invertebrate zoology and is a regular member of the Summer Faculty at the University of New Hampshire and a member of many professional zoological societies.

ACKNOWLEDGMENTS

A book is the product of many people other than the author. My sincere thanks go to the following individuals for their support, helpful criticism, and suggestions in the production of this volume: Charles Baxter of Stanford University, and Judith E. Winston of the American Museum of Natural History, the scientific consultants, read and criticized text and reviewed photographs. Charles Baxter contributed copy for Pacific echinoderms and tunicates, as did Judith Winston for bryozoans and brachiopods. Harald Rehder, of the Smithsonian Institution, reviewed the list of mollusks to be included. Larry G. Harris of the University of New Hampshire provided information on sea anemones and nudibranchs. My wife, Marian Meinkoth, read the introductory chapter and introductions to some of the phyla for clarity. My daughter, Pantip Meinkoth, and W. Lloyd Merritts, typed manuscript. I have enjoyed the support, cooperation, and encouragement of the staff of Chanticleer Press; of Publisher Paul Steiner, Editor-in-Chief Gudrun Buettner, and Senior Editor Milton Rugoff. Special thanks go to Carole Anne Slatkin, who edited the entire text and pointed out my sins of

omission and commission; to Mary Beth Brewer, whose editorial expertise helped me to get started; to John Farrand, Jr., who served as scientific consultant; to Townsend P. Dickinson, for his yeoman efforts in locating photographs; and to Rebecca W. Atwater and Constance Mersel, for their special efforts in the production of this book. My thanks also go to Carol Nehring, who supervised the layout, to Laurie McBarnette, her assistant, and to Helga Lose, Dean Gibson, and John Holliday, who saw the book through production. Finally, I should like to acknowledge and thank the many biologists and naturalists whose books and papers provided the wealth of information from which the text of this book was largely gleaned.
Outstanding among these are the monumental volume by R. H. Morris, D. P. Abbott, and E. C. Haderlie, *Intertidal Invertebrates of California* (Stanford, 1980); R. Tucker Abbott's *American Seashells* (New York, 1974); and Gilbert L. Voss' *Seashore Life of Florida and the Caribbean* (Miami, 1976).

INTRODUCTION

The fascination of the seashore and its creatures probably antedates civilization. Who among us is not stirred by the mystery of the ocean, by a crab scuttling about, a snail or sea star creeping along, or a flowerlike sea anemone with tentacles fully extended? If prehistoric man did not feel this sense of wonder (and who is to say he did not?) he certainly appreciated the sea's largesse, as the remains of edible species dug by archaeologists from kitchen middens testify. The edge of the sea, with its tidepools, rocky shores, sandy beaches, mudflats, marshes, and reefs, is an endless world of creatures that pique our curiosity. It is to satisfy such interest that this guide book has been written. By means of striking color photographs and nontechnical descriptions, it provides an easy way for the reader to identify marine animals and learn something about them. The creatures encountered here are members of the many invertebrate phyla. While certain birds, mammals, reptiles, fishes, and insects are also seashore creatures, they are treated in other volumes of the Audubon Society Field Guide series.

What Is An Invertebrate?

The term "invertebrate" is vaguely familiar but somewhat mysterious to many people. The word actually describes what certain animals are not, that is, vertebrates—animals with a backbone or spinal column. Fishes, amphibians, reptiles, birds, and mammals, including man, are vertebrates. All other animals lack a backbone, and are, by definition, invertebrates. Invertebrates include such diverse and unrelated creatures as sponges, jellyfishes, worms, snails, clams, squids, shrimps, lobsters, crabs, sea stars, and sea urchins. Many of the less familiar kinds of invertebrates such as sponges, sea anemones or hydroids may be mistaken for plants. Closer examination, however, reveals their animal nature.

Plant or Animal? Most plants manufacture their own food from inorganic raw materials through a process called photosynthesis, that combines carbon dioxide, water, and minerals with the help of sunlight and the green pigment chlorophyll. In contrast, animals get energy by metabolizing food obtained from an outside source. In addition, the cells of most plants are surrounded by a tough substance called cellulose; animal cells are not thus enclosed. So while a sponge, hydroid, coral or bryozoan may be as immobile as a seaweed, it still qualifies as an animal on the basis of these and other characteristics such as the nature of reproductive organs, the pattern of development, and the presence of larval stages.

Classifying Invertebrates: The Animal Kingdom is divided into major groups called phyla. Each phylum is divided into classes, each class into orders, each order into families, each family into genera (singular: genus), and each genus into species (singular: species). All members of a phylum share fundamental characteristics such as general body plan, segmentation, and symmetry. All members of a class within that phylum share somewhat more specific characteristics. In the phylum Echinodermata, for example, the Class Stelleroidea includes star-shaped echinoderms, the class Echinoidea includes those with globe-shaped or disk-shaped bodies and moveable spines, and the class Holothuroidea includes long-bodied, spineless echinoderms with a ring of tentacles around the mouth. Orders within a class are separated on the basis of still other similarities and differences, and so on, until we distinguish the species of a genus on the basis of minute features such as the structure of mouth-parts, shell shape, number of bristles, and length of antennae.

A species is a population of organisms capable of interbreeding. While hybridization between species does occur, the offspring are usually sterile or have a lower reproductive potential, and hence are at a disadvantage over the course of generations. Species sometimes show slight but measurable differences between populations geographically separated from each other. Such geographical variants are called subspecies.

One problem in the scientific classification of organisms is the occasional change of names resulting from the continuing study of the many thousands of creatures that are only imperfectly known. For example, early biologists sometimes assumed that a different color indicated a different species. Frequently larval stages, juveniles, or those with large differences between the sexes were given separate species names. Conversely, in some cases a species may originally have included organisms of more than one kind. Redescription and assignment of more than one name was therefore in order. This book does not attempt to judge the validity of any scientific name, but simply to use those that represent the latest authoritative designations.

Geographical Scope: All species included in this guide are commonly found along at least part of the North American Coast, some as far south as South America. Many species were excluded as being too small to be readily noticed or identified by most observers, because they occurred in water too deep for general exploration, or because they were not deemed common enough for general interest. Readers who fail to find a particular species should consider that our choice had to be made from thousands of invertebrate species in two oceans.

The Marine Invertebrate Environment: Anyone spending more than a few hours at the seashore soon becomes aware of the changing character of the sea, its tides, temperatures, currents, and waves.

Tides: Tides are of greatest importance to invertebrates, since many marine animals live near or below the low-tide line. In most areas covered by this book, tides rise and fall twice each day, or more accurately, every 24 hours and 45 minutes. The height to which they rise, the high-tide line, and the level to which they fall, the low-tide line, change each day. Tides result from the gravitational pull of the moon and the sun. Their ebb and flow are greatest at the full moon and the new moon, when the moon, earth, and sun are in line with one another. Such tides are called *spring tides.* When the moon is at the first and third quarters, the tides ebb and flow the least. These are *neap tides,* when the line between the moon and the earth lies at right angles to a line between the earth and the sun, and the moon's pull is thus partially offset. The vertical distance between high- and low-tide lines varies geographically, and is generally greater in the north. Maximal tidal range in eastern Maine is about 23′ (7 m), diminishing to about 10′ (3 m) in Massachusetts Bay. From Cape Cod, Massachusetts, to northern Florida it is less than 3′ (1 m). Much of the coast of the Gulf of Mexico has diurnal tides most of the year, with only one low and one high tide per day. Tidal range there is also less than 3′ (1 m). On the Pacific Coast there is less variation, with maximal tidal ranges in Washington of about 16′ (5 m) decreasing to about 10′ (3 m) in southern California. Since invertebrates are most in evidence at low tide, an observer should check the newspapers that serve coastal areas for the hours of the daily high and low tide. Marinas, fishing supply stores, and diving shops frequently distribute local tide tables. More complete information is given in tide tables printed annually by the National Oceanic and Atmospheric Administration (NOAA) of the U.S. Department of Commerce. Separate volumes covering the entire coastline of North and South America are available for the Atlantic and Pacific oceans.

Water Temperature: Another important aspect of the environment of marine invertebrates is water temperature. This is directly related to ocean currents. In the northern hemisphere, water in both the Atlantic and the Pacific circulates in a clockwise direction. Consequently, our entire Pacific Coast is washed by the cold waters of the California Current flowing south from the north Pacific. Monthly averages are 44°–52° F (7°–11° C) in Washington, 53°–57° F (12°–14° C) in central California, and 59°–72° F (15°–22° C) in southern California. Currents along the Atlantic Coast are somewhat more complex. Warm waters flowing westward as the North Equatorial Current circulate through the West Indies and Gulf of Mexico, and eventually flow northward

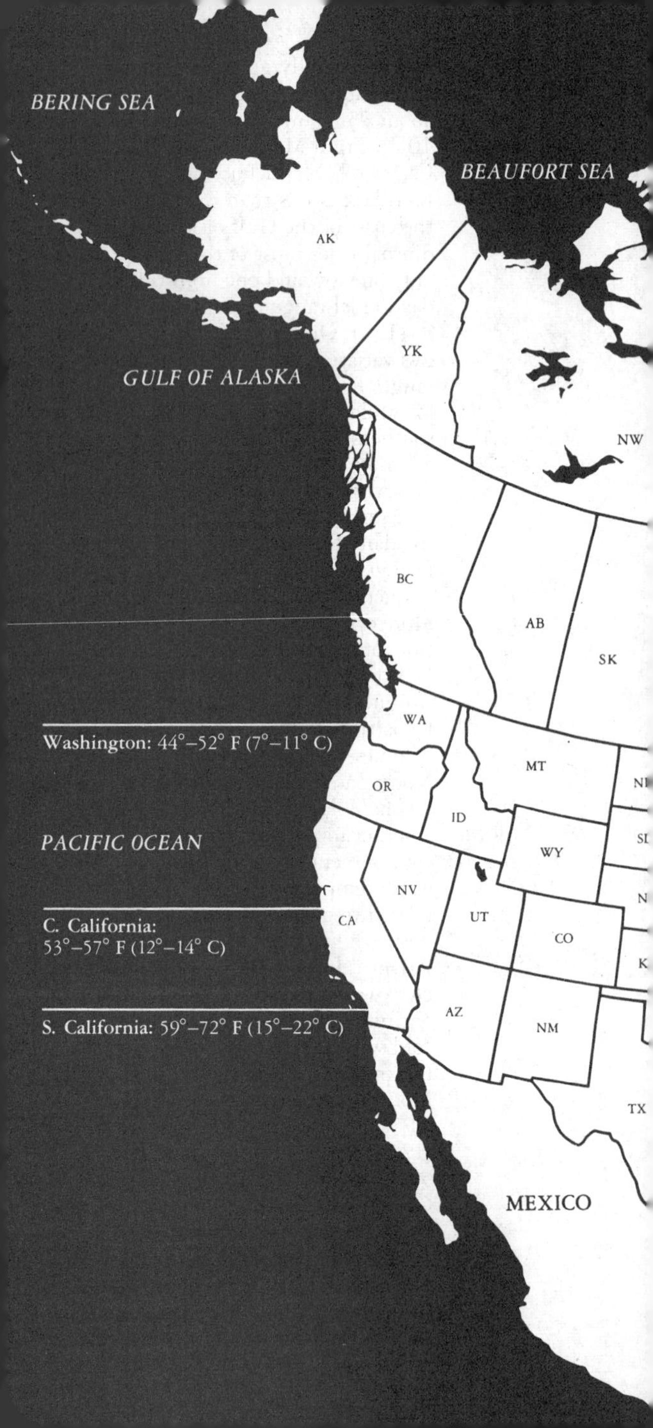
BERING SEA
BEAUFORT SEA
AK
YK
GULF OF ALASKA
NW
BC
AB
SK
WA
Washington: 44°–52° F (7°–11° C)
MT
OR
ID
PACIFIC OCEAN
WY
NV
C. California:
53°–57° F (12°–14° C)
CA
UT
CO
AZ
S. California: 59°–72° F (15°–22° C)
NM
TX
MEXICO

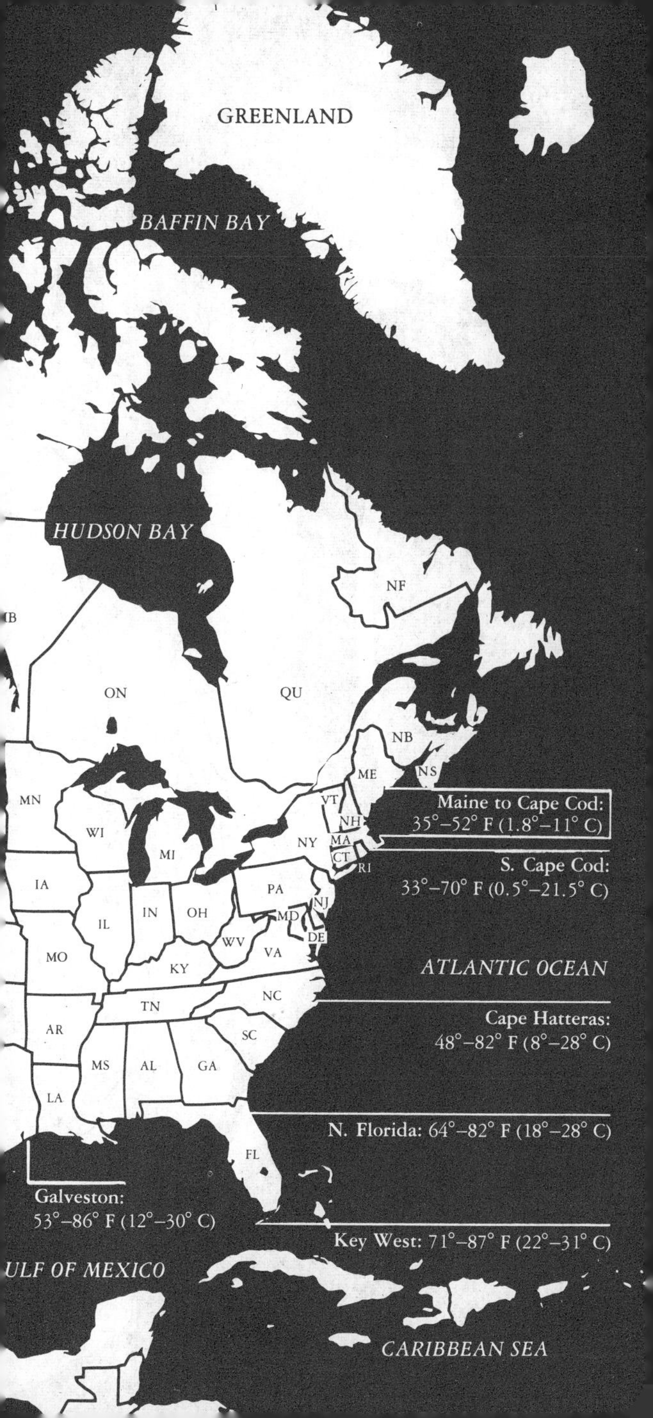

GREENLAND
BAFFIN BAY
HUDSON BAY
NF
MB
ON
QU
NB
NS
ME
VT
NH
MA
CT
RI
NY
MN
WI
MI
IA
PA
NJ
IN
OH
MD
DE
IL
WV
VA
MO
KY
NC
TN
AR
SC
MS
AL
GA
LA
FL
Maine to Cape Cod:
35°–52° F (1.8°–11° C)
S. Cape Cod:
33°–70° F (0.5°–21.5° C)
ATLANTIC OCEAN
Cape Hatteras:
48°–82° F (8°–28° C)
N. Florida: 64°–82° F (18°–28° C)
Galveston:
53°–86° F (12°–30° C)
Key West: 71°–87° F (22°–31° C)
ULF OF MEXICO
CARIBBEAN SEA

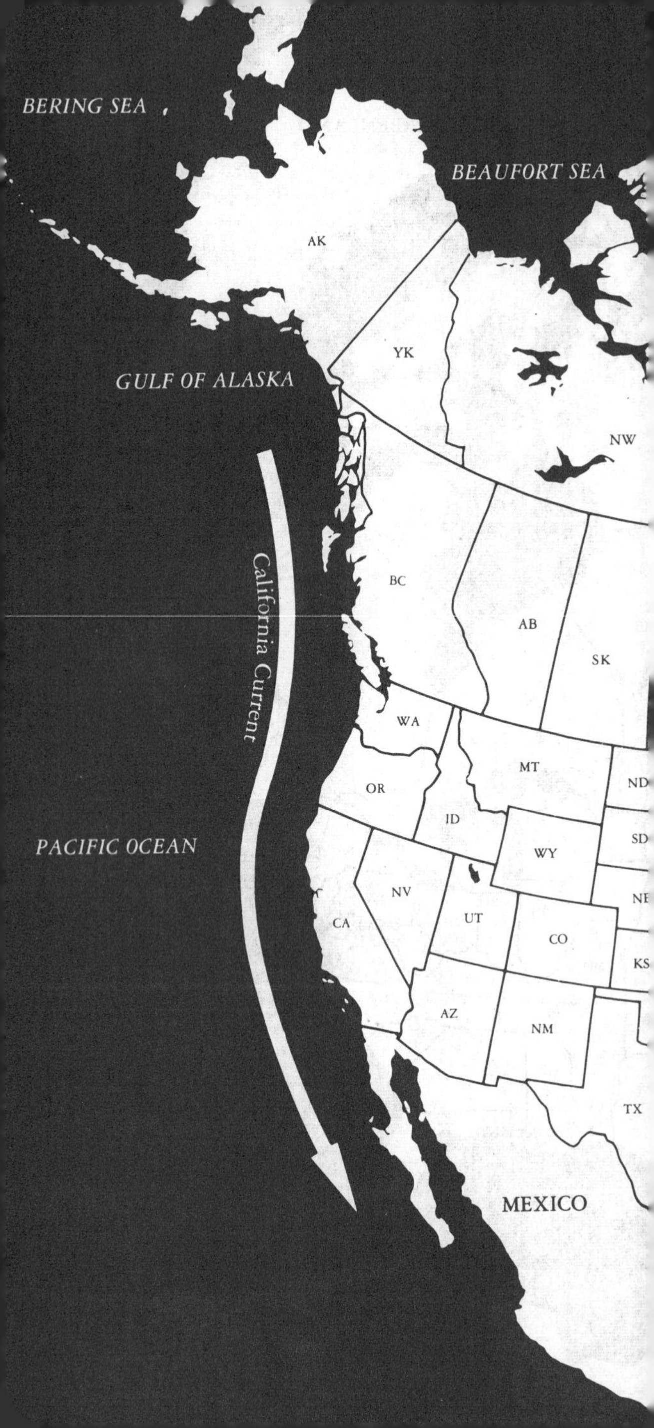
BERING SEA
BEAUFORT SEA
AK
YK
NW
GULF OF ALASKA
California Current
BC
AB
SK
WA
MT
ND
OR
ID
SD
WY
PACIFIC OCEAN
NE
NV
UT
CA
CO
KS
AZ
NM
TX
MEXICO

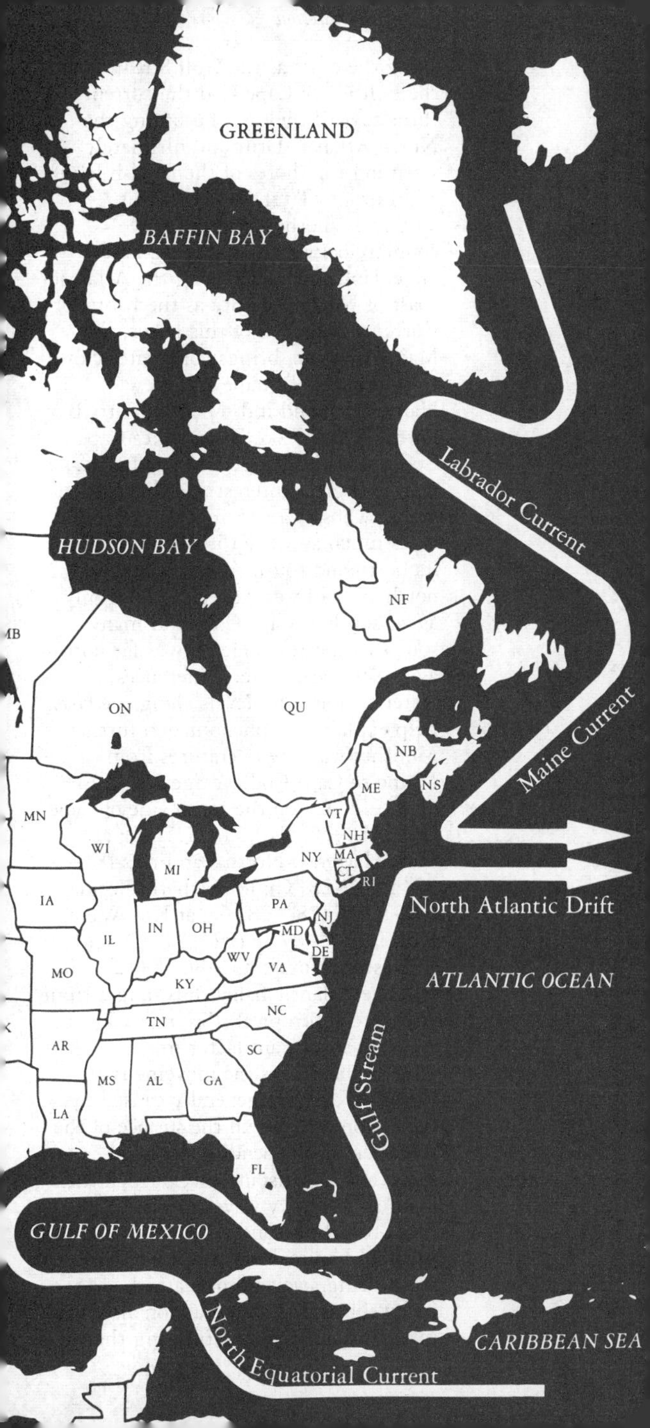
GREENLAND
BAFFIN BAY
HUDSON BAY
Labrador Current
NF
MB
ON
QU
NB
NS
Maine Current
ME
VT
NH
MN
WI
MI
NY
MA
CT
RI
North Atlantic Drift
IA
PA
NJ
IN
OH
IL
MD
DE
WV
VA
MO
KY
ATLANTIC OCEAN
NC
TN
AR
SC
Gulf Stream
MS
AL
GA
LA
FL
GULF OF MEXICO
CARIBBEAN SEA
North Equatorial Current

along the coast as the Gulf Stream. At the latitude of Cape Cod the current turns toward the east, becoming the North Atlantic Drift and ultimately warming the shores of the British Isles and Europe. To the north of this current is another, flowing counterclockwise and moving past Greenland and south along the Atlantic coast of North America as the Labrador Current. A branch of this latter, the Maine Current, brings cold water close to the shores of Maine and New Hampshire, and into Massachusetts Bay north of Cape Cod. This makes Cape Cod a so-called zoogeographic barrier, a region of great interest and diverse fauna, whose water temperatures differ by as much as 10° F (5.5° C) between its north and south shores. Many northern cold water species range only as far south as Cape Cod, and many southern species range only as far north as its southern shore. Differences in water temperature levels along the coast help explain this phenomenon further. Monthly water temperatures from Maine to Cape Cod average 35°–52° F (1.8°–11° C); on the south side of Cape Cod, 33°–70° F (0.5°–21.5° C); at Cape Hatteras, North Carolina, 48°–82° F (8°–28° C); in northern Florida 64°–82° F (18°–28° C); at Key West, Florida, 71°–87° F (22°–31° C); and at Galveston, Texas, 53°–86° F (12°–30° C). The Atlantic fauna thus ranges from Arctic to subtropical. (See maps.)

Waves: Another significant factor in the seashore habitat is the movement of waves. Waves are generally caused by the action of wind on the surface of the sea, their size depending on wind speed and the distance over which the wind acts. Heavy waves grind rocks to pebbles and sand, reducing plant and animal material to detritus, and tearing away animals attached to rocky substrates. Wave action also mixes air and water, saturating the water with oxygen. Many marine creatures not normally found near the low-tide line are washed ashore by waves during storms and thus become accessible to beachcombers.

Salinity: The oceans of the world are remarkably similar in chemical composition and salt concentration. Average sea water is a solution of about 3.5 percent salts—mostly sodium chloride (table salt), along with small amounts of other salts. Salinity or salt concentration is usually expressed in parts per thousand, average salinity being given as 35‰. Water in bays and estuaries has a somewhat higher or lower salinity depending upon the rate of evaporation and the influx of fresh water from land runoff or from rainfall. Water in estuaries and at the mouths of rivers shows, as one moves inland, a gradation from high salinity to brackish water and ultimately to fresh water. Brackish water is a barrier to all echinoderms and marine sponges, most cnidarians, and many other invertebrates that cannot tolerate the change in the osmotic pressure of reduced salinity. Other creatures, such as oysters and blue crabs, can tolerate water of very low salinity. Many invertebrates live in places such as estuaries where salinity increases with each rising tide of seawater and falls as the ebbing tide allows more fresh water from rivers to dilute the seawater. The distribution of many species depends upon their tolerance of such fluctuations.

Food Sources: Like all other living things, invertebrate animals are part of the intricate web of nature. They reproduce, eat, and are eaten. Their ultimate source of energy is the sun, via the photosynthesis of marine plants. Unlike the primary producers, or photosynthesizing plants on land, such as trees, grass, or field crops, those of the ocean are mostly microscopic, and consist of one cell. They are part of the

plankton, an assemblage of plants, animals and bacteria that lives suspended in the water and drifts with its movements. Populations of these unicellular plants, the most common of which are called diatoms and dinoflagellates, vary with season, temperature, and chemical and other factors, but are abundant. Since about 70 percent of the earth's surface is covered by seawater, and since these organisms occur to depths of several hundred feet, their numbers and the quantity of their photosynthetic product stagger the imagination.
The harvesters of these unicellular plants include invertebrates of nearly all kinds and sizes living on the bottom, near shore or in mid-ocean, in waters both shallow and deep. By far the most important of these live in the plankton: most are small crustaceans called copepods. These, in turn, are the principal food of herring and many other fishes, of some whales, and of various other predators.
Closer to the shore are other, larger food sources, mainly algae and some seed plants, such as eelgrass, turtle grass, and marsh grasses, on which many invertebrates browse. In addition, the sea receives a certain amount of terrestrial plant material from rivers. Much of the larger matter is reduced to detritus by the grinding action of waves on shorelines. But no matter what the source, organic matter—dead or alive, intact or disintegrated—is food for some invertebrate, which in turn becomes prey for yet another animal. In the case of invertebrates such as clams, oysters, shrimps, lobsters, and crabs, that animal is man.

How to Identify Marine Invertebrates:

Because marine invertebrates are so varied and abundant, their identification may at first seem forbiddingly difficult. The uninitiated frequently find the line drawings of most field guides confusing, since they tend to stress unfamiliar anatomical details often inaccessible to the novice observer. For that reason this book employs a new approach: taking advantage of the remarkable progress recently achieved in modern underwater camera work, it uses outstanding color photographs of seashore creatures, alive (with the exception of a few species of shells) and in their natural habitat. The pictures are arranged not by scientific groups, which usually emphasize unfamiliar structural features, but by readily distinguishable categories of body pattern and color. Close-up photographs are included to illustrate the individuals within a colony; pictures taken at a greater distance demonstrate the shape and structure of colonies as a whole. These illustrations also capture the animals' characteristic positions and emphasize features most readily visible in the field. The reader thus has immediate access to all visible elements needed to identify a species or its close relative, and is able, at the same time, to enjoy its natural beauty. At this point it should be noted that invertebrates quickly lose their color after death. Creatures stranded by high tide, washed up during storms, or placed in a bucket may not be nearly as richly hued as they were a short time before. The skeletal remains of sponges, corals, snails, bivalves, and crabs may be faded or bleached, little resembling the photograph of the living animal. The photographs in this guide illustrate similar-looking animals side by side to enable the reader to note their differences and thus quickly identify an animal in the field. Whenever possible, photographs are identified as to species, but in a few cases, animals are so similar that they confuse even the experts. In such instances, identifying the genus of the animal should satisfy most observers.

Observing Marine Invertebrates: Though often hidden from sight, seashore creatures are readily discovered by the careful beachcomber, especially at low tide or after storms when many animals are washed ashore. The search should be made by turning over small rocks (beware of sharp-edged barnacles), studying tidepools for the tentacles of tube-dwellers and burrowers, and sorting through seaweeds. A beach walker may also come upon crab burrows, holes left by clams that have withdrawn beneath the sand, or worm tubes extending above the sand's surface. A low, muddy cone surrounding a hole, or sandy or muddy castings, may indicate the presence of a nearby beach denizen. A shovel or clamfork and a coarse sieve are useful beachcombing tools (a clamming permit may be required in some areas), and the sharp-eyed wader with a dip-net or strainer can catch many species including jellyfishes, comb jellies, hermit crabs, snails, and an occasional squid. A bucket and several small plastic containers will keep valuable specimens separate. The snorkeler or scuba diver will naturally gain access to a large number of species beneath the surface of the water; a still richer fauna may be found farther offshore, but is outside the scope of this guide.

Organization of the Color Plates: We have arranged the photographs of marine invertebrates according to the features an observer sees in the field: shape and color. For example, the section called Periwinkles and Other Snails contains, in addition to members of that class, a worm whose tube resembles the shells of certain snails. Similarly, the section called Plantlike Animals compares, among others, the yellowish, highly-branched forms of both a cnidarian and a bryozoan. For species with many variations, more than one photograph has often been used. The photographs have been

juxtaposed so that similar-looking animals are close together, allowing the reader to determine at a glance the differences between them.
The color plates are arranged in the following order:

Sponges and Corals
Plantlike Animals
Sea Squirts
Encrusting Animals
Flowerlike Animals
Sea Anemones
Flatworms, Nudibranchs, and Sea Hares
Wormlike Animals
Barnacles
Oysters, Mussels, and Clams
Scallops and Cockles
Chitons
Limpets and Abalones
Whelks, Conchs, and Augers
Olives and Doves
Periwinkles and Other Snails
Octopods and Squids
Jellylike Animals
Urchins and Sand Dollars
Sea Stars
Lobsters and Shrimps
Crabs

Thumb Tab Guide: The grouping of the color plates is explained in a table preceding that section. A silhouette of a typical member of each group appears on the left. Silhouettes of marine invertebrates within that group are shown on the right. For example, the silhouette of a lobster represents the group Lobsters and Shrimps. This representative silhouette is also inset on a thumb tab at the left edge of each double page of color plates devoted to that group of invertebrates.

Captions: The caption under each photograph gives the plate number, common name, maximum recorded measurement, and page number of the text description. The measurement in inches indicates the largest dimension of the animal. The color plate number is repeated at the beginning of each text description.

Organization of the Text: This guide describes members of 18 phyla of invertebrates. Each phylum is introduced by a discussion of its distinctive features, the kinds of animals it includes, aspects of their life history, and other information. Classes of the larger phyla are treated in greater detail, while those of smaller phyla are more briefly characterized. In a few instances, where a phylum or a class is large, we may include a discussion of subgroups within the phylum. This general description is followed by a species account for each individual animal treated in the book.

All groups, from phyla through species, are presented in the commonly accepted phylogenetic order, beginning with the most primitive and ending with the most advanced. The sequence of species within each phylum depends on the nearest relationships, and in general, is that employed in such standard reference works as *Lights Manual, Intertidal Invertebrates of the Central California Coast* by R. I. Smith and J. T. Carlton (Berkeley: University of California Press, 1975), *Field Book of Seashore Life* by R. W. Miner (New York: G. P. Putnam's Sons, 1950), and *Guide to Identification of Marine and Estuarine Invertebrates, Cape Hatteras to the Bay of Fundy,* by K. L. Gosner (New York: John Wiley and Sons, Inc., 1971).

Plate Number: Each species account begins with the number of the color plate or plates. Where no outstanding photograph was available, line drawings appear in the margins of the text description.

Common Name: The common name of the species opens the text account, and is sometimes followed by other English names (designated by quotation marks) used locally. Since the common names of marine invertebrates have never been standardized, those given here reflect prevailing usage. Where there is no widely accepted common name, a new name has been introduced.

Scientific Name: The scientific names of species and genera are frequently revised by experts. In this guide, the scientific names conform with the most recently accepted terminology. Where confusion is possible, an earlier name is cited in the Comments section. The scientific name of a species consists of two words, usually derived from Latin or Greek. The first, always capitalized, is the name of the genus; the second, always in lower case (even though it may be formed from a proper name) indicates the species within the genus and is latinized if it is derived from a non-Latin source such as the name of a place or person. Thus, the scientific name of the California Mussel is *Mytilus californianus,* and Forbes' Sea Star is *Asterias forbesi.* Of the several barnacles in the genus *Balanus,* the species whose common name is the Ivory Barnacle is designated *Balanus eburneus* from the genus *Balanus* and the species *eburneus.* A related species in the same genus, the Giant Acorn Barnacle, has the same genus name, *Balanus,* but the species name *nubilis.* This system of scientific designation was devised by an 18th century Swedish naturalist, Karl von Linné, also known by his latinized name, Carolus Linnaeus. Common names vary greatly from country to country, and even within a country. In contrast, scientific names signify the same organism to scientists everywhere. When a phylum description includes a discussion of subgroups within the phylum, such as classes, the subgroup to which each species belongs is indicated in the text entry under the animal's scientific name.

Description: Each description begins with the maximum measurements recorded for the species described, the dimensions listed depending on the shape and symmetry of the animal. Many animals found in the field will be considerably smaller, but will have the same body proportions. In some groups (*e.g.* sponges, hydroid colonies, corals), there is no absolute maximum size, and specimens even larger than the sizes cited may occasionally be found. For convenience, measurements are given in fractions of inches as marked on a ruler. Approximate metric equivalents of inches are given in millimeters up to 6″, in centimeters up to 39″, and then in meters. Measurements greater than 60″ are given in feet. The account then describes the animal's general shape and color in life. (Beached animals may be considerably paler and may have lost some of their body parts, and regional variation may produce alteration in features.) The description continues with information on anatomical features visible in the field without magnification (in a few instances a hand lens may be necessary), and texture where applicable. Diagnostic or distinguishing features are italicized for quick reference. These details should be sufficient to verify an initial identification made on the basis of the color photograph.

If the sexes are different, both are described. We have avoided description of technical details of structure which cannot be readily detected in the field. Wherever possible, familiar terms are used; technical terms are defined in the glossary or illustrated with labelled drawings in the introduction. Where the exact species of an animal can be determined only by laboratory examination, the text gives the name and description of the genus in which the species is included. (For example, the Bougainvillia Hydroids are described as *Bougainvillia* spp., with "spp." the plural abbreviation for "species".)

Habitat: Many marine invertebrates frequent only certain kinds of environments or substrata. Habitats are therefore described as specifically as possible. Although all animals included here may be found above low-tide line, or in water no more than knee-deep, the greatest depth to which any animal is known to occur is given, in both feet and meters, where such information is available. While a knowledge of likely habitats should help you to find certain species, you may not always find them in a specific habitat. Abundance varies with geographical range, season, temperature, pollution, and human disturbance of the habitat.

Range: The range descriptions are given from north to south for North America and adjacent islands, and for South America as far as it has been determined, with West Coast ranges preceded by those on the East Coast. No attempt is made to cite wider ranges, although many species are much more widely distributed.

Comments: The species accounts conclude with comments on the animal's behavior, food, reproduction and life cycle, edibility, commercial value, ecological role, or conservation status. If an animal is poisonous or dangerous to human beings, that fact is indicated in bold face in the Comments section. In many cases, closely related species are mentioned and described in order to increase the scope and usefulness of the guide. While no photograph is included for such species, it may be assumed that they are very similar to the related illustrated species.

Part I
Color Plates

HOW TO USE THIS GUIDE

Example 1
A round, spiny creature on a Gulf Coast beach

Walking near the low-tide line on a Florida beach, you notice a round, reddish creature covered with longitudinally-grooved spines.

1. Turn to the Thumb Tab guide preceding the color plates, and look for the silhouette that most resembles the creature you have seen. In the group called urchins and sand dollars you find it: the silhouette for urchins, color plates 517–534.
2. Check the color plates. Two creatures have the right shape and coloring: the Atlantic Purple Sea Urchin and the Rock-boring Urchin, color plates 518 and 519. The captions indicate the sizes and the text pages 689 and 692.
3. Reading the text, you find that only the Atlantic Purple Sea Urchin has spines with grooves. The Rock-boring Urchin has ungrooved spines. Your creature is an Atlantic Purple Sea Urchin.

Example 2
A shelled, many-legged animal in a Cape Cod estuary

Paddling a canoe down a brackish estuary on Cape Cod, you observe a small animal with two dark eyes on stalks, and numerous tan legs protruding from a shell. The end of the large, right front leg is broad and rounded, with the tip forming a blunt angle on one edge.

1. In the Thumb Tab Guide you find the silhouette of an animal with many

segmented legs and discover that it is a crab.

2. Turning to the color plates you narrow your choice to 2 photographs—the Flat-clawed Hermit Crab and the Long-clawed Hermit Crab, color plates 676 and 677. The captions refer you to text pages 631 and 632.
3. Reading the text you discover that the hand of the right pincer of the Long-clawed Hermit Crab is nearly cylindrical and three times longer than it is wide—rather than broad and rounded. You have found a Flat-clawed Hermit Crab.

Example 3
An orange, oval-shaped animal on a rock near San Diego

Wading in the coast near San Diego you spot a bright orange, oval shape, covered with granular scales, attached to a rock. From its elongated neck grows a cluster of frondlike tentacles.

1. Among the silhouettes preceding the color plates you select one of an oval-shaped animal with a feathery tentacle-cluster: a sea cucumber, color plates 150–157.
2. Turning to the color plates you note two very similar sea cucumbers: the Scarlet Psolus, color plates 153 and 155, and the Slipper Sea Cucumber, color plate 154.
3. Checking the text on page 703 you eliminate the Scarlet Psolus because it occurs only on the East Coast. You have seen the Slipper Sea Cucumber.

Key to the Color Plates

The color plates on the following pages are divided into 22 groups:

Sponges and Corals
Plantlike Animals
Sea Squirts
Encrusting Animals
Flowerlike Animals
Anemones
Flatworms, Nudibranchs, and Sea Hares
Worms
Barnacles
Clams
Scallops and Cockles
Chitons
Limpets and Abalones
Whelks, Conchs, and Augers
Olives and Doves
Periwinkles and Other Snails
Octopods and Squids
Jellylike Animals
Urchins and Sand Dollars
Sea Stars
Lobsters and Shrimps
Crabs

Thumb Tab Guide: To help you find the correct group, a table of silhouettes precedes the color plates. On the left side of the table, each group is represented by a silhouette of a typical member of that group. On the right, you will find the silhouettes of animals found within that group.

The representative silhouette for each group is repeated as a thumb tab at the left edge of each double page of color plates, providing a quick and convenient index to the color section.

Thumb Tab	Group	Plate Numbers
	Sponges and Corals	1–33
	Plantlike Animals	34–90

Typical Shapes		Plate Numbers
	brain corals, Large Flower Coral	1–5
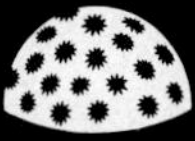	corals, sponges, Knobbed Zoanthidean, Mat Anemone, Elephant Ear Tunicate	6–16, 20, 27, 30
	sponges, Encrusted Tunicate	17–19, 21, 22, 24, 26
	sponges, corals	23, 25, 31, 32
	sponges, Lettuce Coral	28, 29, 33
	hydroids, bryozoans, corals, Green Encrusting Tunicate, Lacy Tube Worm, Clapper Hydromedusa, Thick-based Entoproct	34–38, 65–88
	stalked jellyfishes, stalked tunicates, sponges, Red Soft Coral, Green Phoronid Worm, Gurney's Sea Pen, Solitary Hydroid	39–46, 89, 90

Thumb Tab	Group	Plate Numbers
	Plantlike Animals	34–90
	Sea Squirts	91–111
	Encrusting Animals	112–129

Typical Shapes		Plate Numbers
	bryozoans	47–51
	sponges, corals	52–63
	Sea Fan	64
	tunicates	91–94, 97, 99, 105–108
	tunicates	95, 96, 98, 100–104
	Heath's Sponge, Bristly Tunicate, Cactus Tunicate	109–111
	bryozoans	112–114, 116, 117
	sponges, white crusts, Purple Stylasterine	115, 118, 123–129
	tunicates	119–122

Thumb Tab	Group	Plate Numbers
	Flowerlike Animals	130–165
	Sea Anemones	166–198
	Flatworms, Nudibranchs, and Sea Hares	199–234
	Wormlike Animals	235–273

Typical Shapes		Plate Numbers
	feather dusters, tube worms, Ragged Sea Hare	130–149
	sea cucumbers	150–157
	terebellid worms, fringed worms, Johnston's Ornate Worm	158–165
	anemones, corals	166–198
	nudibranchs, sea slugs	199–204, 208, 212
	nudibranchs, Sea Lemon, Navanax	205, 206, 220–231, 233, 234
	sea hares, California Stichopus, Warty Sea Cat	207, 209–211
	flatworms, Oyster Leech	213–219, 232
	sea cucumbers	235–238

Thumb Tab	Group	Plate Numbers
	Wormlike Animals	235–273
	Barnacles	274–288
	Oysters, Mussels, and Clams	289–345

Typical Shapes		Plate Numbers
	scale worms	239, 240, 245
	fire worms, paddle worms, clam worms, and other worms, Sticky-skin Sea Cucumber	241–244, 246–250, 252
	nemerteans and other worms, Common White Synapta, Limulus Leech, Silky Sea Cucumber	251, 253–263, 265–267
	worms	264, 268–273
	barnacles	274–286
	Leaf Barnacle, Common Goose Barnacle	287, 288
	oysters	289–291
	mussels, pen shells	292–295, 298, 299

Thumb Tab	Group	Plate Numbers
	Oysters, Mussels, and Clams	289–345
	Scallops and Cockles	346–369

Typical Shapes		Plate Numbers
	razor clams and other bivalves, Tongue-shell Brachiopod	296, 297, 300–309
	piddocks	310, 313, 314
	shipworms	311, 312
	clams, macomas, tellins, lucines, cockles, astartes, quahogs, and other bivalves	315–343
	jingle shells	344, 345
	oysters, jewel boxes	346–349
	scallops, oysters, and other bivalves, brachiopods	350–360
	cockles, arks, and other bivalves	361–369

Thumb Tab	Group	Plate Numbers
	Chitons	370–381
	Limpets and Abalones	382–393
	Whelks, Conchs, and Augers	394–438

Typical Shapes		Plate Numbers
	chitons	370–381
	limpets	382–391
	Red Abalone	392, 393
	wentletraps, horn snails, augers, and other snails	394–405
	whelks, rock snails, tritons, and other snails	406–420
	tulip snails, whelks, Junonia, Florida Horse Conch	421–426
	whelks, conchs, murexes, and other snails	427–429, 433–438
	cones	430–432

Thumb Tab	Group	Plate Numbers
	Olives and Doves	439–456
	Periwinkles and Other Snails	457–477
	Octopods and Squids	478–486
	Jellylike Animals	487–516

	olives, Salt-marsh Snail, and other snails	439–450
	snails, Unicorn, Scotch Bonnet	451–456
	top snails	457–459, 468
	snails	460–467, 469, 470
	periwinkles, Chink Snail	471–474
	worm snails, Sinistral Spiral Tube Worm	475–477
	octopods	478–483
	squids	484–486
	comb jellies, salps, Chain Siphonophore, Common Doliolid	487–493

Thumb Tab	Group	Plate Numbers
	Jellylike Animals	487–516
	Urchins and Sand Dollars	517–534
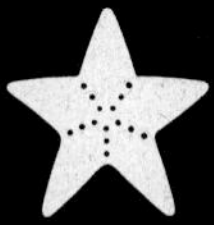	Sea Stars	535–579

Typical Shapes		Plate Numbers
	jellyfishes, hydromedusae, Sea Gooseberry, Blue Buttons, Sea Nettle, Lion's Mane	494–511, 514
	Portuguese man-of-war, By-the-wind Sailor	512, 513, 515, 516
	sea urchins, Sea Egg, Long-spined Sea Biscuit	517–528
	sand dollars, sea urchins, West Indian Sea Biscuit	529–534
	sea stars	535–541, 547–564
	sun stars, Sunflower Star	542–546
	brittle and basket stars	565–573
	sea spiders, Arrow Crab	574–579

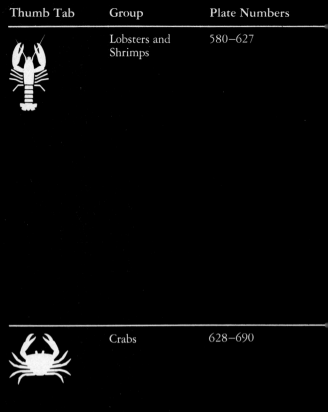

Thumb Tab	Group	Plate Numbers
	Lobsters and Shrimps	580–627
	Crabs	628–690

Typical Shapes		Plate Numbers
	isopods, sea roaches	580–585, 594
	amphipods, beach fleas, Scud, Mottled Tube-maker	586–591, 598
	shrimps, squillas, Horned Krill	592, 593, 595–597, 599–622
	lobsters	623–627
	fiddler crabs	628–630
	crabs	631–665, 667–675
	Horseshoe Crab	666
	hermit crabs, Sharp-nosed Crab, Fuzzy Crab	676–687
	mole crabs	688–690

The color plates on the following pages correspond with the numbers preceding the text descriptions. The caption under each photograph gives the plate number, common name, measurement, and page number of the text description. The measurement in inches indicates the animal's maximum dimension, which may be length (*l.*), width (*w.*), or height (*h.*). The color plate number is repeated at the beginning of each text description.

Sponges and Corals

These large and colorful animals are frequently found in tropical seas where their highly-branched, boulderlike, or vase-shaped communities vivify the underwater world. Red, purple, and blue sponges intermingle with yellow, brown and green corals and such similar marine inhabitants as the Knobbed Zoanthidean, Mat Anemone and Elephant Ear Tunicate to provide an impressive spectacle. But certain sponges and corals can sting severely, and should not be handled.

1 Large Flower Coral, *w.* 36″, *p.* 394

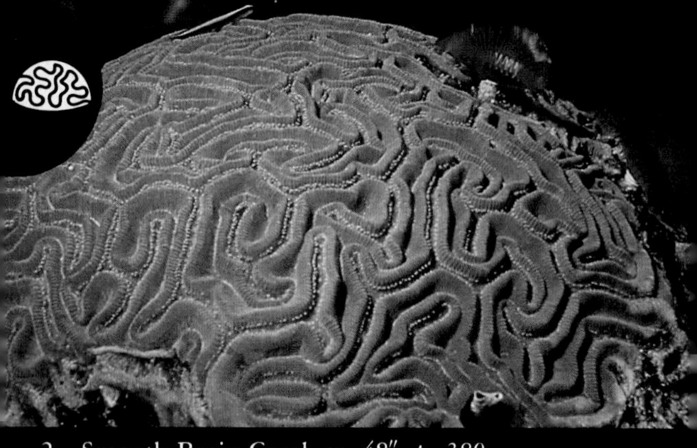

2 Smooth Brain Coral, *w.* 48″, *p.* 389

4 Knobbed Brain Coral, *w.* 48″, *p. 388*

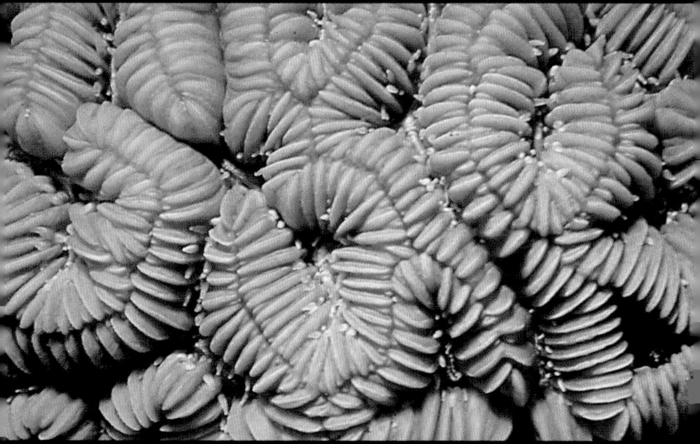

5 Meandrine Brain Coral, *h.* 12″, *p. 392*

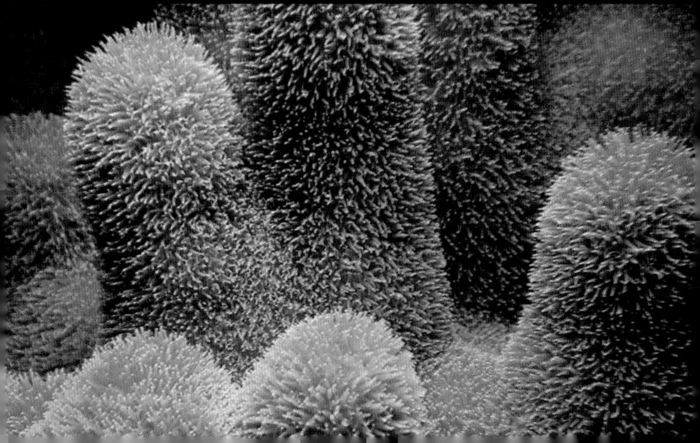

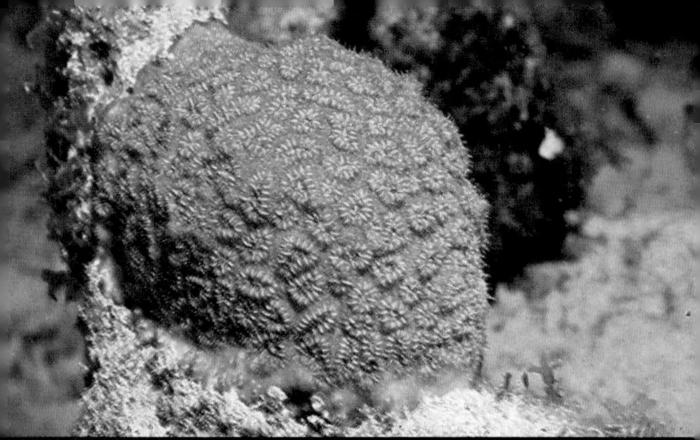

7 Stokes' Star Coral, *h.* 12", *p. 393*

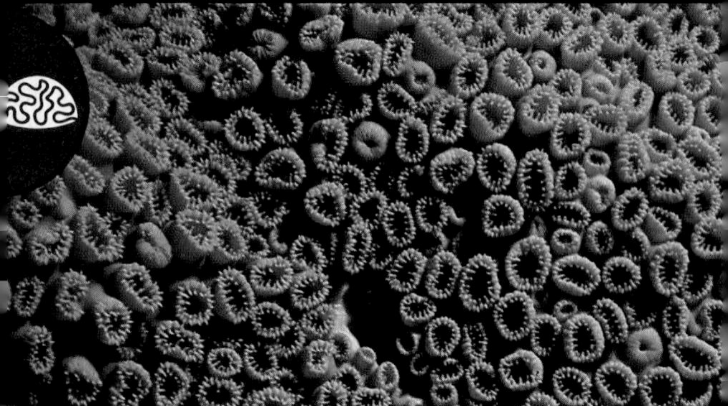

8 Knobbed Zoanthidean, *w.* 24", *p. 371*

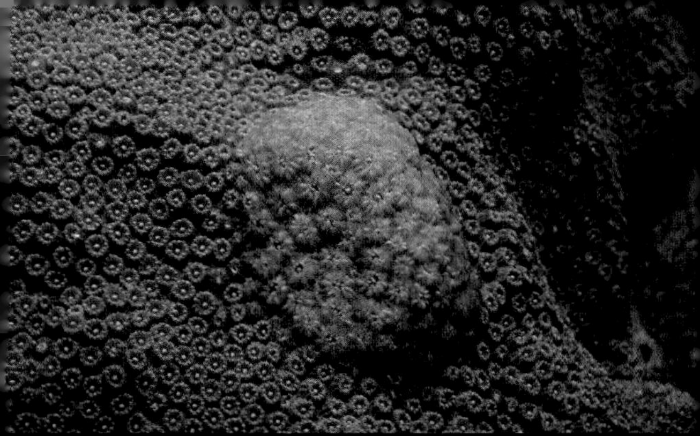

10 Common Star Coral, *w.* 48″, *p. 391*

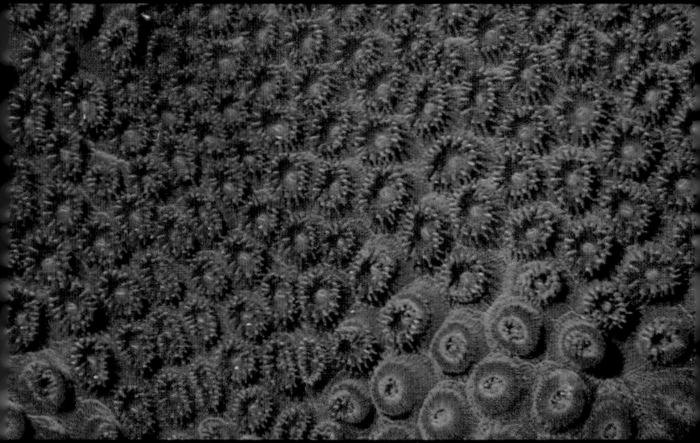

11 Large Star Coral, *w.* 60″, *p. 390*

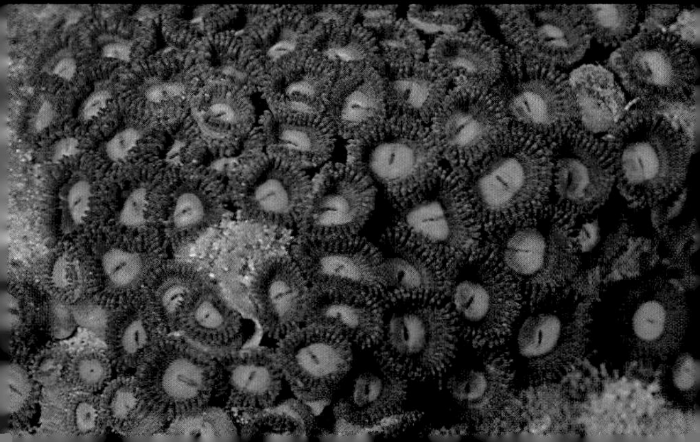

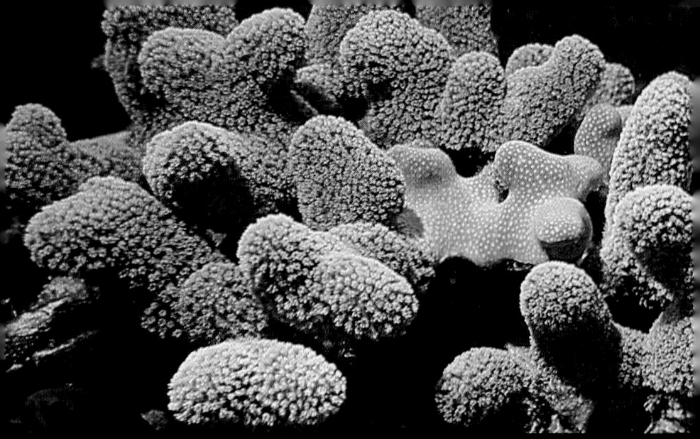

13 Clubbed Finger Coral, *l.* 12″, *p.* 388

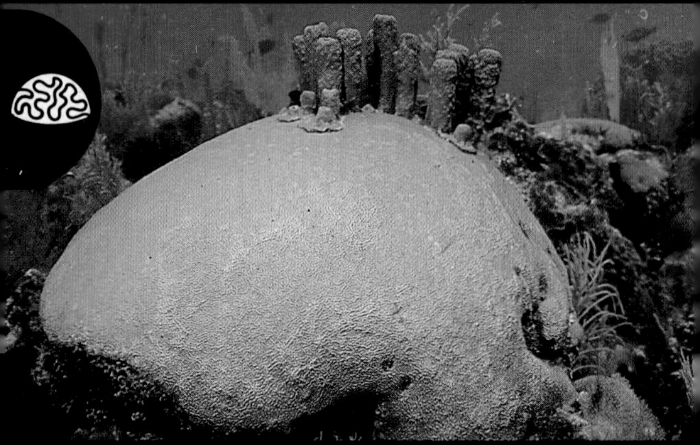

14 Reef Starlet Coral, *w.* 36″, *p.* 387

16 Starlet Coral, *w.* 12″, *p. 386*

17 Loggerhead Sponge, *w.* 36″, *p. 329*

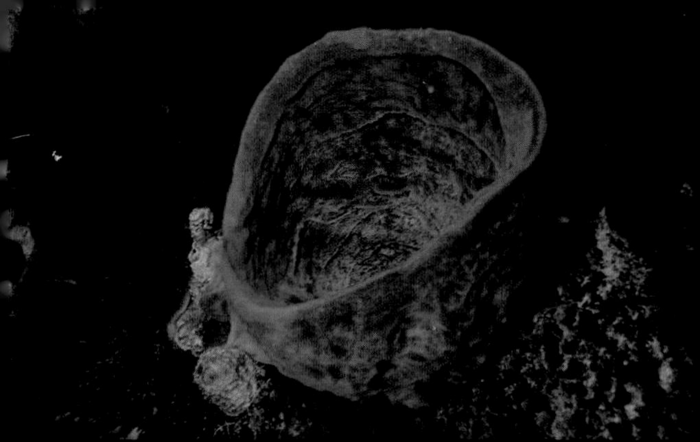

19 Vase Sponge, *h.* 36″, *p.* 332

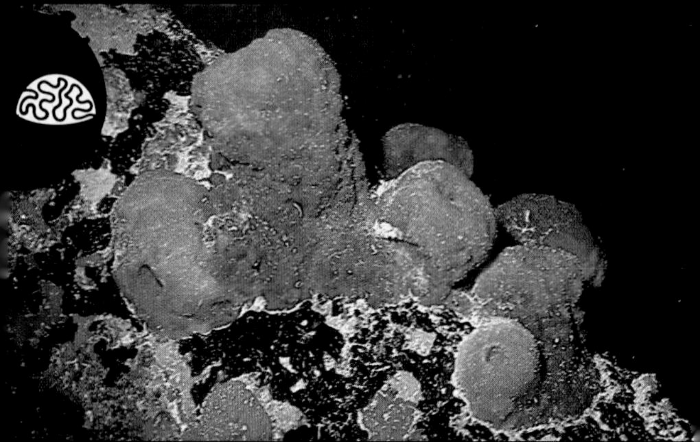

20 Do-not-touch-me Sponge, *h.* 12″, *p.* 332

22 Stinker Sponge, *h.* 12″, *p. 331*

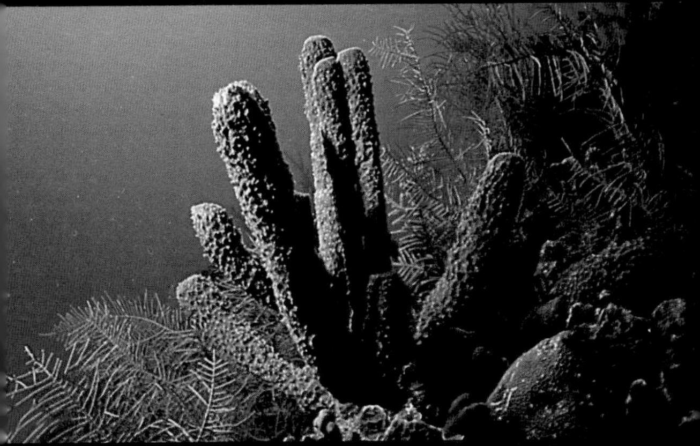

23 Tube Sponge, *h.* 36″, *p. 333*

25 Fire Coral, *h.* 24", *p.* 358

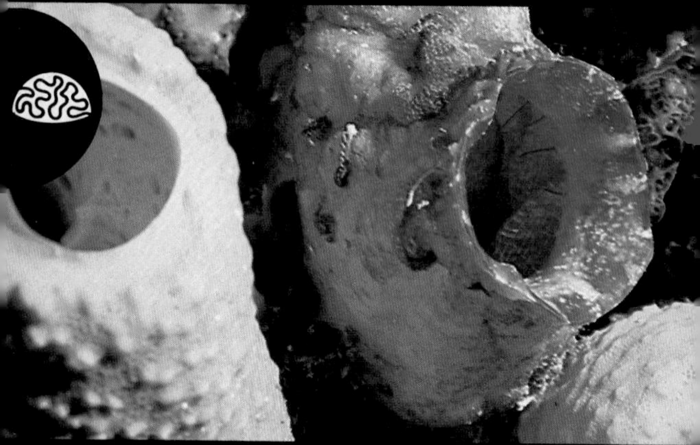

26 Encrusted Tunicate, *l.* 2", *p.* 740

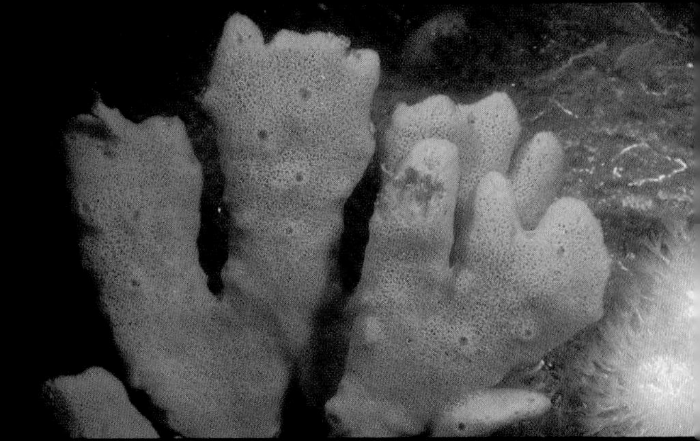

28 Common Palmate Sponge, *h.* 12″, *p. 328*

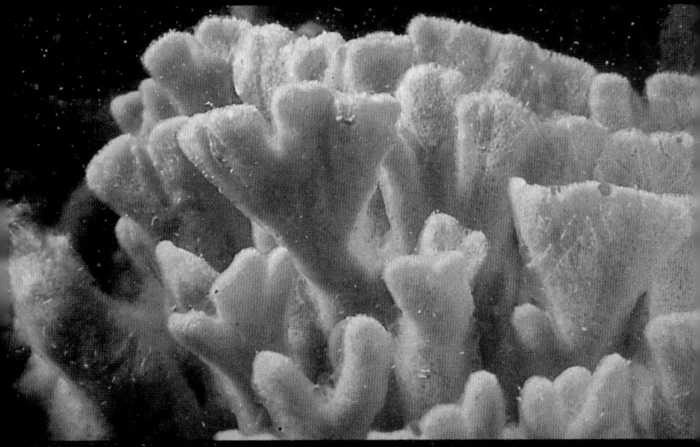

29 Red Beard Sponge, *w.* 8″, *p. 327*

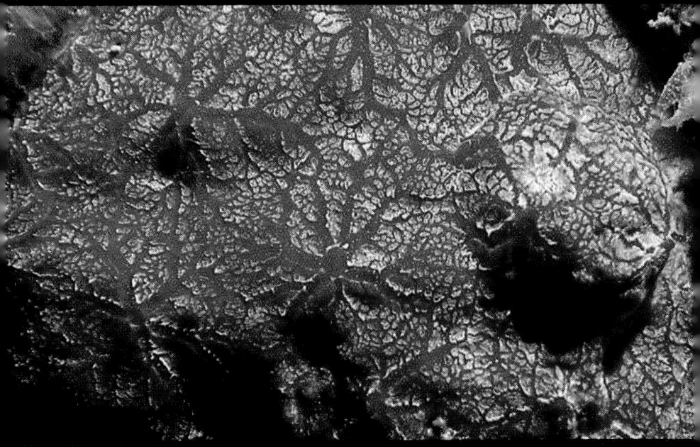

30 Smooth Red Sponge, *w.* 10″, *p. 328*

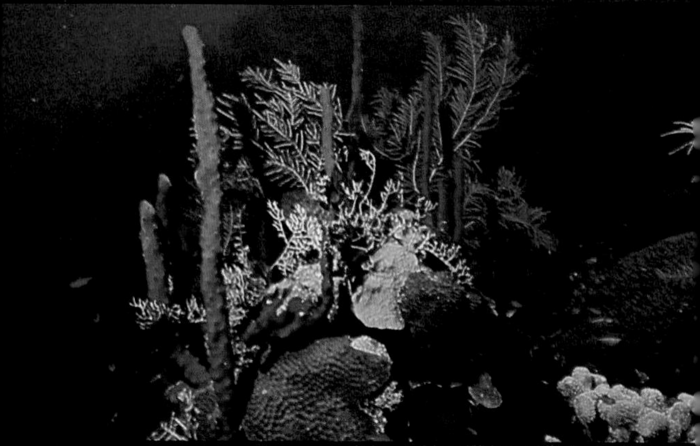

31 Red Sponge, *w.* 16″, *p. 326*

32 Elkhorn Coral, *h.* 120″, *p. 385*

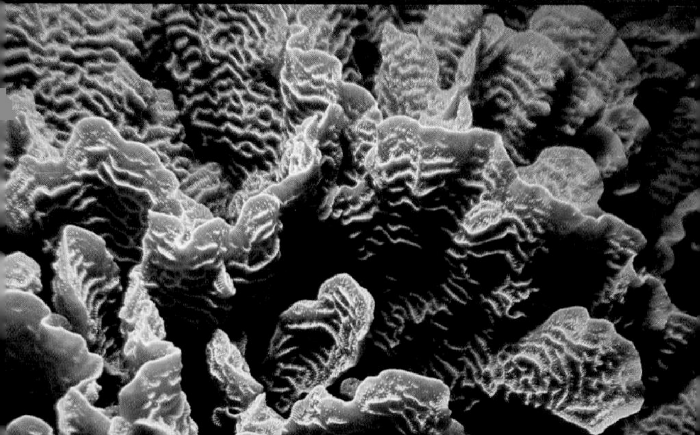

33 Lettuce Coral, *l.* 6″, *p. 386*

Plantlike Animals

All these animals display a certain resemblance to members of the plant kingdom. They vary in size from ¼″ (6 mm) to 36″ (91 cm), and range in appearance from the fuzzy or fernlike growths of hydroids, some bryozoans and the Lacy Tube Worm, to the upright stalks of sponges, sea pens, the Red Soft Coral, the Green Phoronid Worm, and the stalked tunicates and jellyfish. Some bryozoans form dense, irregular colonies which look much like a moss or lichen, and most soft corals, as well as the Staghorn Coral, the Sea Fan and the tube sponges, branch upward.

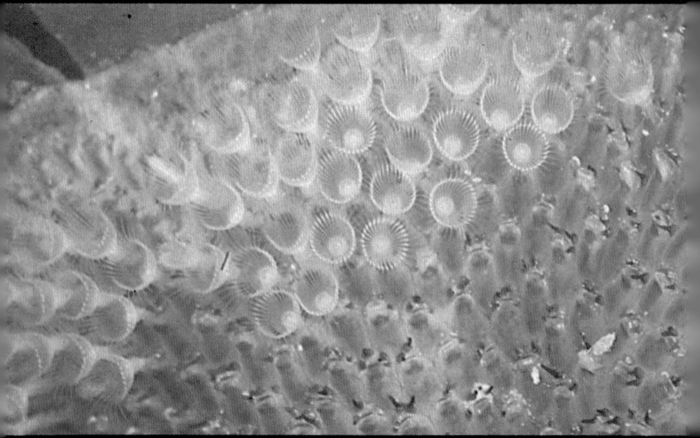

34 Porcupine Bryozoan, *l.* 4", *p.* 708

35 Ivory Bush Coral, *h.* 12", *p.* 392

36 Green Encrusting Tunicate, *l.* 3", *p.* 740

37 Red Soft Coral, *h.* 6″, *p.* 364

38 Snail Fur, *h.* ⅛″, *p.* 343

39 Red Soft Coral, *h.* 6″, *p.* 364

40 Trumpet Stalked Jellyfish, *h.* 1″, *p. 359*

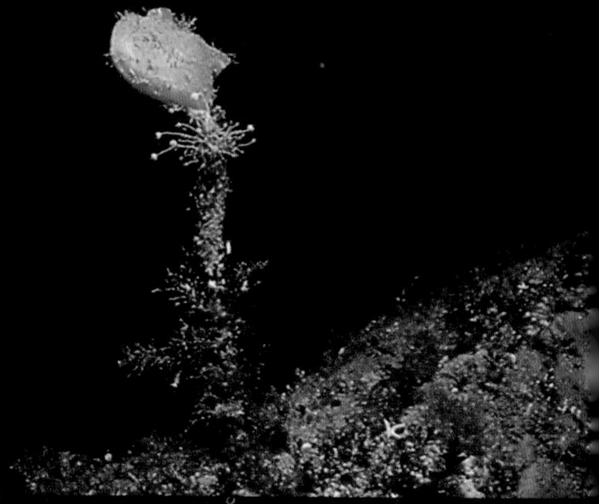

41 Stalked Tunicate, *l.* 3″, *p. 741*

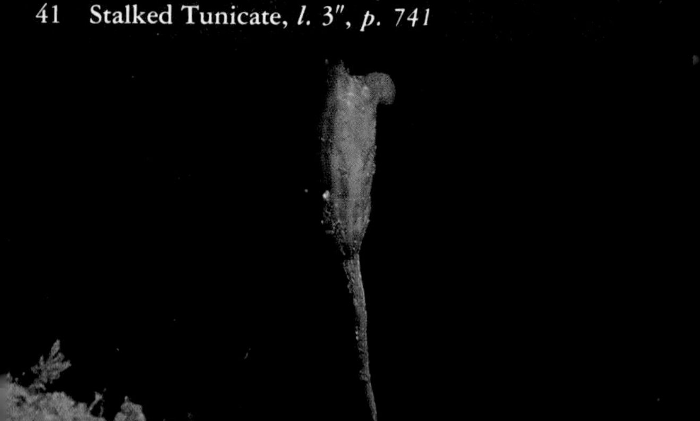

42 Monterey Stalked Tunicate, *h.* 10″, *p. 736*

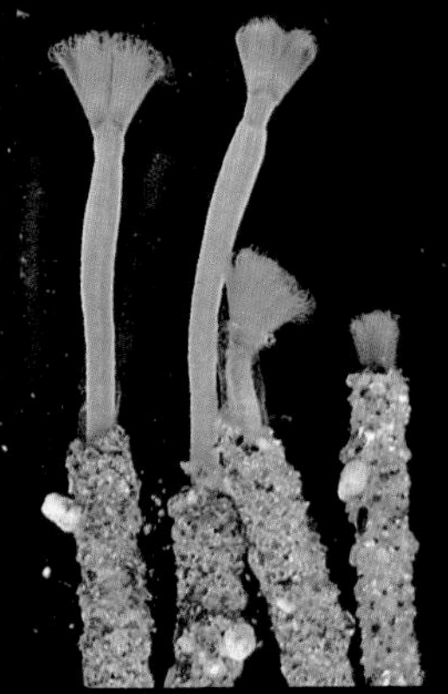

43 Green Phoronid Worm, *l.* 5″, *p.* 719

44 Gurney's Sea Pen, *h.* 18″, *p.* 370

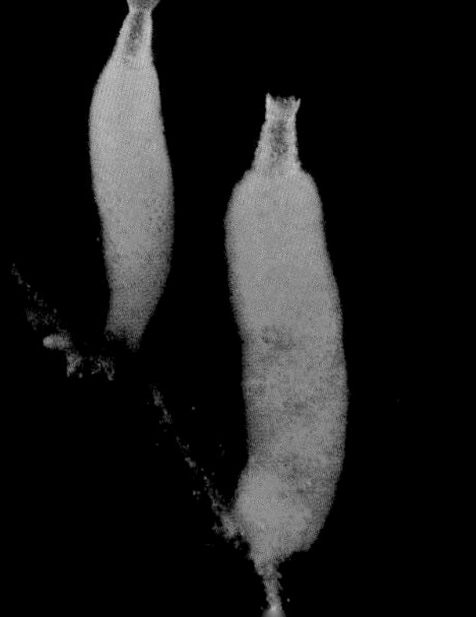

46 Nutting's Sponge, *h.* 2", *p.* 324

47 Ellis' Bryozoan, *w.* 2", *p.* 714

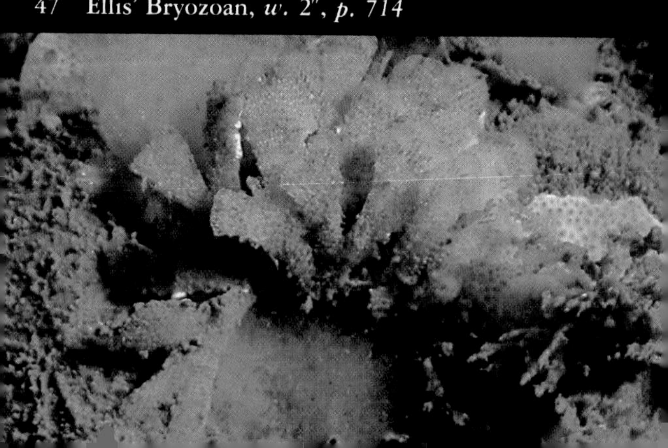

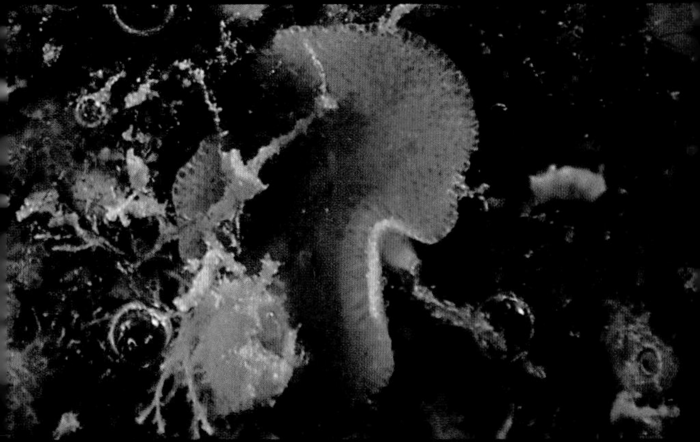

49 Rubbery Bryozoan, *w.* 3″, *p.* 708

50 Lattice-work Bryozoan, *w.* 8½″, *p.* 716

51 Coralline Bryozoan, *w.* 10″, *p.* 710

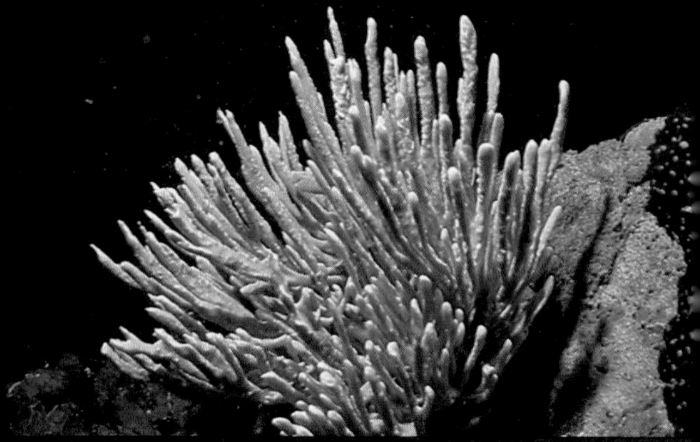

52 Finger Sponge, *h.* 18", *p. 325*

53 Nipple Sponge, *w.* 4", *p. 329*

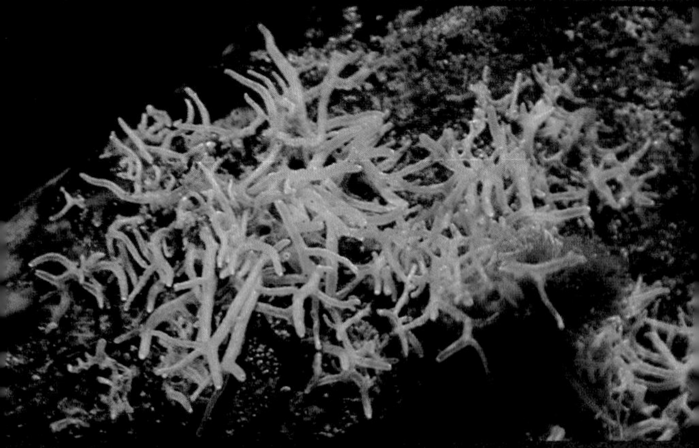

54 Organ-pipe Sponge, *h.* ½", *p. 323*

55 Staghorn Coral, *h.* 10′, *p.* 385

56 Eunicea Sea Rod, *h.* 36″, *p.* 368

57 Black Sea Rod, *h.* 24″, *p.* 368

58 Double-forked Plexaurella, *h.* 24″, *p. 369*

59 Yellow Sea Whip, *h.* 24″, *p. 367*

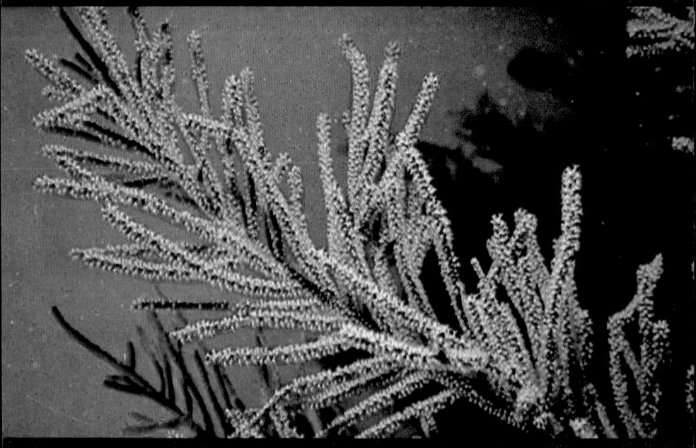

60 Sea Plume, *h.* 36″, *p. 366*

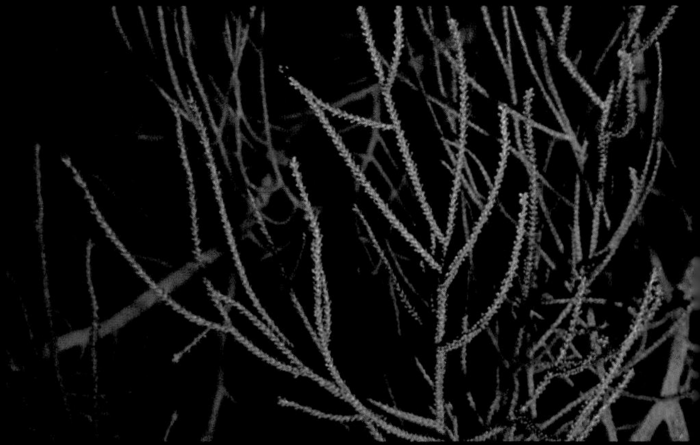

61 Spiny Muricea, *h.* 24″, *p.* 367

62 Yellow Sea Whip, *h.* 24″, *p.* 367

63 Sea Whip, *h.* 36″, *p.* 365

64 Sea Fan, *h.* 36″, *p. 366*

65 Articulated Bryozoan, *l.* 1″, *p. 710*

66 Sea Plume, *h.* 36″, *p. 366*

67 Feathery Hydroid, *h.* 24″, *p.* 354

68 Corky Sea Fingers, *h.* 12″, *p.* 365

69 Feathery Hydroid, *h.* 24″, *p.* 354

70 Tropical Garland Hydroid, *h.* 8″, *p. 354*

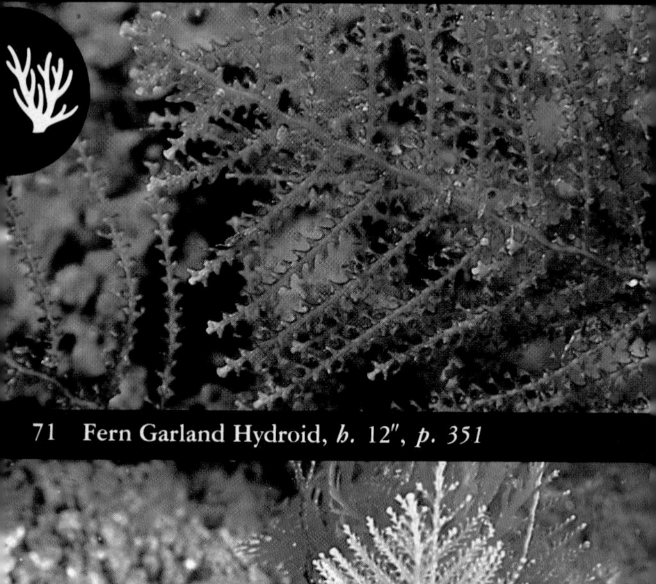

71 Fern Garland Hydroid, *h.* 12″, *p. 351*

72 Halecium Hydroid, *h.* 3″, *p. 351*

73 Spiral-tufted Bryozoan, *l.* 12″, *p. 713*

74 Silvery Hydroid, *h.* 12″, *p. 353*

75 Wine-glass Hydroid, *h.* 10″, *p. 347*

76 Bushy Twinned Bryozoan, *h.* 10″, *p. 711*

77 Wine-glass Hydroid, *h.* 10″, *p. 347*

78 Zig-zag Wine-glass Hydroid, *w.* 12″, *p. 348*

79 Bushy Wine-glass Hydroid, *h.* 8″, *p. 348*

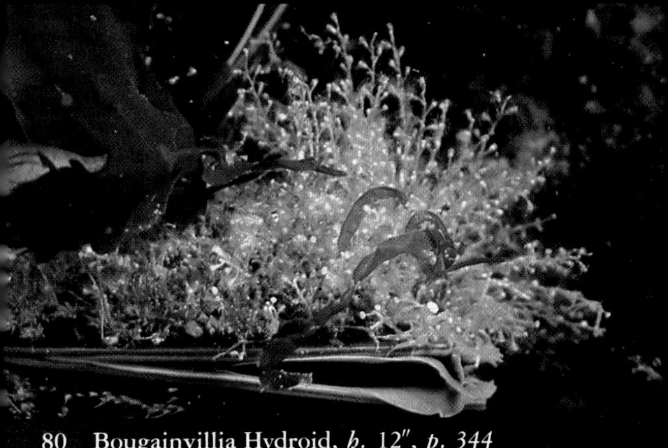

80 Bougainvillia Hydroid, *h.* 12″, *p. 344*

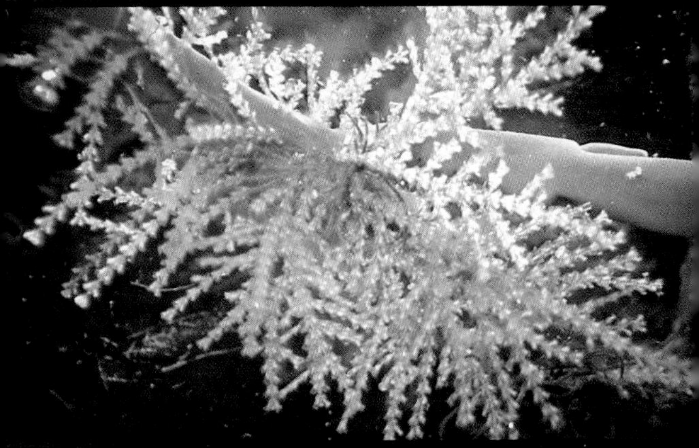

81 Garland Hydroid, *h.* 2″, *p. 352*

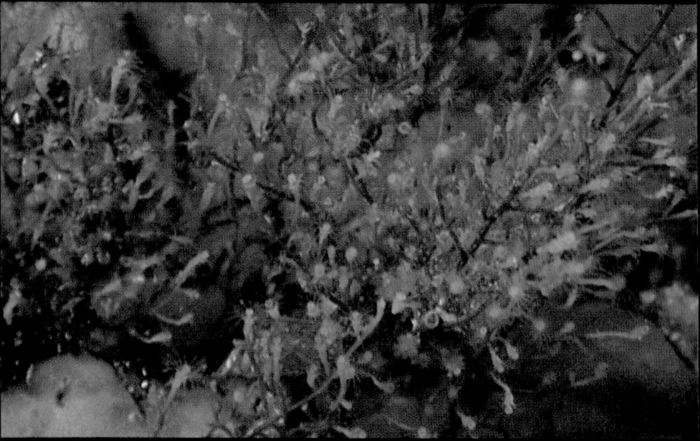

82 Stick Hydroid, *h.* 6″, *p. 345*

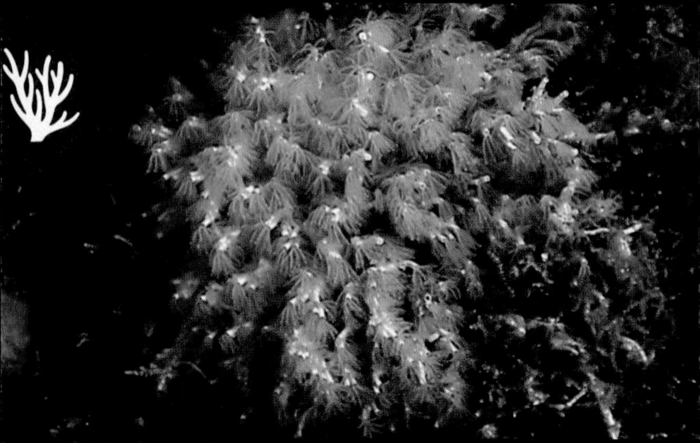

83 Lacy Tube Worm, *l.* ¼″, *p. 441*

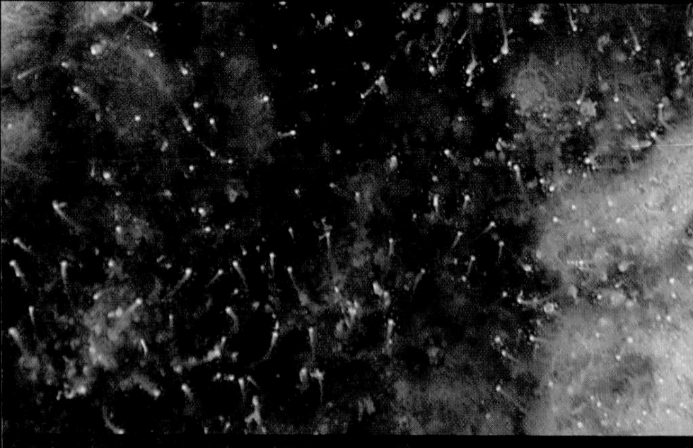

84 Clapper Hydromedusa, *w.* 4″, *p. 342*

85 Thick-based Entoproct, *w.* 1¼″, *p. 718*

86 Feathered Hydroid, *h.* 6″, *p. 342*

87 Club Hydroid, *w.* 1″, *p. 343*

88 Tubularian Hydroid, *w.* 12", *p. 341*

89 Solitary Hydroid, *h.* 4", *p. 340*

90 Horned Stalked Jellyfish, *h.* 3", *p. 359*

Sea Squirts

Both sea squirts and colonial tunicates are enclosed by a thin, flexible tunic perforated by two siphons at the top. These animals are rounded or vase-shaped, and range in color from solid tones of orange, red and yellow to transparent or translucent blue or green. Also vase-shaped, and thus included in this group, is Heath's Sponge.

91 Mushroom Tunicate, *w.* ¾", *p.* 732

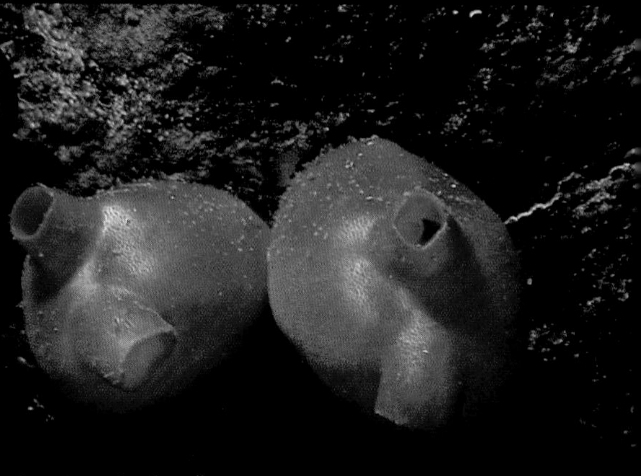

92 Sea Peach, *h.* 5", *p.* 742

93 Blood Drop Tunicate, *w.* ½", *p.* 736

94 Taylor's Colonial Tunicate, *w.* 8″, *p.* 739

95 Mangrove Tunicate, *l.* 1″, *p.* 734

96 Mangrove Tunicate, *l.* 1″, *p.* 734

97 Orange Sea Grape, *l.* ⅝", *p.* 742

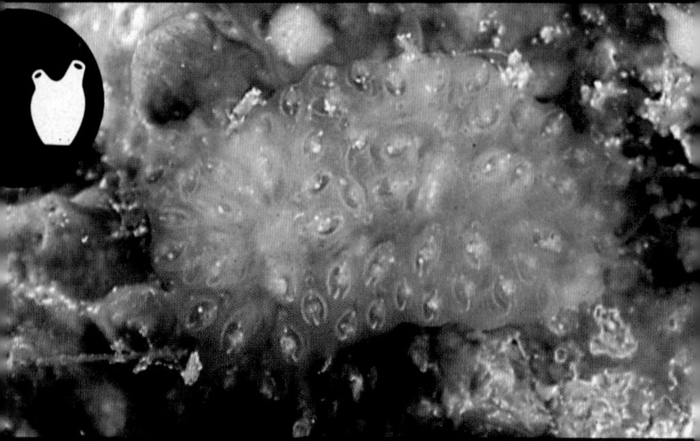

98 Northern Sea Pork, *w.* 3", *p.* 730

99 Green Encrusting Tunicate, *l.* 3", *p.* 740

100 Painted Tunicate, *w.* 12″, *p.* 729

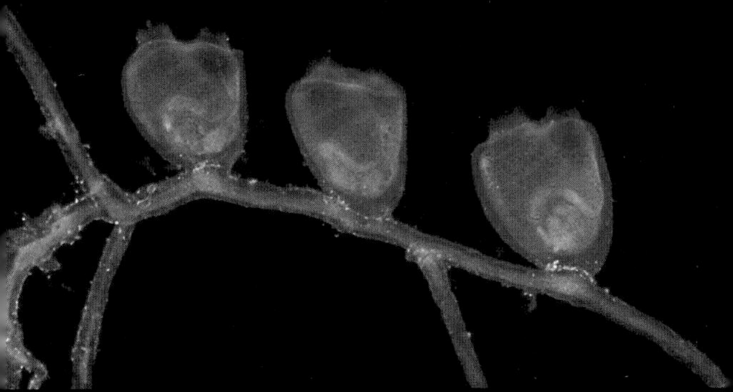

101 Creeping Tunicate, *w.* 3″, *p.* 734

102 Light Bulb Tunicate, *l.* 2″, *p.* 729

103 Light Bulb Tunicate, *l.* 2″, *p.* 729

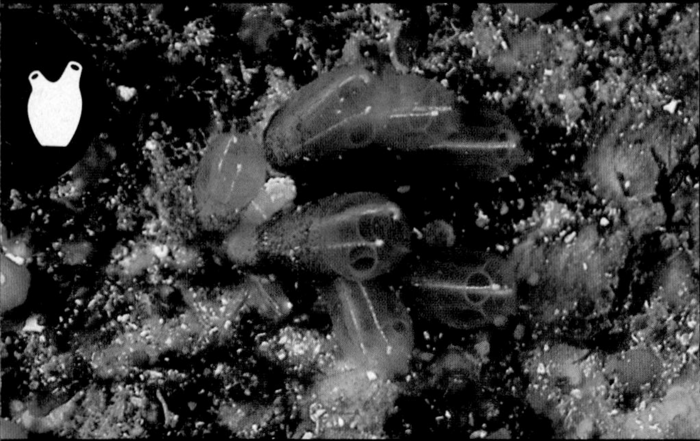

104 Light Bulb Tunicate, *l.* 2″, *p.* 729

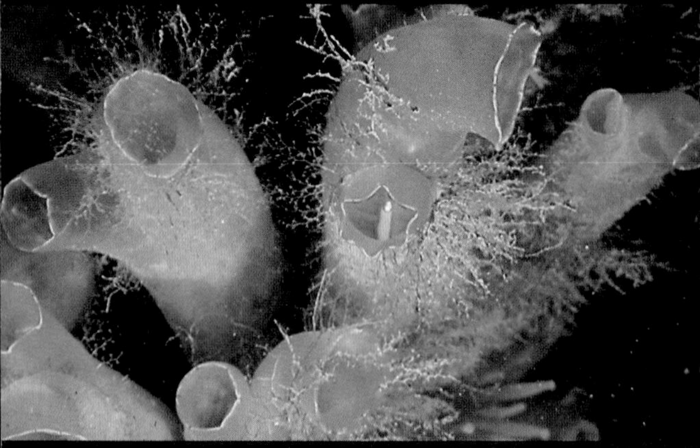

105 Sea Vase, *h.* 6″, *p.* 733

106 Striped Tunicate, *l.* 3⅝″, *p.* 737

107 Striped Tunicate, *l.* 3⅝″, *p.* 737

108 Club Tunicate, *h.* 6″, *p.* 736

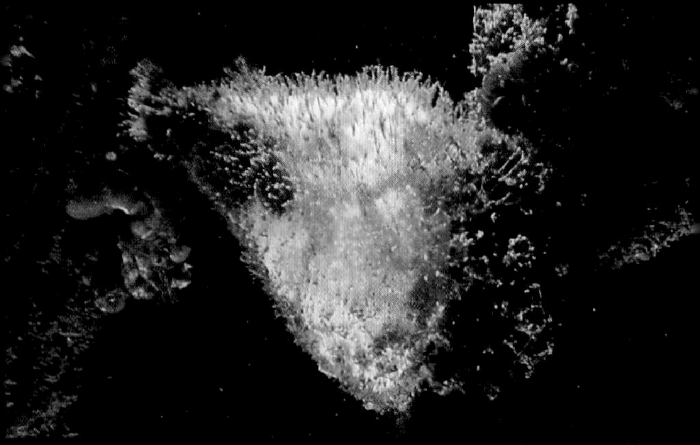

109 Heath's Sponge, *h.* 4⅜", *p. 324*

110 Bristly Tunicate, *h.* 1⅝", *p. 742*

111 Cactus Tunicate, *w.* 1⅜", *p. 741*

Encrusting Animals

The shapes of the sponges, bryozoans and tunicates grouped here vary with the substrate. Encrusting sponges grow in an irregular mat, have numerous pores, and are often richly colored. Encrusting bryozoans are clusters of individuals joined by a gelatinous, membranous, rubbery or limy covering that ranges in color from pure white to pink or red. Compound tunicates are also composed of colonial individuals, and sometimes form a star-shaped pattern. Included with these is the Boring Sponge, which penetrates its host. Certain red sponges in this group are powerful stingers, and should not be handled.

112 Lacy-crust Bryozoan, *w.* 3″, *p. 711*

113 Hairy Bryozoan, *w.* 2″, *p. 712*

114 Common Red Crust Bryozoan, *w.* 3″, *p. 715*

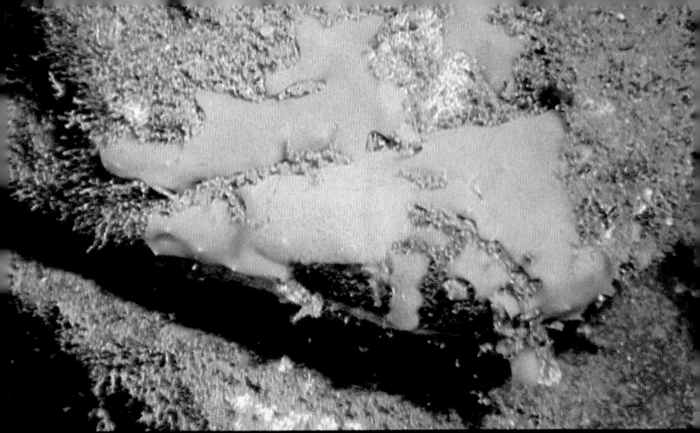

115 Dujardin's Slime Sponge, *w.* 4", *p.* 325

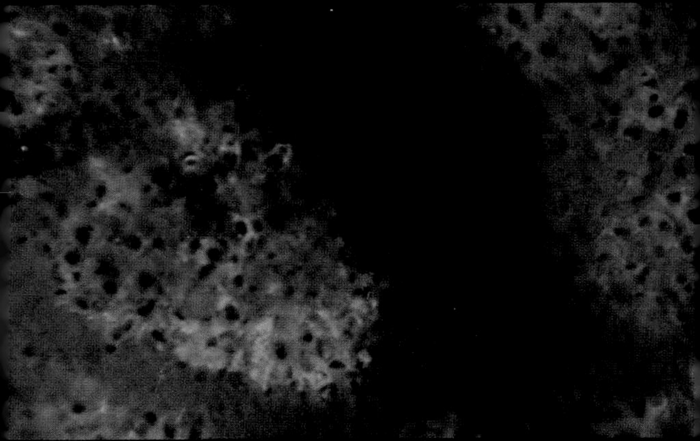

116 Black-speckled Bryozoan, *w.* 2½", *p.* 714

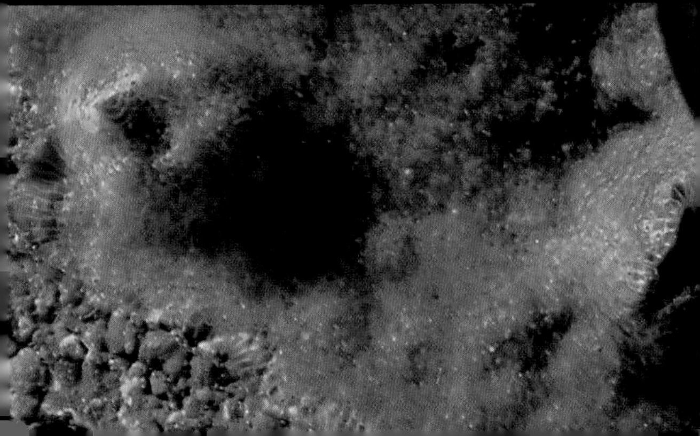

118 Pacific White Crust, *w.* 4¾″, *p.* 731

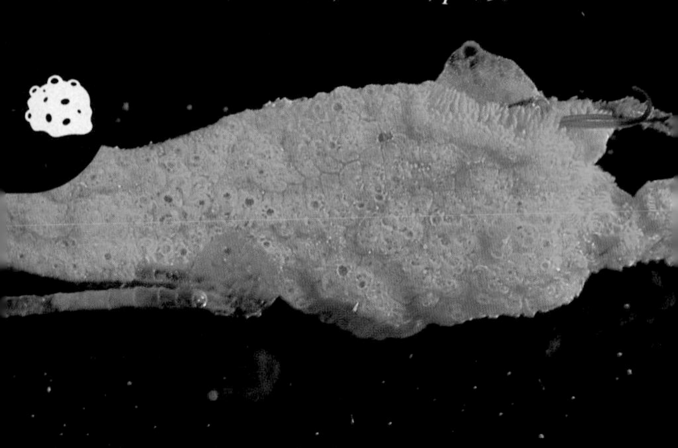

119 Orange Sheath Tunicate, *w.* 2″, *p.* 738

121 Golden Star Tunicate, *w.* 4″, *p.* 738

122 Disk-top Tunicate, *h.* 2⅜″, *p.* 733

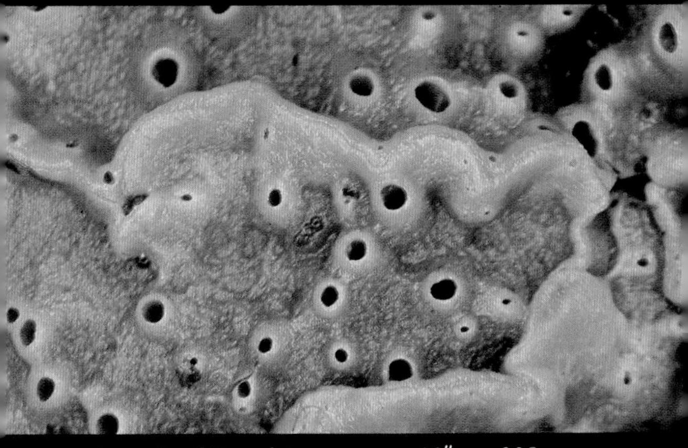

123 Crumb of Bread Sponge, *w.* 12″, *p.* 328

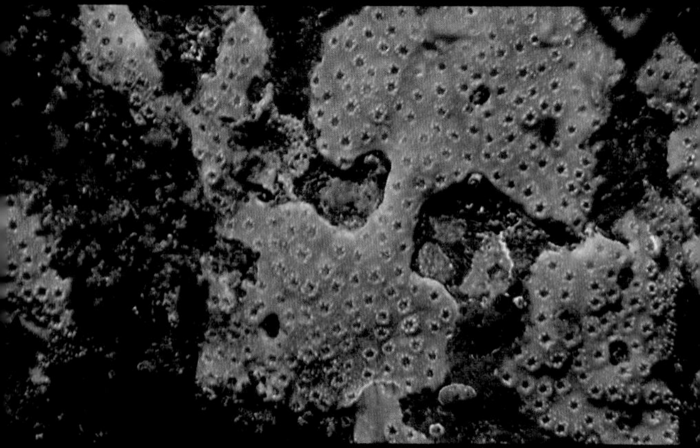

124 Purple Stylasterine, *w.* 6″, *p.* 358

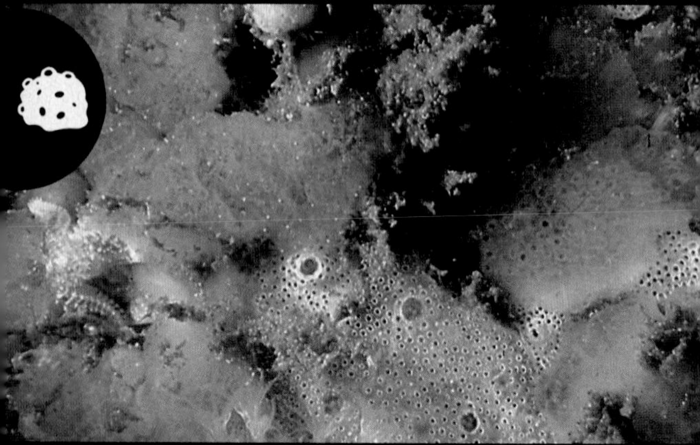

125 Northern White Crust, *w.* 4″, *p.* 731

126 Purple Sponge, *w.* 36″, *p.* 326

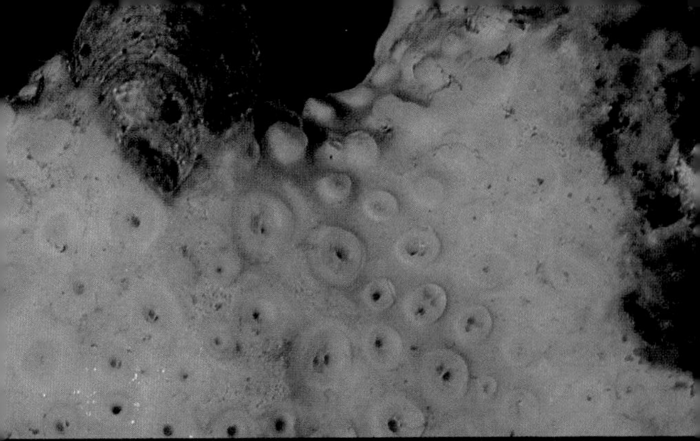

127 Boring Sponge, *w.* ⅛", *p. 330*

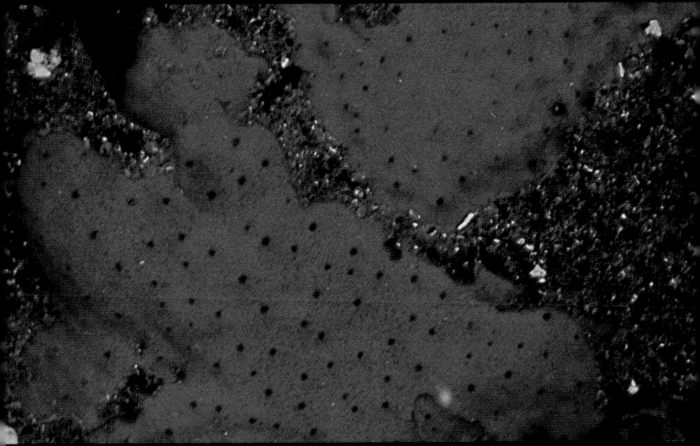

128 Velvety Red Sponge, *w.* 36", *p. 327*

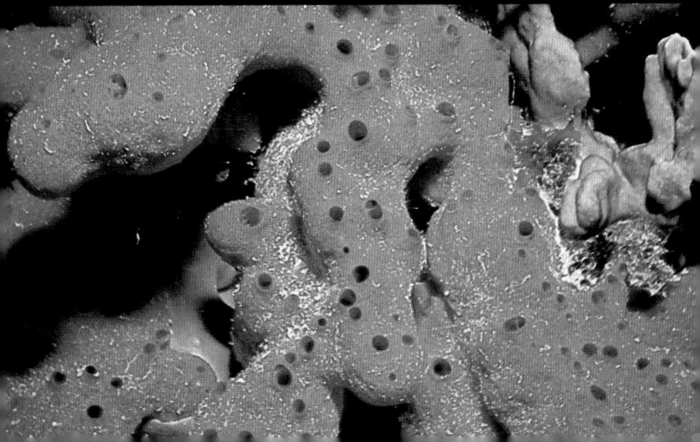

Flowerlike Animals

Brilliantly-colored feeding tentacles and gills lend a flowerlike appearance to these sea cucumbers and polychaete worms. Sometimes white or brown, often rich red or purple, these showy mucus-covered tufts surround the animal's mouth, and trap floating microscopic organisms. The Ragged Sea Hare also has mucus-covered filaments.

130 Spiral-gilled Tube Worm, *l.* 4″, *p.* 443

131 Spiral-gilled Tube Worm, *l.* 4″, *p.* 443

133 Spiral-gilled Tube Worm, *l.* 4″, *p.* 443

134 Spiral-gilled Tube Worm, *l.* 4″, *p.* 443

135 Spiral-gilled Tube Worm, *l.* 4″, *p.* 443

136 Star Tube Worm, *l.* 4", *p.* 442

137 Red Tube Worm, *l.* 4", *p.* 443

138 Large-eyed Feather Duster, *l.* 4", *p.* 439

139 Large-eyed Feather Duster, *l.* 4″, *p. 439*

140 Slime Feather Duster, *l.* 8″, *p. 439*

142 Red Tube Worm, *l.* 4″, *p. 443*

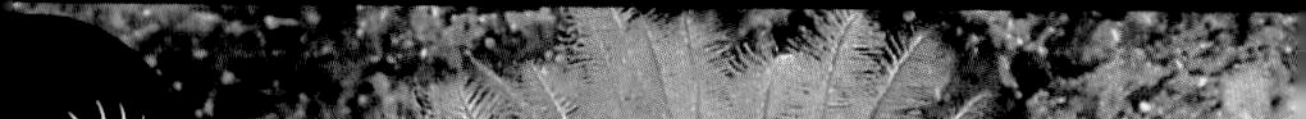

143 Giant Feather Duster, *l.* 10″, *p. 438*

145 Red Tube Worm, *l.* 4″, *p.* 443

146 Giant Feather Duster, *l.* 10″, *p.* 438

148 Ragged Sea Hare, *l.* 4″, *p.* 520

149 Lacy Tube Worm, *l.* ¼″, *p.* 441

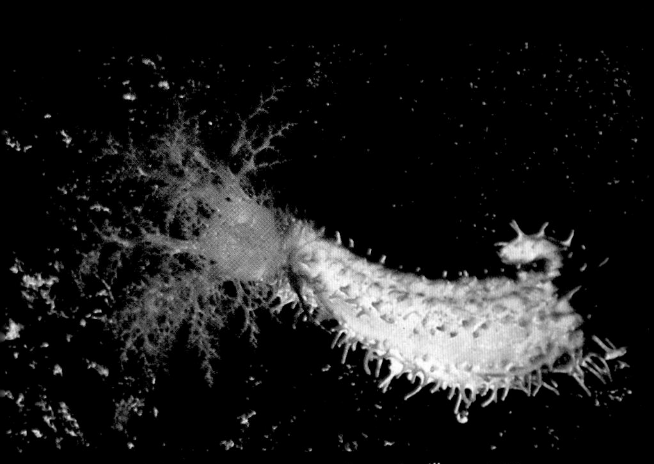

151 Stiff-footed Sea Cucumber, *l.* 4″, *p.* 702

152 Orange-footed Sea Cucumber, *l.* 19″, *p.* 701

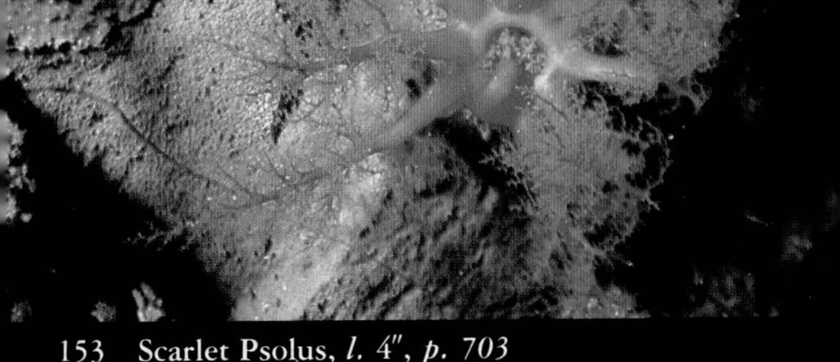

153 Scarlet Psolus, *l.* 4″, *p.* 703

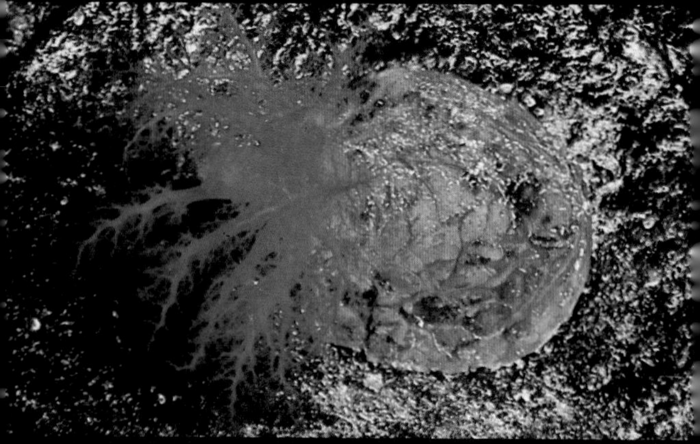

154 Slipper Sea Cucumber, *l.* 4¾", *p. 704*

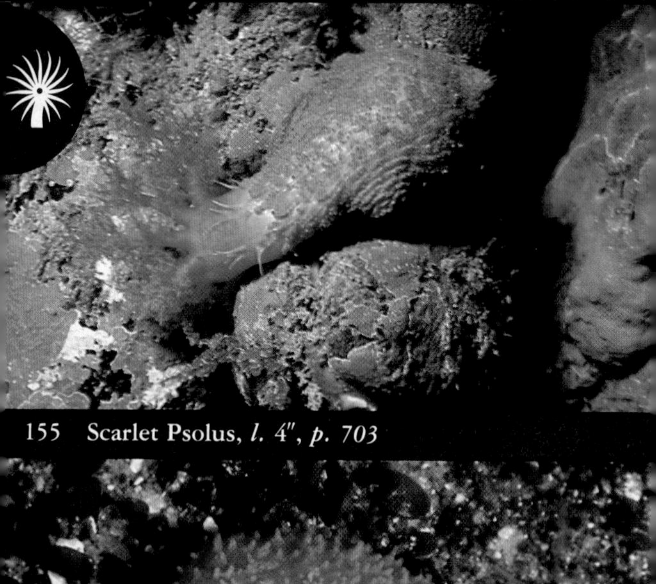

155 Scarlet Psolus, *l.* 4", *p. 703*

156 Dwarf Sea Cucumber, *l.* ¾", *p. 704*

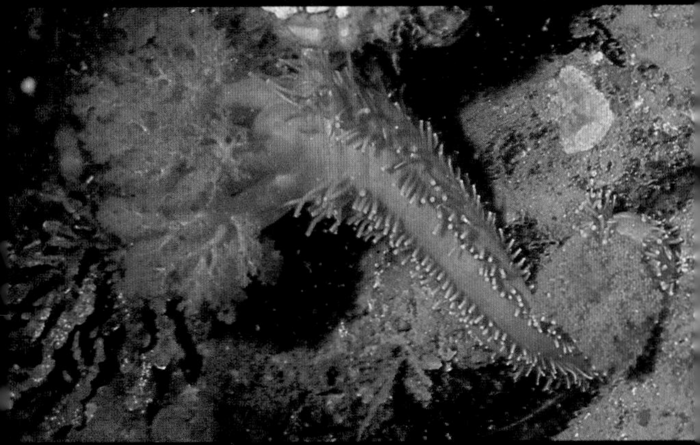

157 Red Sea Cucumber, *l.* 10″, *p. 701*

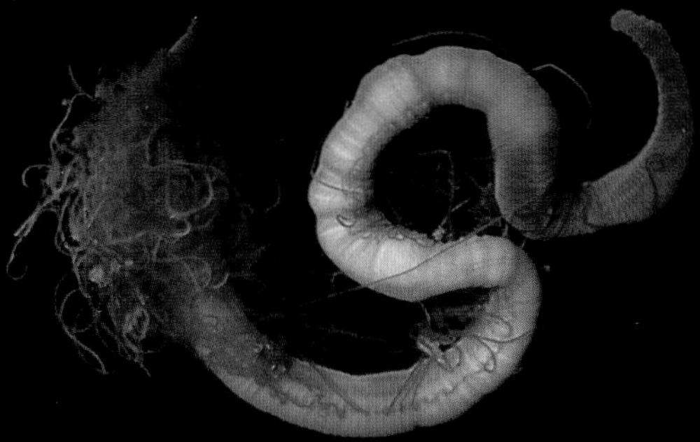

158 Large Fringed Worm, *l.* 6″, *p. 431*

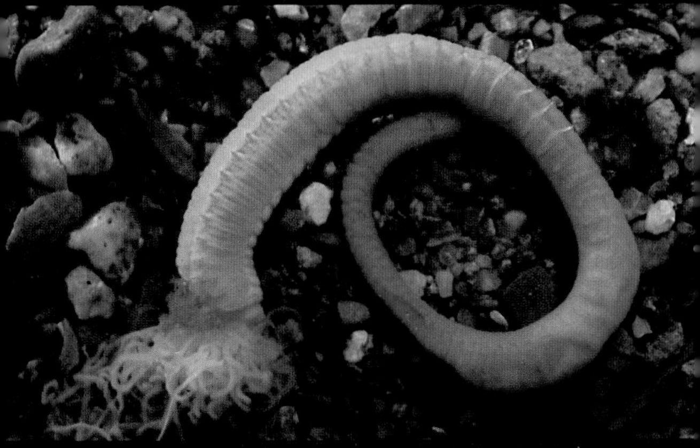

159 Johnston's Ornate Worm, *l.* 10″, *p. 435*

160 Luxurious Fringed Worm, *l.* 6″, *p. 432*

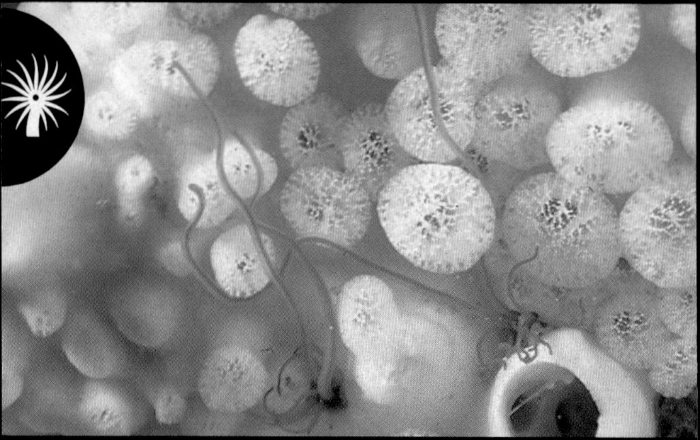

161 Eyed Fringed Worm, *l.* 4¾″, *p. 431*

162 Curly Terebellid Worm, *l.* 11¼″, *p. 437*

163 Pale Terebellid Worm, *l.* 8″, *p. 438*

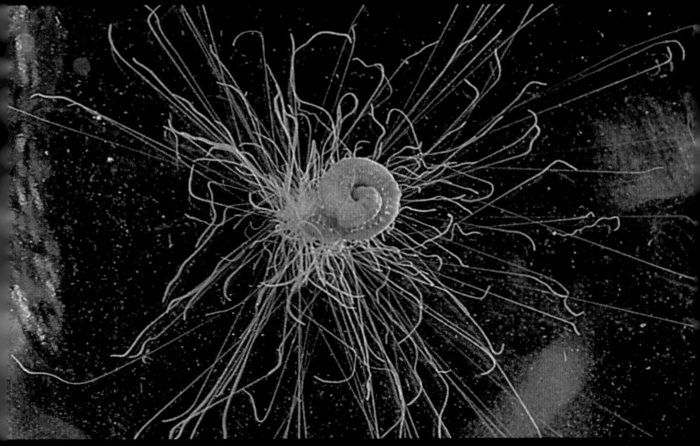

164 Red Terebellid Worm, *l.* 2¾″, *p.* 437

165 Curly Terebellid Worm, *l.* 11¼″, *p.* 437

Sea Anemones

Among the most visually appealing of invertebrates, sea anemones are immediately recognizable by their unique form: a tube with a mouth at the top surrounded by unbranched tentacles. While many anemones live attached to a solid object, burrowing anemones excavate a hole in the ocean floor from which their feeding tentacles extend. Certain corals, most of which secrete a calcium carbonate base, are also included in this section because their polyp shape is similar to that of the anemones.

166 Lined Anemone, *h.* 1⅜″, *p.* 372

167 Pale Anemone, *h.* 2″, *p.* 380

169 Ghost Anemone, *h.* 1½", *p.* 383

170 Northern Cerianthid, *h.* 18", *p.* 383

172 Smooth Burrowing Anemone, *h.* 2½″, *p. 381*

173 Flower Coral, *w.* 12″, *p. 394*

175 Rose Coral, *h.* 4", *p.* 390

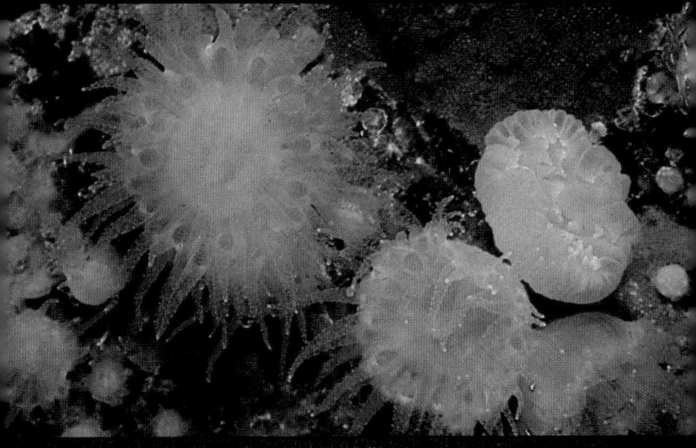

176 Orange Cup Coral, *h.* ⅜", *p.* 384

177 Orange Cup Coral, *h.* ⅜", *p.* 384

178 Red Stomphia, *h.* 2″, *p. 378*

179 Buried Sea Anemone, *h.* 10″, *p. 375*

180 Club-tipped Anemone, *h.* 1¼″, *p. 384*

181 Proliferating Anemone, *w.* 2″, *p. 377*

182 Northern Red Anemone, *h.* 5″, *p. 373*

183 Strawberry Anemone, *h.* 6″, *p. 374*

184 Leathery Anemone, *h.* 5⅝″, *p.* 374

185 Aggregating Anemone, *h.* 20″, *p.* 376

186 Giant Green Anemone, *h.* 12″, *p.* 376

187 Pink-tipped Anemone, *w.* 12″, *p.* 377

188 Pink-tipped Anemone, *w.* 12″, *p.* 377

190 Ringed Anemone, *h.* 2″, *p. 381*

191 Elegant Burrowing Anemone, *h.* 2″, *p. 372*

193 Warty Sea Anemone, *h.* 4″, *p.* 378

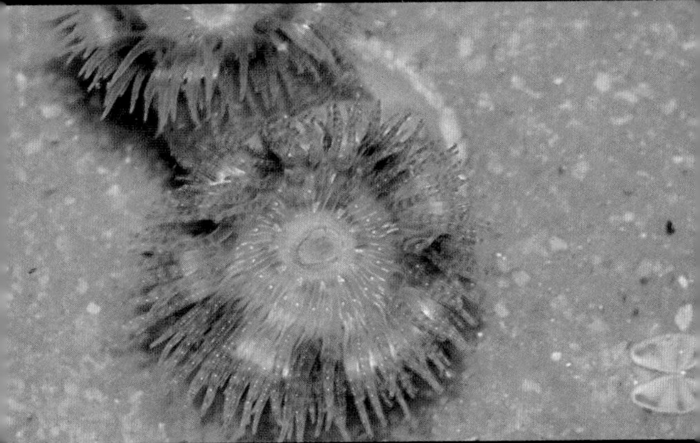

194 Tricolor Anemone, *h.* 3″, *p.* 379

196 Silver-spotted Anemone, *w.* 2″, *p.* 375

197 Proliferating Anemone, *w.* 2″, *p.* 377

Flatworms, Nudibranchs, and Sea Hares

This exotic group includes wonderfully colored, mobile animals that crawl or swim through the marine environment. Flatworms have an extremely compressed body that allows them to undulate through the water and creep into crevices between rocks or shells. Nudibranchs and sea hares glide gracefully over the ocean floor, the nudibranchs seeking invertebrate prey while the sea hares graze on algae in shallow waters. Also in this group is the California Stichopus.

199 Hopkins' Rose, *l.* 1¼", *p.* 526

200 Elegant Eolid, *l.* 3⅝", *p.* 530

201 Hermissenda Nudibranch, *l.* 3¼", *p.* 531

202 Red-gilled Nudibranch, *l.* 1¼", *p.* 529

203 Salmon-gilled Nudibranch, *l.* 1¾", *p.* 529

204 Salmon-gilled Nudibranch, *l.* 1¾", *p.* 529

205 Atlantic Ancula, *l.* ½", *p.* 522

206 Sea Clown Nudibranch, *l.* 6", *p.* 527

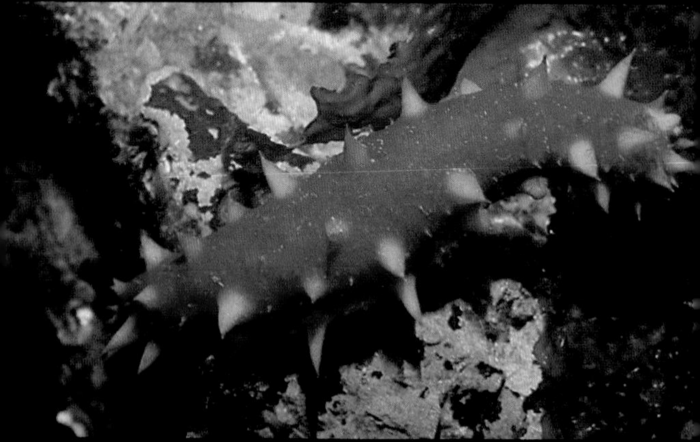

207 California Stichopus, *l.* 16", *p.* 699

208 Bushy-backed Sea Slug, *l.* 4⅝", *p.* 528

209 California Sea Hare, *l.* 16", *p.* 519

210 Spotted Sea Hare, *l.* 5", *p.* 518

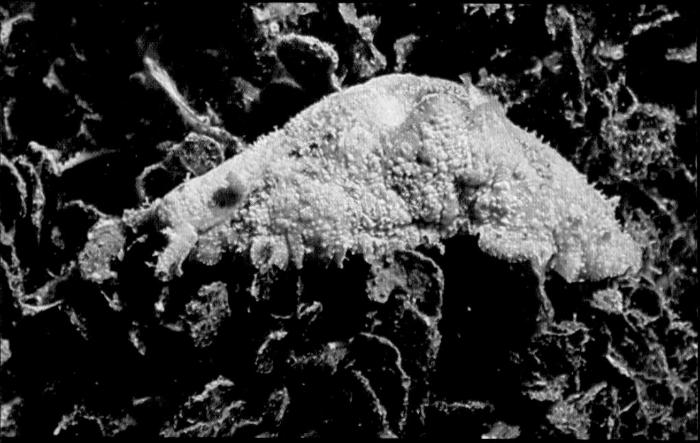

211 Warty Sea Cat, *l.* 6″, *p. 518*

212 Common Lettuce Slug, *l.* 1½″, *p. 521*

213 Monterey Flatworm, *l.* 2″, *p. 403*

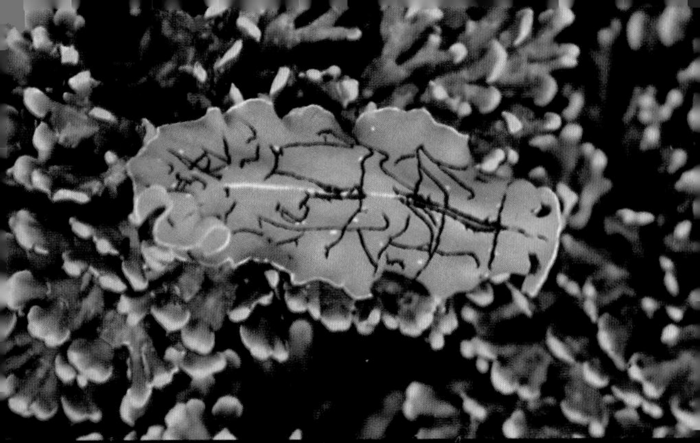

214 Horned Flatworm, *l.* 1¼", *p. 403*

215 Crozier's Flatworm, *l.* 2", *p. 404*

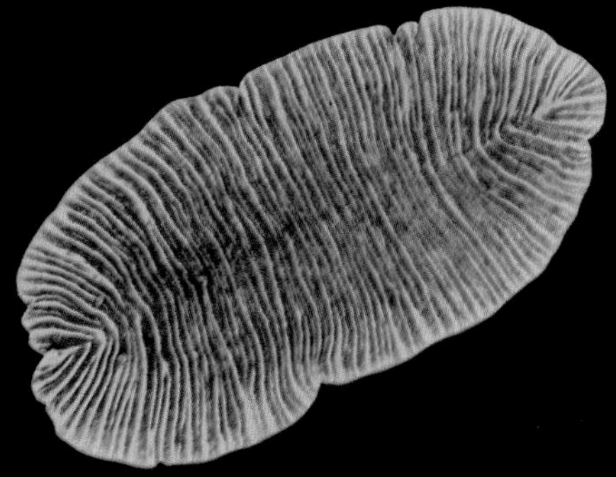

216 Zebra Flatworm, *l.* 1½", *p. 402*

217 Oval Flatworm, *l.* 1½", *p. 401*

218 Oyster Leech, *l.* 1", *p. 401*

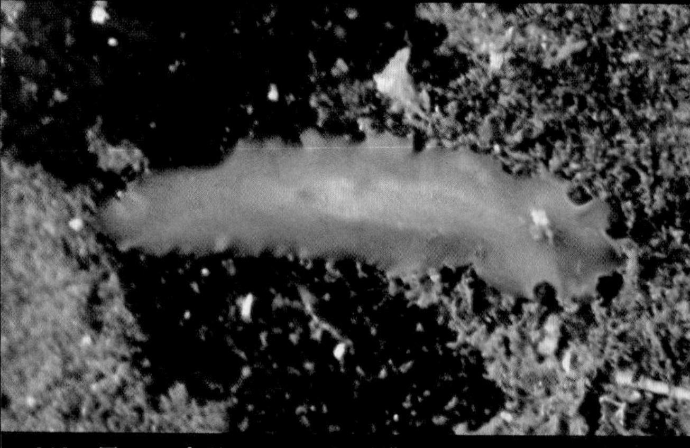

219 Tapered Flatworm, *l.* 2⅜", *p. 402*

220 Lion Nudibranch, *l.* 4", *p.* 531

221 Ringed Doris, *l.* 3⅝", *p.* 525

222 Hairy Doris, *l.* 1¼", *p.* 521

223 White Atlantic Cadlina, *l.* 1″, *p.* 524

224 White Knight Doris, *l.* 8″, *p.* 523

225 Yellow-edged Cadlina, *l.* 3¼″, *p.* 524

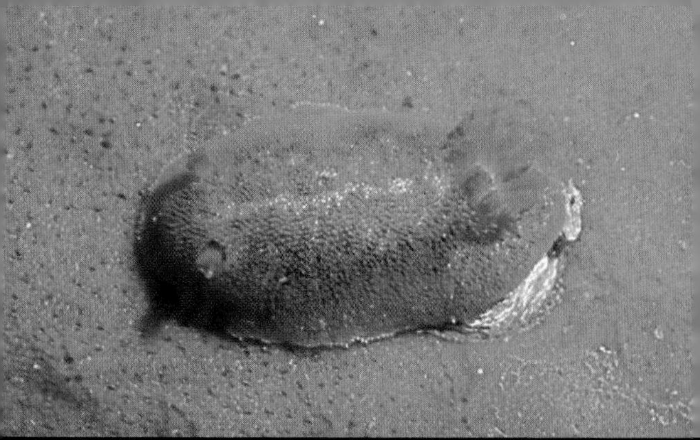

226 Crimson Doris, *l.* 1¼", *p.* 527

227 Salted Doris, *l.* 2¾", *p.* 525

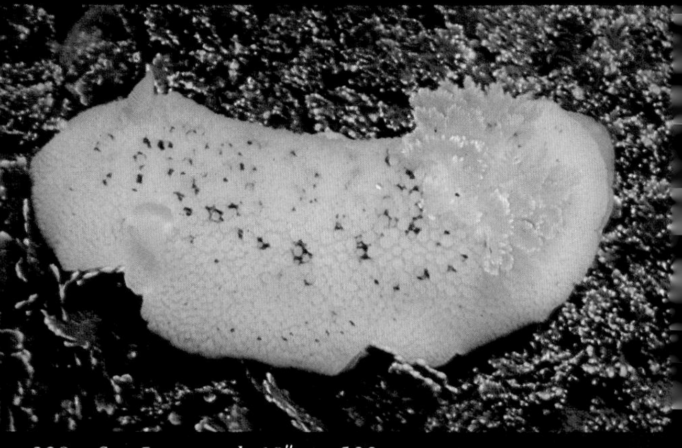

228 Sea Lemon, *l.* 10", *p.* 522

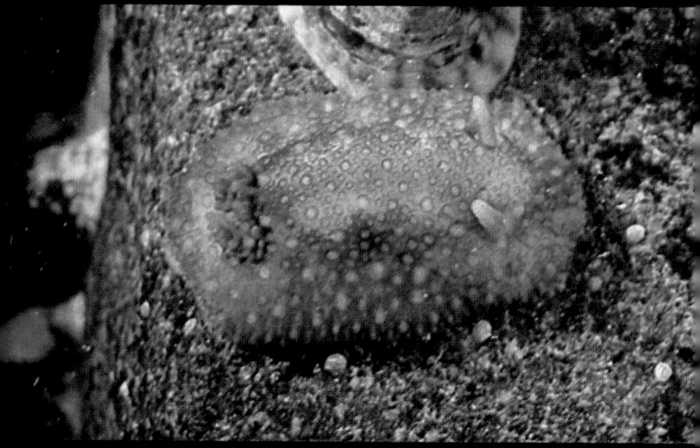

229 Rough-mantled Doris, *l.* 1", *p.* 526

230 Rough-mantled Doris, *l.* 1", *p.* 526

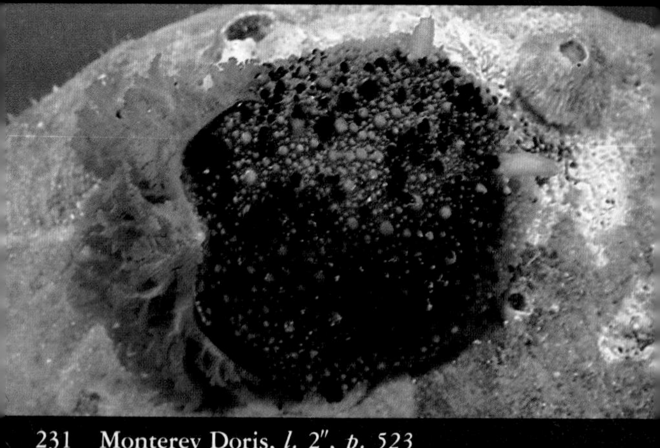

231 Monterey Doris, *l.* 2", *p.* 523

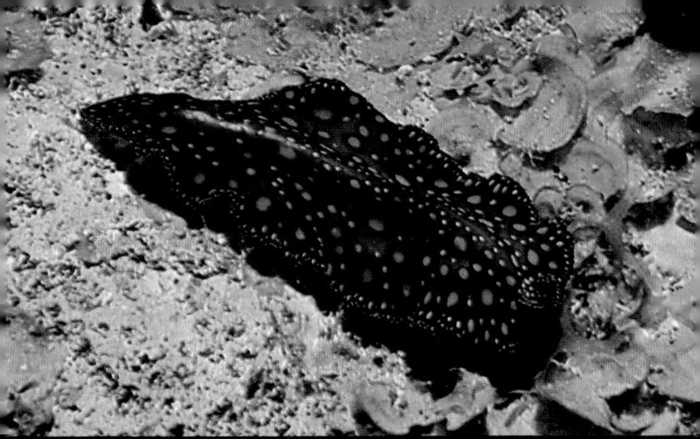

232 Leopard Flatworm, *l.* $1\frac{1}{2}''$, *p. 404*

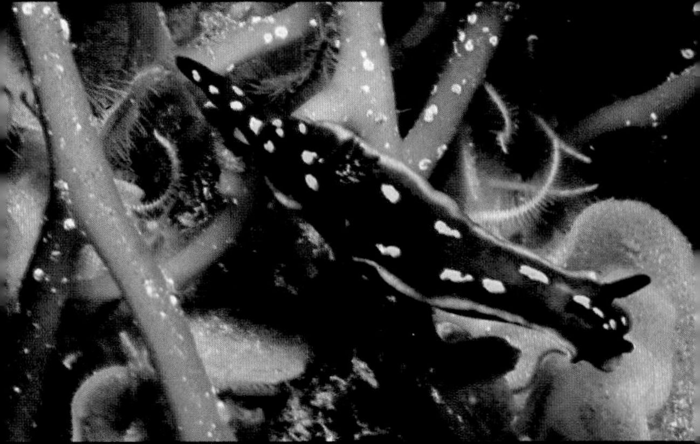

233 Blue-and-gold Nudibranch, *l.* $2\frac{5}{8}''$, *p. 526*

Wormlike Animals

A variety of wormlike animals of different sizes and shapes lives on or in the ocean floor. Some, such as the paddle worms, clam worms, and Sticky-skin Sea Cucumber have protruding legs or spines; others like the nemerteans, Limulus Leech, and Silky Sea Cucumber have smooth, flattened or rounded bodies; and still others, like the scale worms, are covered with protective plates. Fire worms can inflict a painful sting and should not be handled.

235 Agassiz's Sea Cucumber, *l.* 12″, *p. 700*

236 Florida Sea Cucumber, *l.* 10″, *p. 700*

237 Fissured Sea Cucumber, *l.* 18″, *p. 699*

238 Four-sided Sea Cucumber, *l.* 14″, *p.* 698

239 Fifteen-scaled Worm, *l.* 2½″, *p.* 417

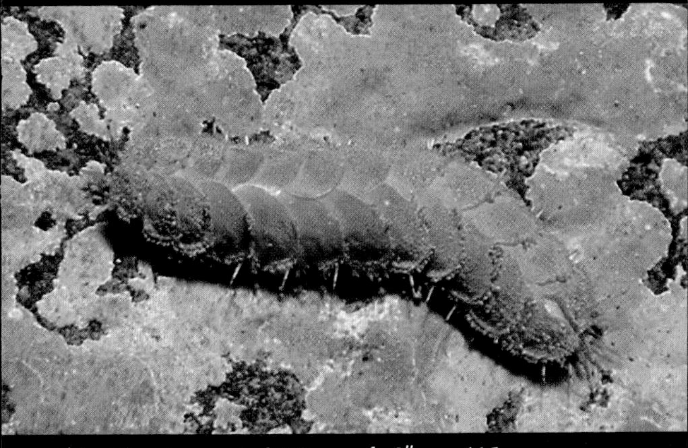

240 Twelve-scaled Worm, *l.* 2″, *p.* 415

241 Green Fire Worm, *l.* 10″, *p.* 424

242 Orange Fire Worm, *l.* 6″, *p.* 424

243 Red-tipped Fire Worm, *l.* 5″, *p.* 425

244 Sticky-skin Sea Cucumber, *l.* 39″, *p.* 706

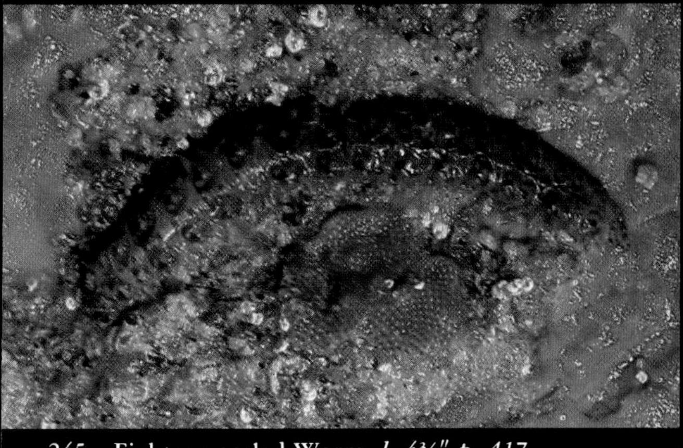

245 Eighteen-scaled Worm, *l.* 4⅜″ *p.* 417

246 Lug Worm, *l.* 12″, *p.* 425

247 Leafy Paddle Worm, *l.* 18", *p. 413*

248 Green Paddle Worm, *l.* 6", *p. 414*

249 Clam Worm, *l.* 36", *p. 420*

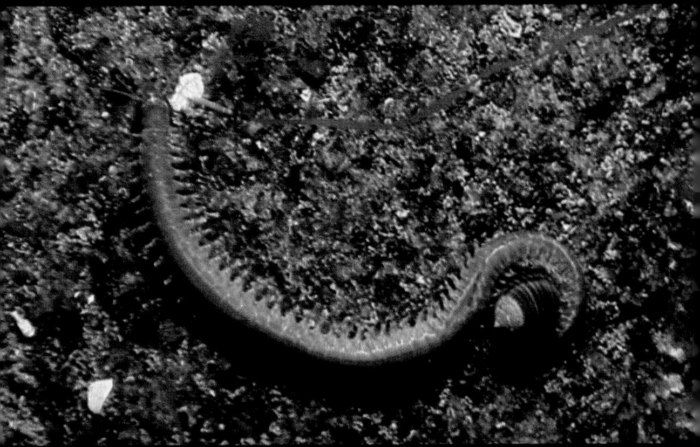

250 Pelagic Clam Worm, *l.* 6⅛", *p. 421*

251 Two-gilled Blood Worm, *l.* 15⅜", *p. 418*

252 Opal Worm, *l.* 24", *p. 423*

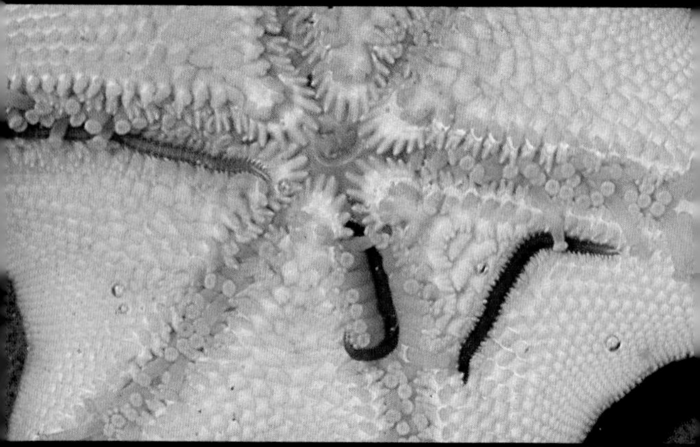

253 Bat Star Worm, *l.* 1½", *p. 420*

254 Six-lined Nemertean, *l.* 8", *p. 406*

255 Chevron Amphiporus, *l.* 6", *p. 409*

256 Chevron Amphiporus, *l.* 6″, *p.* 409

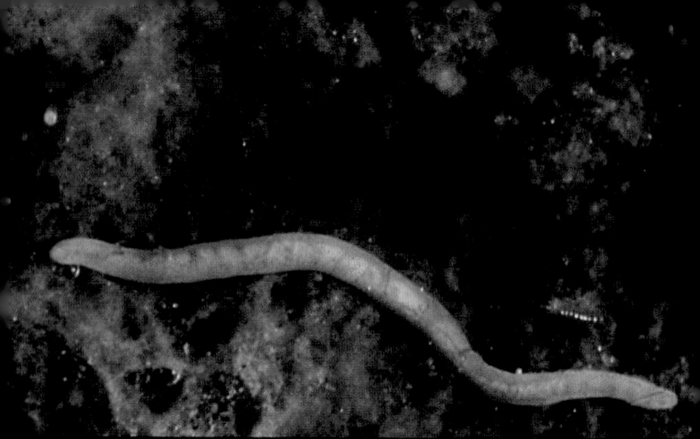

259 Red Lineus, *l.* 8″, *p.* 407

262 Tailed Priapulid Worm, *l.* 3¼″, *p.* 445

263 Gould's Peanut Worm, *l.* 12″, *p.* 447

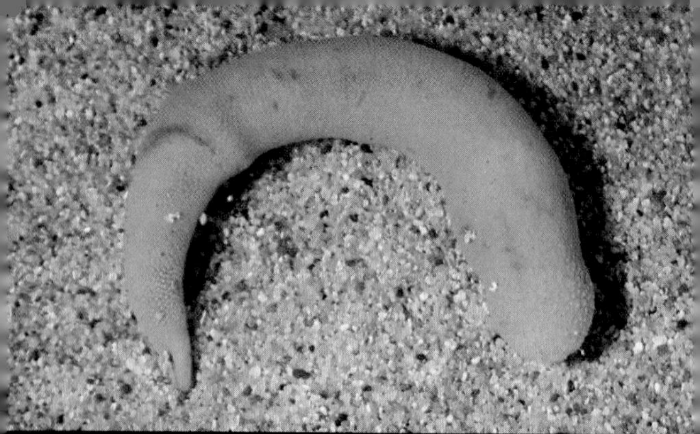

265 Innkeeper Worm, *l.* 7¼", *p.* 452

266 Agassiz's Peanut Worm, *l.* 4¾", *p.* 448

268 Ice Cream Cone Worm, *l.* 1⅝″, *p. 433*

269 Polydora Mud Worm, *l.* 1″, *p. 428*

270 Plumed Worm, *l.* 12″, *p. 422*

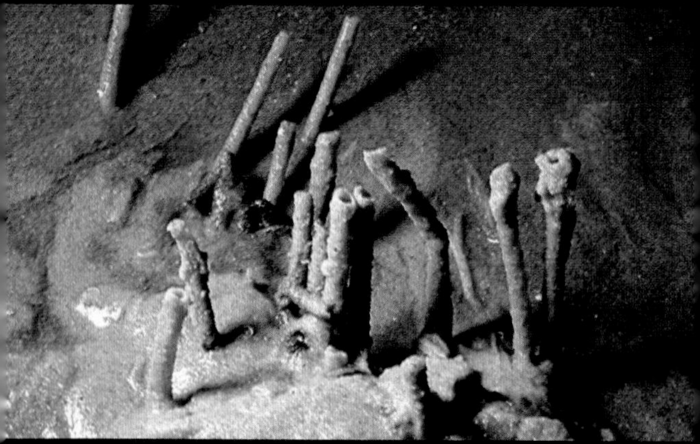

271 Bamboo Worm, *l.* 6", *p.* 426

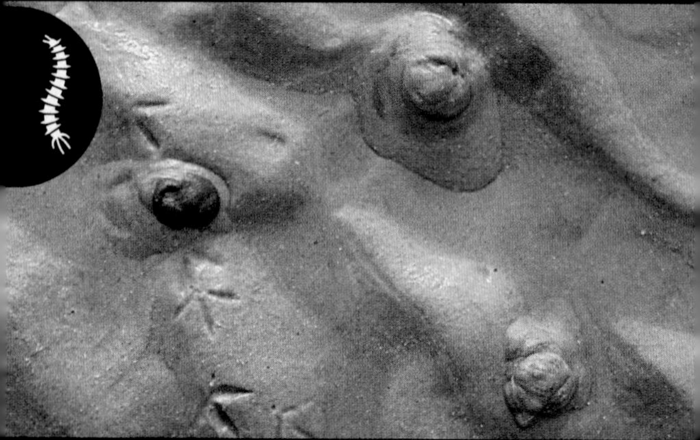

272 Golden Acorn Worm, *l.* 6", *p.* 726

273 Ornate Worm, *l.* 15", *p.* 434

Barnacles

The white barnacles seen at low tide covering rocks and pilings on the seacoast are sedentary crustaceans. They secrete limy shells composed of many interlocking plates, often with a trapdoor opening at the top that can be closed for protection. In the Common Goose Barnacle and Leaf Barnacle the plates are reduced and the body extended on a long stalk.

274 Bay Barnacle, *w.* ½", *p.* 596

275 Ivory Barnacle, *w.* 1", *p.* 594

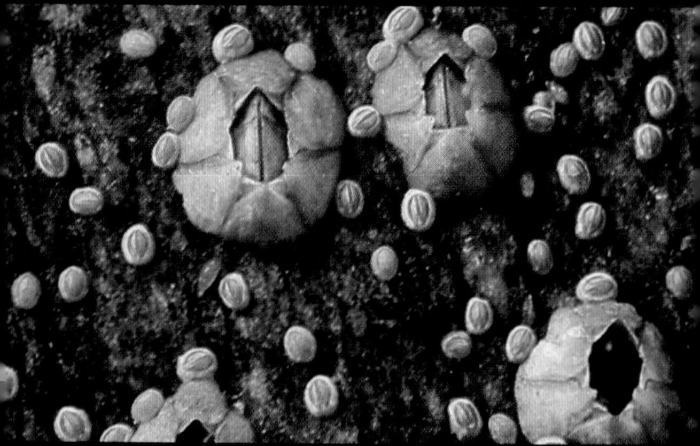

276 Little Gray Barnacle, *w.* ⅜", *p.* 593

277 Leaf Barnacle, *l.* 3¼″, *p.* 592

278 Northern Rock Barnacle, *h.* 1″, *p.* 593

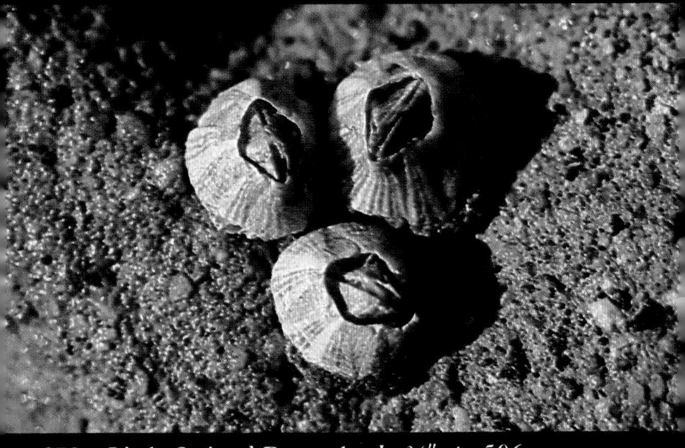

279 Little Striped Barnacle, *h.* ¾″, *p.* 596

280 Giant Acorn Barnacle, *w.* 4⅜″, *p.* 595

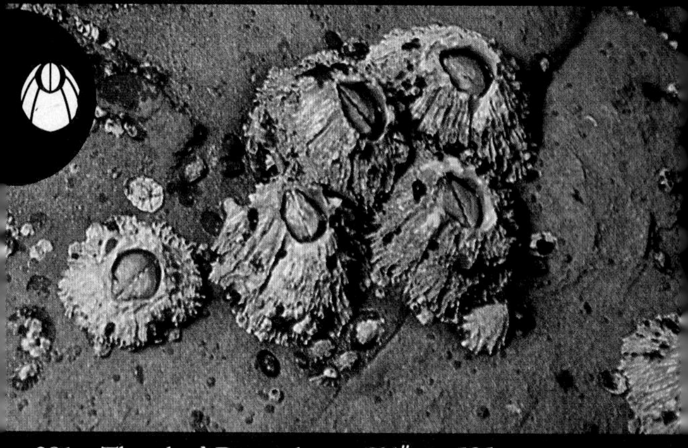

281 Thatched Barnacle, *w.* 2⅜″, *p.* 595

282 Volcano Barnacle, *h.* 2″, *p.* 597

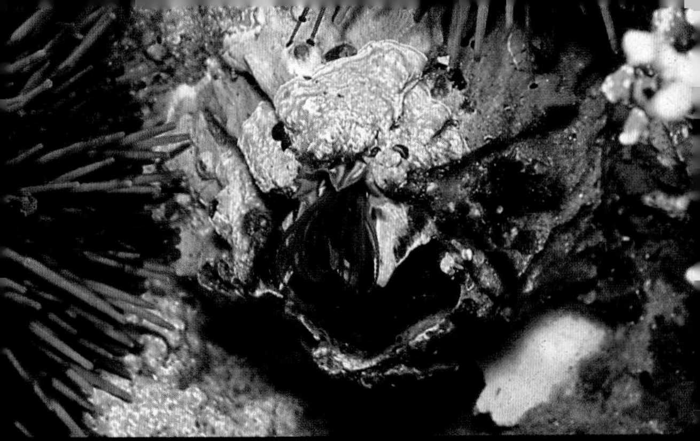

283 Giant Acorn Barnacle, *w.* 4⅜″, *p.* 595

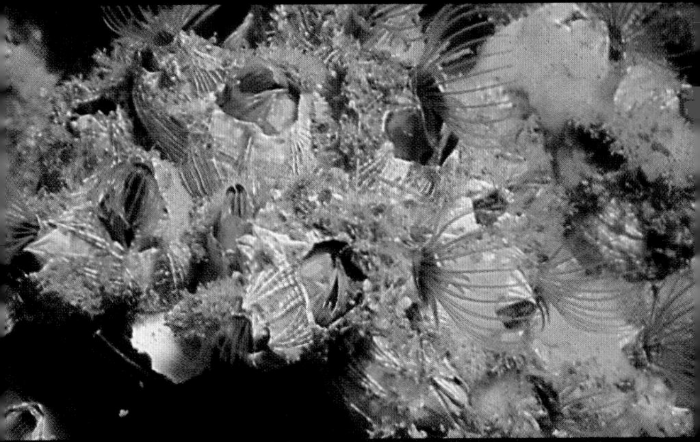

284 Red-striped Acorn Barnacle, *w.* 2⅜″, *p.* 597

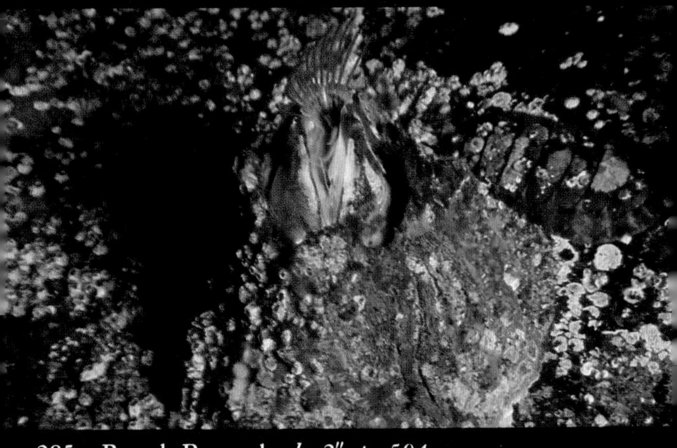

285 Rough Barnacle, *h.* 2″, *p.* 594

286 Northern Rock Barnacle, *h.* 1″, *p.* 593

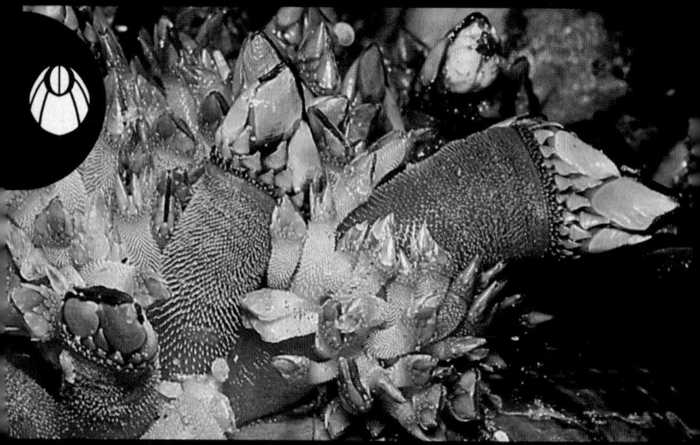

287 Leaf Barnacle, *l.* 3¼″, *p.* 592

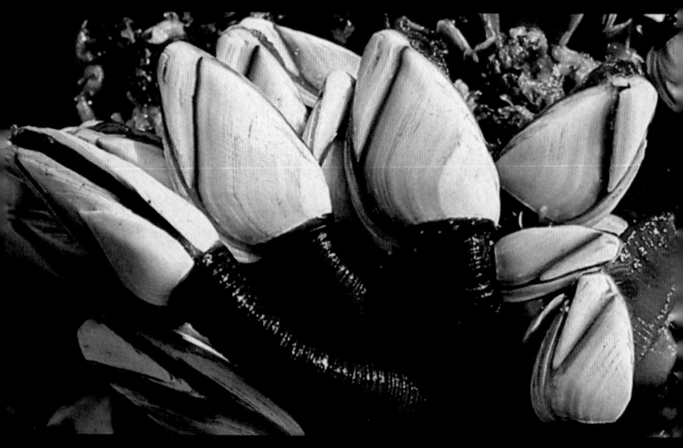

288 Common Goose Barnacle, *l.* 6″, *p.* 591

Oysters, Mussels, and Clams

In this group are many of those bivalve mollusks well-known to both beachcombers and seafood fanciers: oysters, mussels, razor clams, and quahogs. Also included are some unusual bivalves such as piddocks, with bulging bodies and tiny valves, jingle shells with fragile transparent valves, and destructive shipworms whose valves are modified for burrowing through wood. The Tongue-shell Brachiopod is also found here.

289 Eastern Oyster, *l.* 10″, *p.* 547

290 Giant Pacific Oyster, *l.* 12″, *p.* 548

291 Native Pacific Oyster, *l.* 3½″, *p.* 549

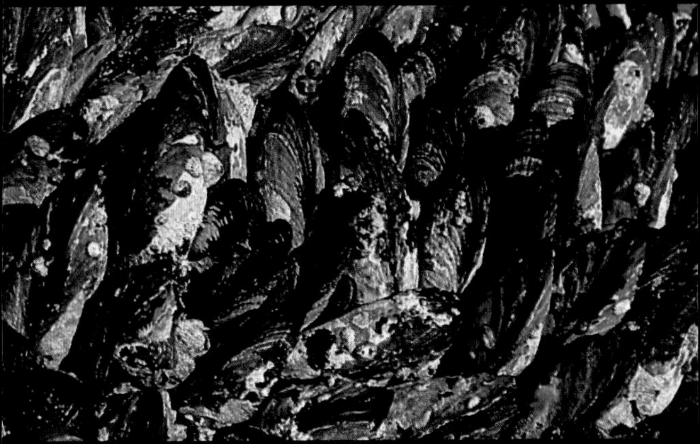

292 California Mussel, *l.* 10″, *p.* 539

293 Blue Mussel, *l.* 4″, *p.* 538

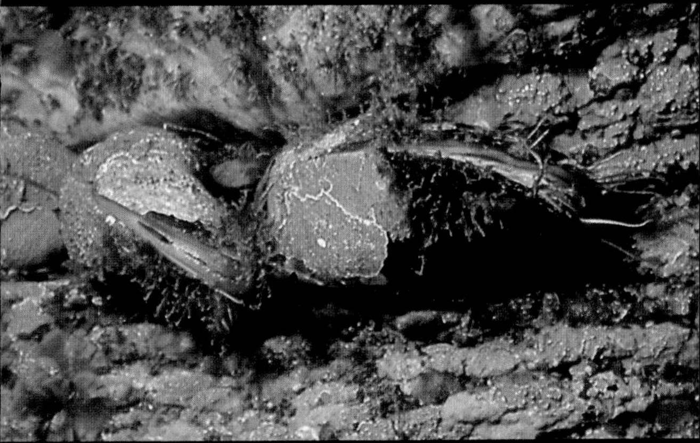

294 Horse Mussel, *l.* 6″, *p.* 537

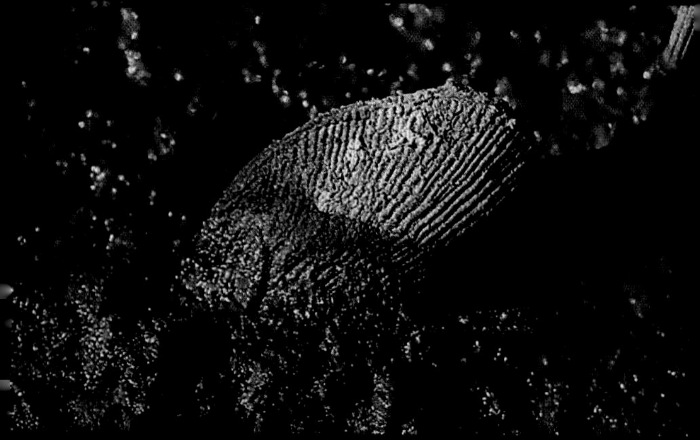

295 Ribbed Mussel, *l.* 4″, *p.* 538

296 Angel Wing, *l.* 8″, *p.* 573

297 False Angel Wing, *l.* 2″, *p.* 562

298 Stiff Pen Shell, *l.* 12″, *p.* 540

299 Saw-toothed Pen Shell, *l.* 10″, *p.* 541

300 File Yoldia, *l.* 2½″, *p.* 534

301 Thin Nut Clam, *l.* ¾″, *p.* 533

302 Sunray Venus, *l.* 5″, *p.* 558

303 Pacific Razor Clam, *l.* 6¾″, *p.* 568

304 Atlantic Razor Clam, *l.* 2½″, *p.* 568

305 Jackknife Clam, *l.* 4″, *p.* 567

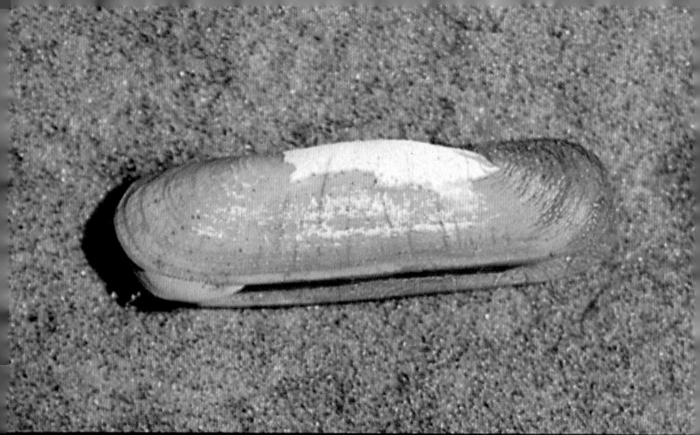

307 California Jackknife Clam, *l.* 4″, *p.* 567

308 Common Razor Clam, *l.* 10″, *p.* 569

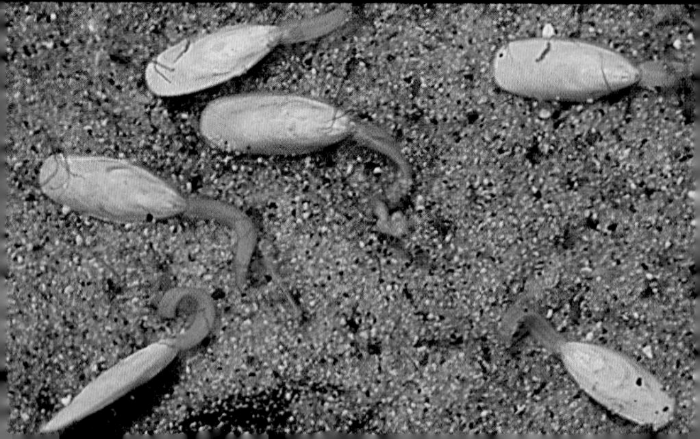

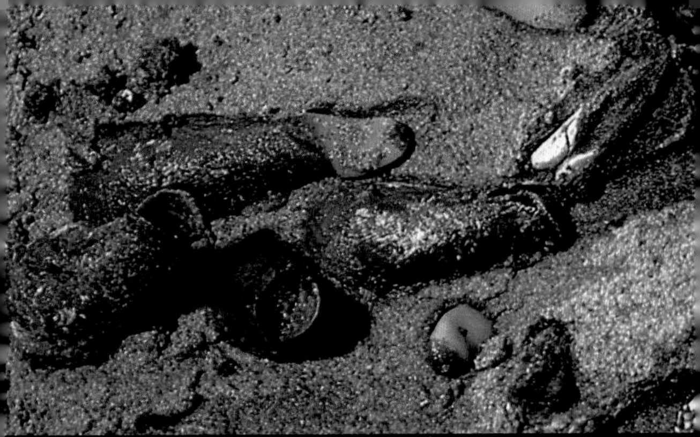

310 Flat-tipped Piddock, *l.* 3″, *p.* 574

311 Common Shipworm, *l.* 12″, *p.* 576

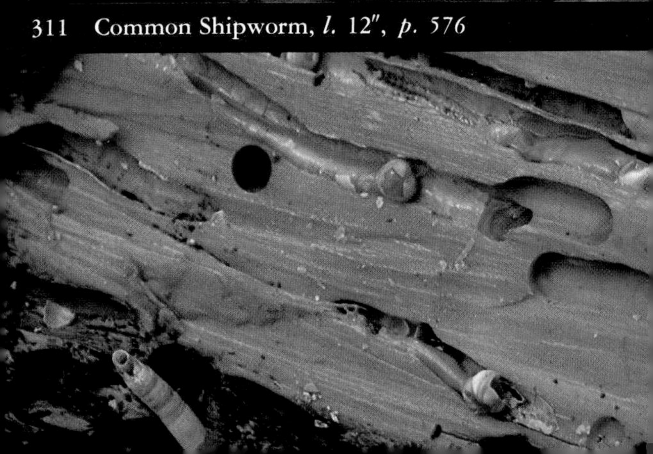

313 Striated Wood Piddock, *l.* 2″, *p.* 573

314 Great Piddock, *l.* 3″, *p.* 574

316 Soft-shelled Clam, *l.* 6″, *p.* 572

317 White Sand Macoma, *l.* 3⅝″, *p.* 566

318 Gaper Clam, *l.* 8″, *p.* 570

319 Surf Clam, *l.* 7″, *p.* 569

320 Bent-nosed Macoma, *l.* 4⅜″, *p.* 565

321 Coquina, *l.* ¾″, *p.* 566

322 Pismo Clam, *l.* 6″, *p.* 561

323 White-bearded Ark, *l.* 3″, *p.* 535

325 Antillean File Shell, *h.* 2¼", *p.* 546

326 Dwarf Tellin, *l.* ½", *p.* 563

328 Carpenter's Tellin, *l.* ⅜″, *p.* 564

329 Atlantic Nut Clam, *l.* ⅜″, *p.* 533

331 Cross-hatched Lucine, *l.* 1″, *p.* 553

332 Tiger Lucine, *l.* 3¾″, *p.* 553

334 Disk Dosinia, *l.* 3", *p.* 558

335 Common Washington Clam, *l.* 6", *p.* 561

336 Southern Quahog, *l.* 6", *p.* 560

337 Carolina Marsh Clam, *l.* 1½″, *p.* 551

338 Quahog, *l.* 5″, *p.* 559

339 Black Clam, *l.* 5″, *p.* 552

340 Wavy Astarte, *l.* 1¼″, *p. 550*

341 Boreal Astarte, *l.* 2″, *p. 550*

342 Morton's Egg Cockle, *h.* 1⅛″, *p. 557*

343 Baltic Macoma, *l.* 1½″, *p.* 564

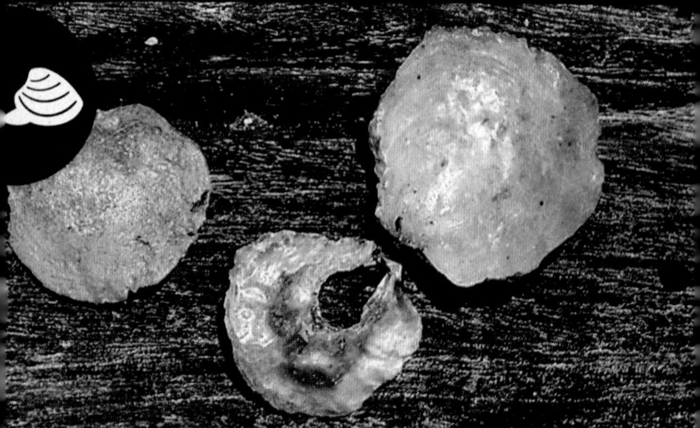

344 False Pacific Jingle Shell, *l.* 4″, *p.* 547

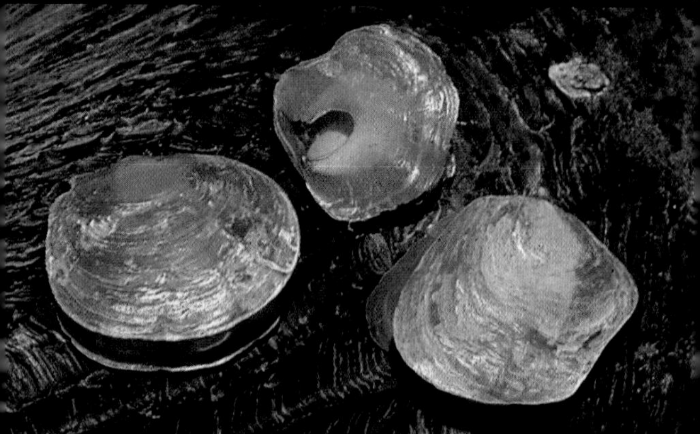

345 Common Jingle Shell, *l.* 2¼″, *p.* 546

Scallops and Cockles

This group includes bivalves with singular shapes: the fan of the scallop, the heart pattern of the cockle, and the spiny silhouette of the jewel box, Atlantic Pearl Oyster and Atlantic Thorny Oyster. Other unusually shaped bivalves and two brachiopods that resemble them are also in this group.

346 Atlantic Pearl Oyster, *l.* 3", *p.* 540

347 Leafy Jewel Box, *l.* 3½", *p.* 554

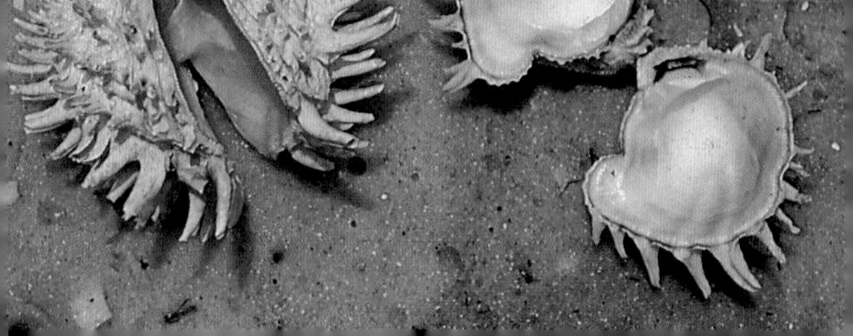

349 Atlantic Thorny Oyster, *l.* 5½", *p.* 545

350 Rough File Shell, *h.* 3¾", *p.* 545

351 Giant Rock Scallop, *l.* 10", *p.* 543

352 Lion's Paw, *l.* 6″, *p.* 544

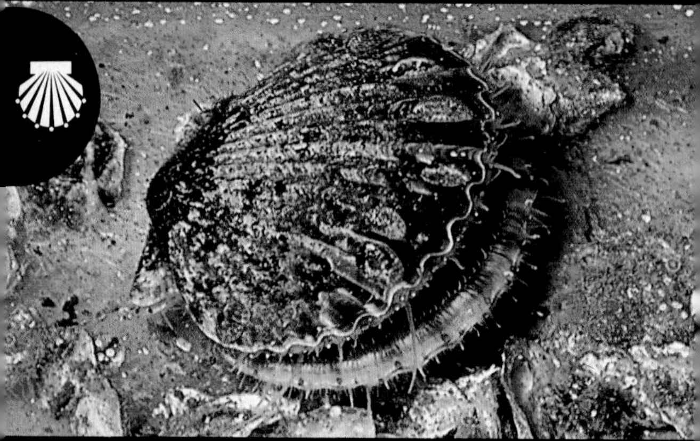

353 Atlantic Bay Scallop, *w.* 3″, *p.* 542

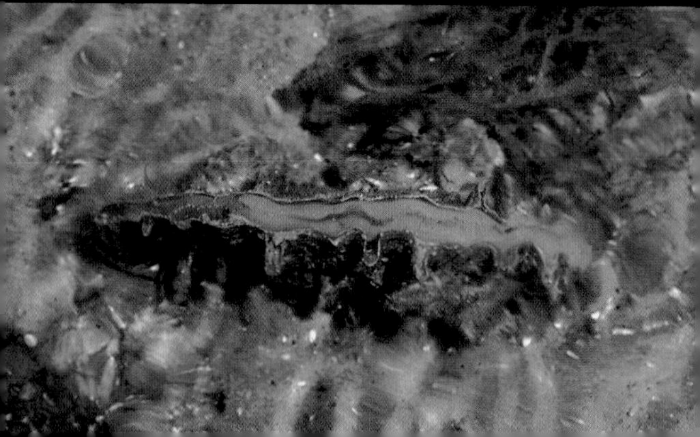

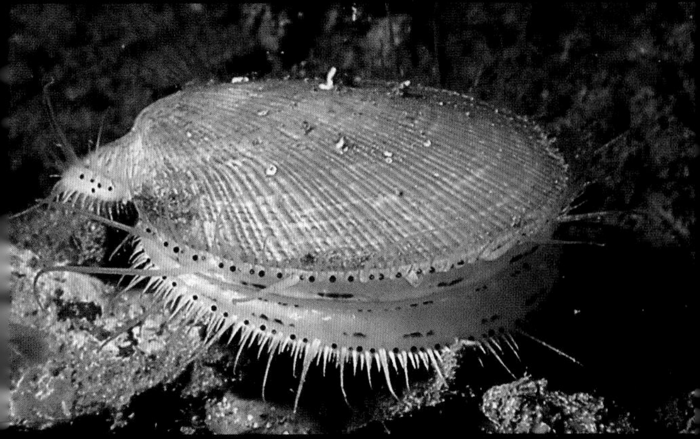

355 Iceland Scallop, *l.* 4″, *p.* 542

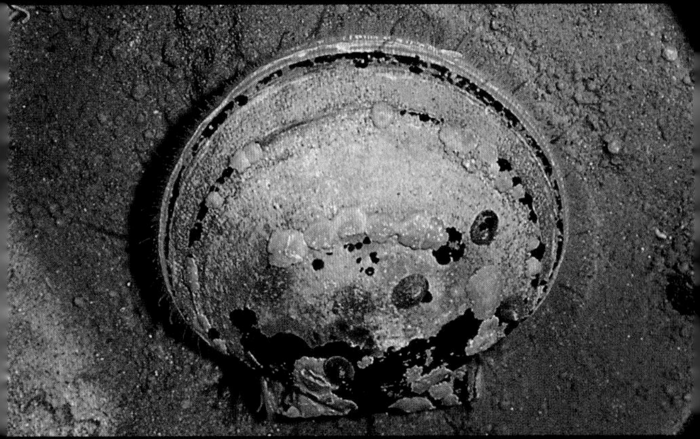

356 Atlantic Deep-sea Scallop, *l.* 8″, *p.* 544

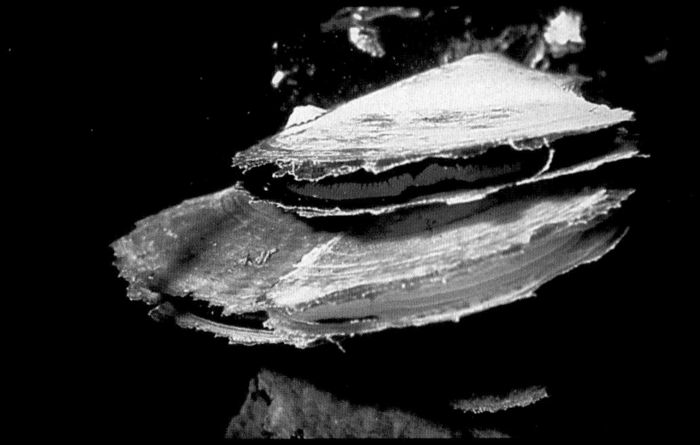

357 Flat Tree Oyster, *l.* 3″, *p.* 539

358 Northern Lampshell, *l.* 1¼", *p.* 723

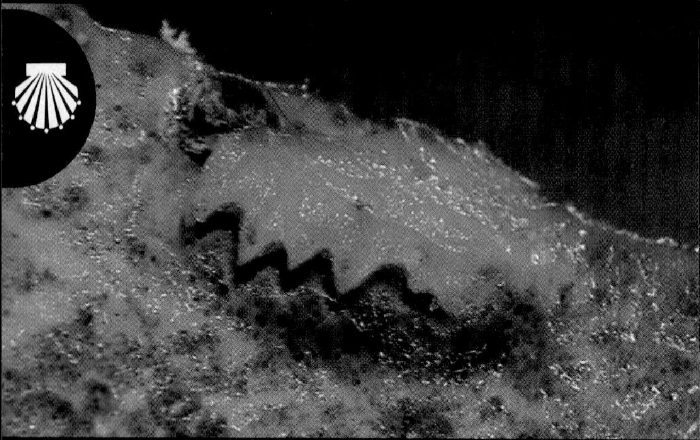

359 Coon Oyster, *l.* 2¾", *p.* 549

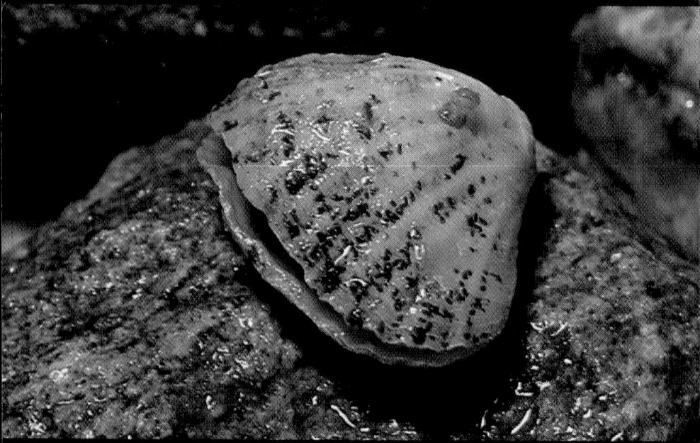

360 Common Pacific Brachiopod, *w.* 2¼", *p.* 723

361 Kitten's Paw, *w.* 1″, *p.* 541

362 Atlantic Strawberry Cockle, *h.* 2″, *p.* 555

363 Yellow Cockle, *h.* 2½″, *p.* 557

364 Nuttall's Cockle, *l.* 6″, *p.* 555

365 Comb Bittersweet, *l.* 1¼″, *p.* 536

366 Giant Atlantic Cockle, *l.* 5¼″, *p.* 556

367 Ponderous Ark, *l.* 2½″, *p.* 536

368 Common Pacific Littleneck Clam, *l.* 2¾″, *p.* 560

369 Blood Ark, *l.* 2¼″, *p.* 535

Chitons

Chitons are flattened, lozenge-shaped mollusks most often found tightly clamped to a rock. The 8 shells running down a chiton's back are held in place by a girdle that completely obscures the shells in some species. Chitons vary widely in color from black, browns, reds, and greens to vivid pinks, aquamarines and violets.

370 Mottled Red Chiton, *l.* 1½", *p. 461*

371 Lined Chiton, *l.* 2", *p. 462*

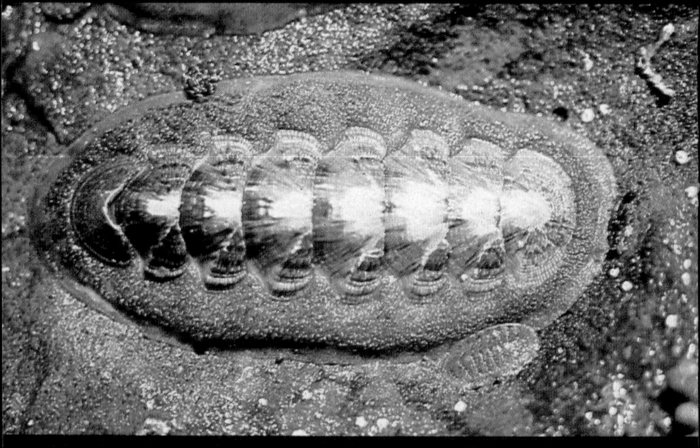

372 California Nuttall's Chiton, *l.* 2", *p. 464*

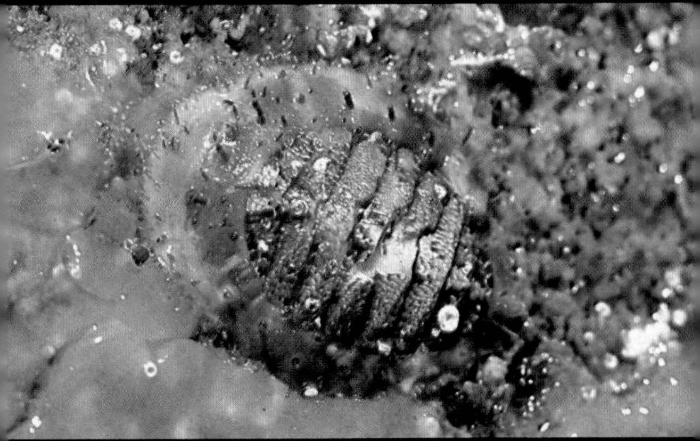

373 Veiled Chiton, *l.* 2″, *p. 467*

374 Mottled Red Chiton, *l.* 1½″, *p. 461*

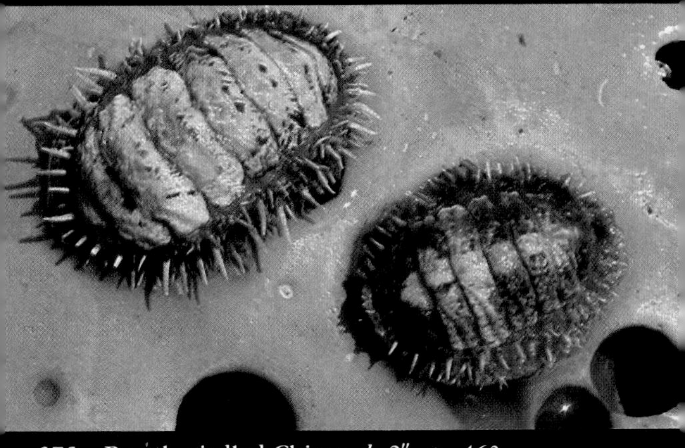

375 Rough-girdled Chiton, *l.* 2″, *p. 463*

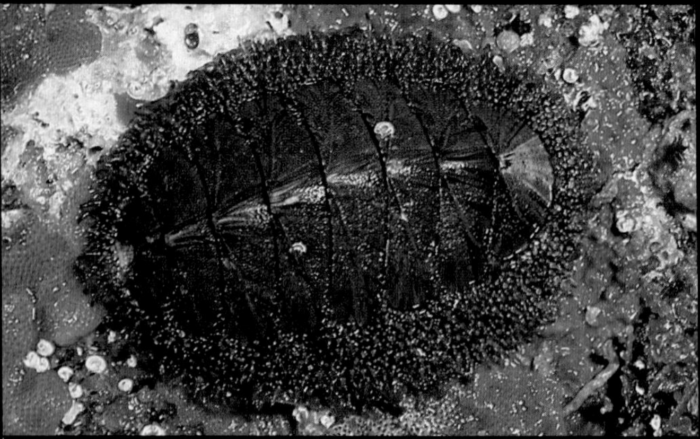

376 Mossy Chiton, *l.* 3⅝"; *p. 466*

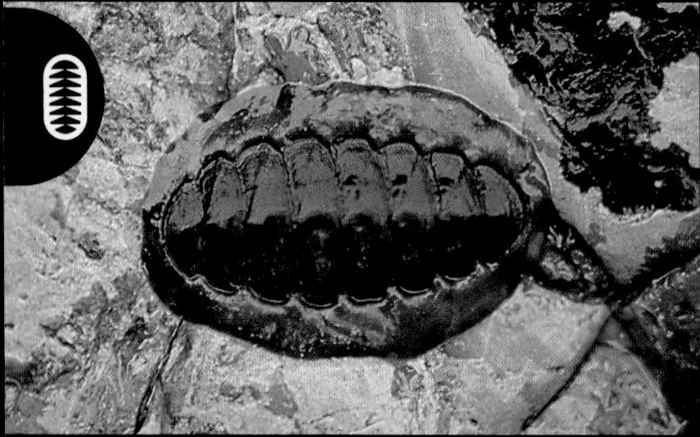

377 Hartweg's Chiton, *l.* 2", *p. 465*

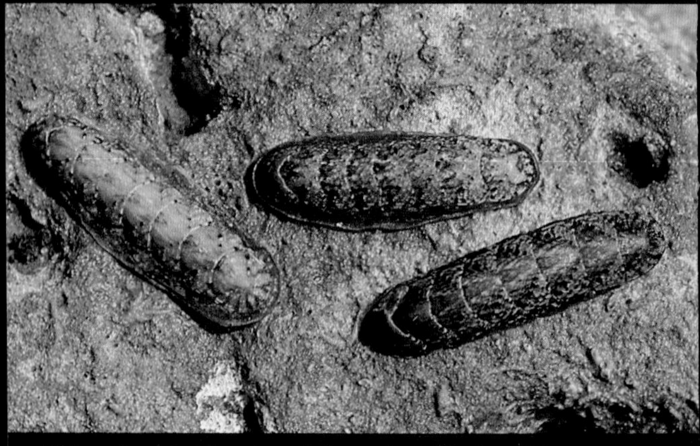

378 Florida Slender Chiton, *l.* 1½", *p. 465*

379 Gum Boot Chiton, *l.* 13″, *p.* 464

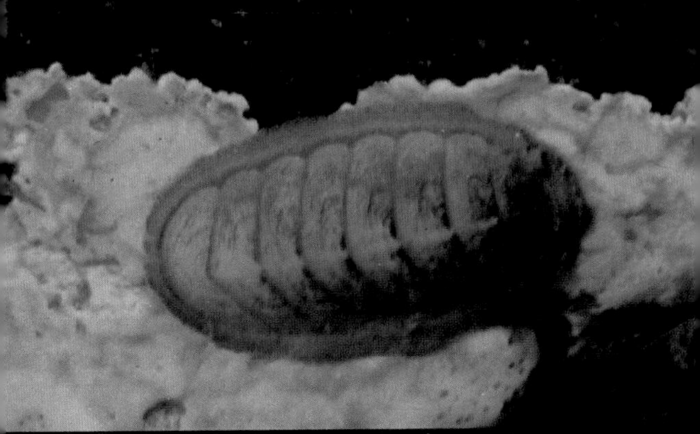

380 White Chiton, *l.* ½″, *p.* 463

Limpets and Abalones

Both these snail species have flattened shells, often pierced with holes that are used by the animal to drain water from its gills. Limpets sometimes have beaded, streaked or ribbed shells; abalones have disk-shaped shells, with the exterior often covered by marine growth. The iridescent interior is coveted by shell collectors.

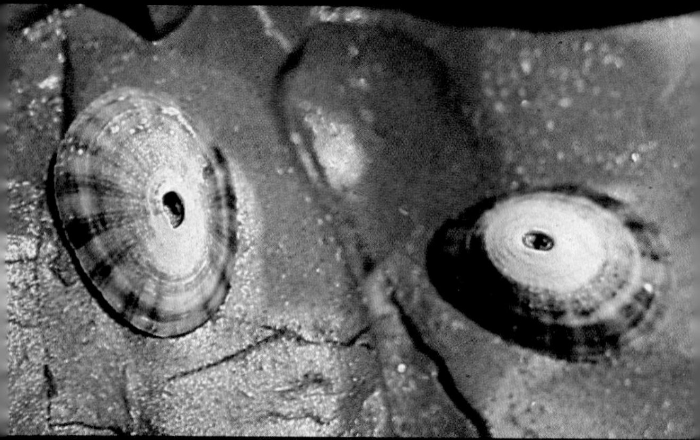

382 Volcano Limpet, *l.* 1⅜″, *p. 470*

383 Tortoise-shell Limpet, *l.* 1½″, *p. 471*

384 Cayenne Keyhole Limpet, *l.* 2″, *p. 469*

385 Plate Limpet, *l.* 2½", *p.* 472

386 Plate Limpet, *l.* 2½", *p.* 472

387 Shield Limpet, *l.* 1⅝", *p.* 473

388 Giant Keyhole Limpet, *l.* 5⅛", *p. 471*

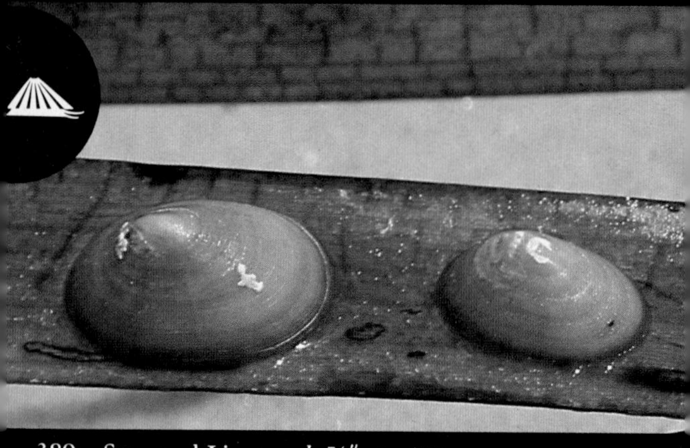

389 Seaweed Limpet, *l.* ⅞", *p. 472*

390 Owl Limpet, *l.* 3½", *p. 473*

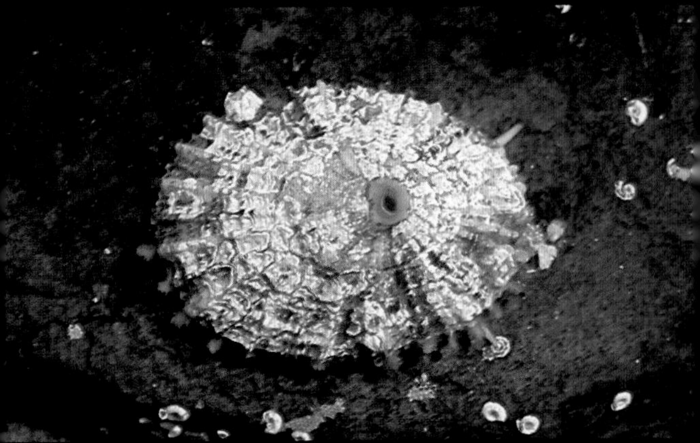

391 Rough Keyhole Limpet, *l.* 2¾″, *p. 469*

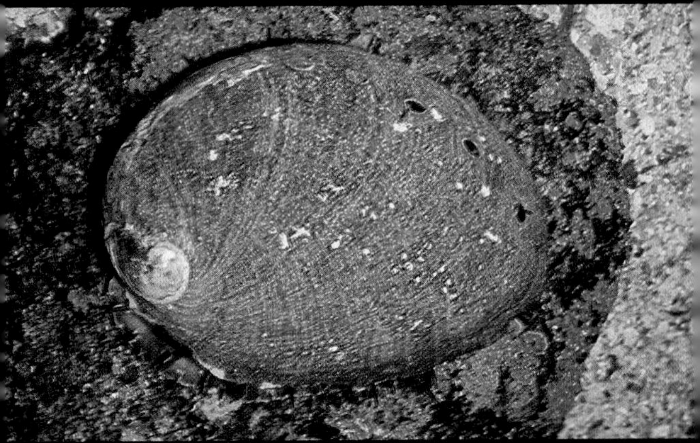

392 Red Abalone, *l.* 12″, *p. 468*

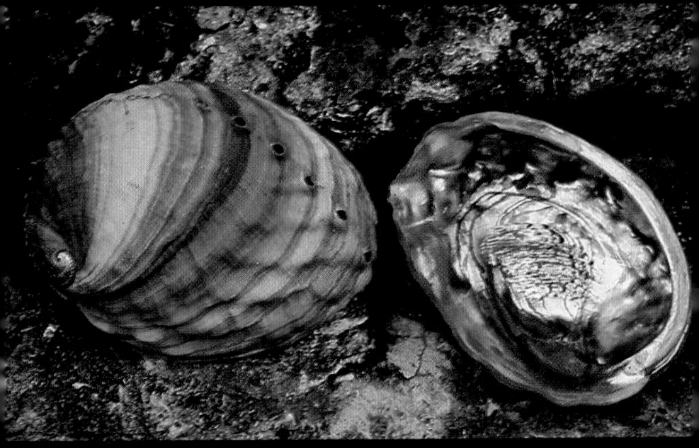

393 Red Abalone, *l.* 12″, *p. 468*

Whelks, Conchs, and Augers

Many of these gastropods have shells prized by collectors for their shape and design: the long, narrow spires of wentletraps, augers and miter shells with their intricate sculpture; the thickened body whorls of whelks, rock snails and the Emarginate Dogwinkle tapering to a high spire; the spindle shapes of tulip snails, other whelks and the Florida Horse Conch; the massive and showy sculpture of some of the larger whelks, conchs and murexes. Also in this group are the distinctive cone shells.

394 Greenland Wentletrap, *l.* 2″, *p.* 487

395 Angulate Wentletrap, *l.* 1″, *p.* 486

396 Costate Horn Snail, *l.* ½″, *p.* 483

397 Concave Auger, *l.* 1″, *p.* 515

398 Common Atlantic Auger, *l.* 2″, *p.* 515

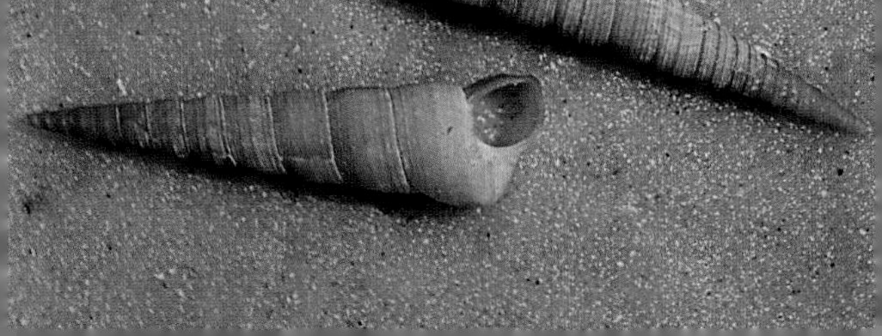

400 Ida's Miter, *l.* 3¼", *p. 511*

401 Black Horn Snail, *l.* ¾", *p. 483*

402 Florida Cerith, *l.* 1½", *p. 484*

403 Beaded Miter, *l.* 1″, *p. 510*

404 ♀ Oyster Turret, *l.* 1″, *p. 516*

405 Alternate Bittium, *l.* ¼″, *p. 485*

406 Greedy Dove Snail, *l.* ½", *p. 501*

407 Atlantic Oyster Drill, *l.* 1¼", *p. 499*

408 Waved Whelk, *l.* 4", *p. 502*

409 Spotted Thorn Drupe, *l.* 1⅝", *p. 495*

410 Poulson's Rock Snail, *l.* 2⅜", *p. 498*

411 Giant Western Nassa, *l.* 2", *p. 506*

412 New England Dog Whelk, *l.* ¾", *p.* 506

413 Mottled Dog Whelk, *l.* ½", *p.* 507

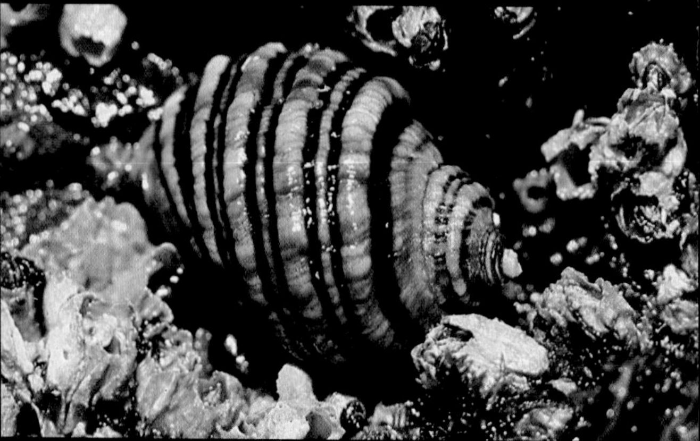

414 Emarginate Dogwinkle, *l.* 1⅝", *p.* 497

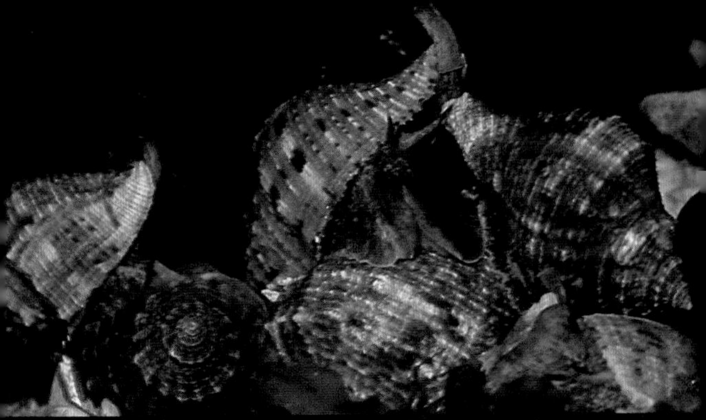

415 Rock Snail, *l.* 5″, *p.* 498

416 Common Nutmeg, *l.* 1¾″, *p.* 512

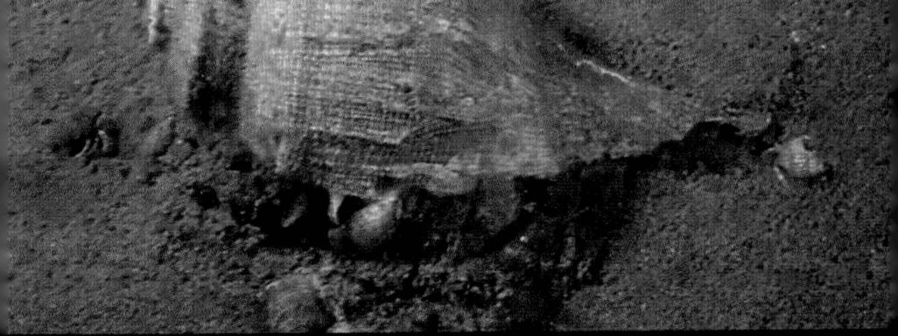

417 Channeled Whelk, *l.* 7″, *p.* 503

418 Oregon Hairy Triton, *l.* 5″, *p. 494*

419 Atlantic Hairy Triton, *l.* 5½″, *p. 493*

420 Corded Neptune, *l.* 5″, *p. 503*

421 True Tulip Snail, *l.* 10″, *p.* 508

422 Banded Tulip Snail, *l.* 3″, *p.* 508

424 Stimpson's Whelk, *l.* 4″, *p.* 502

425 Corded Neptune, *l.* 5″, *p.* 503

426 Florida Horse Conch, *l.* 24″, *p.* 507

427 Lightning Whelk, *l.* 16″, *p.* 504

428 Lightning Whelk, *l.* 16″, *p.* 504

430 Alphabet Cone, *l.* 3″, *p. 513*

431 Mouse Cone, *l.* 1½″, *p. 514*

432 Stearns' Cone, *l.* 1″, *p. 514*

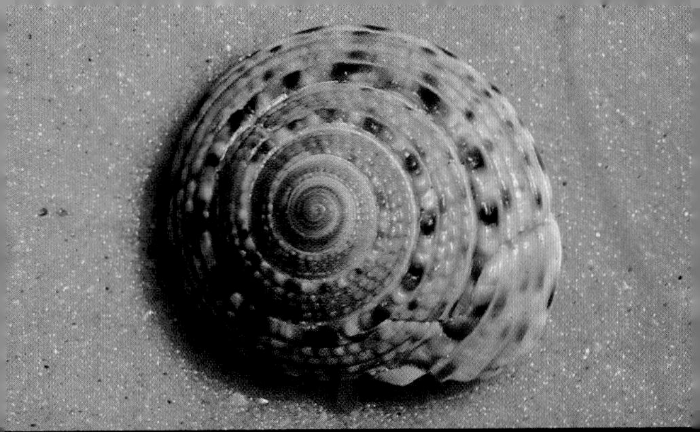

433 Common Sundial, *w.* 2″, *p. 482*

434 Emperor Helmet, *l.* 9″, *p. 492*

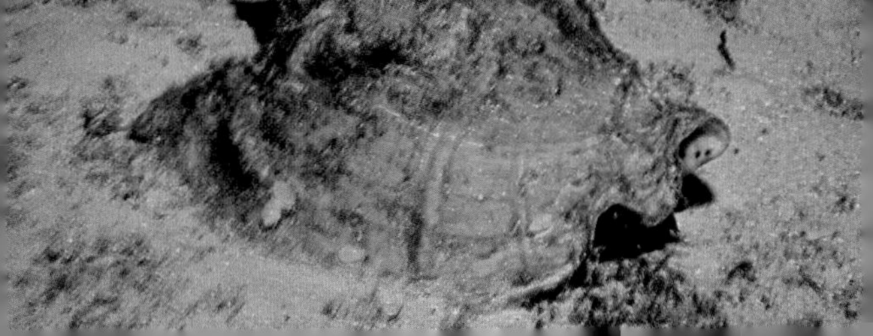

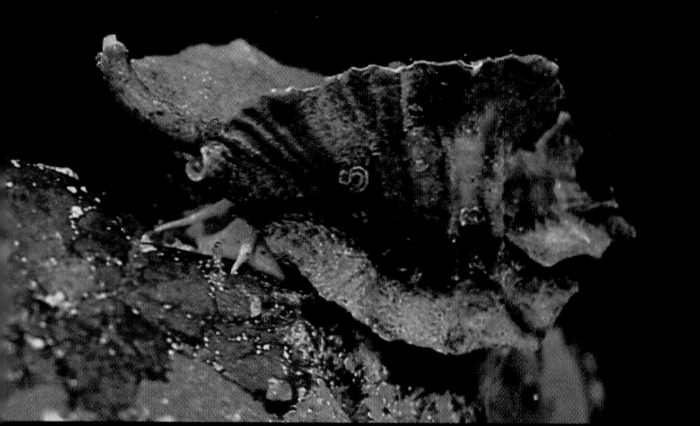

436 Leafy Hornmouth, *l.* 4″, *p.* 495

437 Apple Murex, *l.* 4½″, *p.* 496

438 Lace Murex, *l.* 3″, *p.* 496

Olives and Doves

These gastropods are distinguished by oblong-to-squat shells with rounded or tapered ends. Those of the olives, Salt-marsh Snail, Common Marginella, Common West Indian Bubble, and Flamingo Tongue are smooth, rounded oblongs patterned in a variety of colors. The Four-spotted Trivia's shell has numerous fine riblets, and those of the snails, Unicorn, Mud Dog Whelk, Scotch Bonnet and Atlantic Dogwinkle, often subtly colored, taper sharply to pointed ends, and bear fine sculpture.

439 Common Marginella, *l.* ½", *p.* 512

440 Netted Olive, *l.* 1½", *p.* 509

441 Lettered Olive, *l.* 2½", *p.* 509

442 Lettered Olive, *l.* 2½″, *p.* 509

443 Common West Indian Bubble, *l.* 1″, *p.* 516

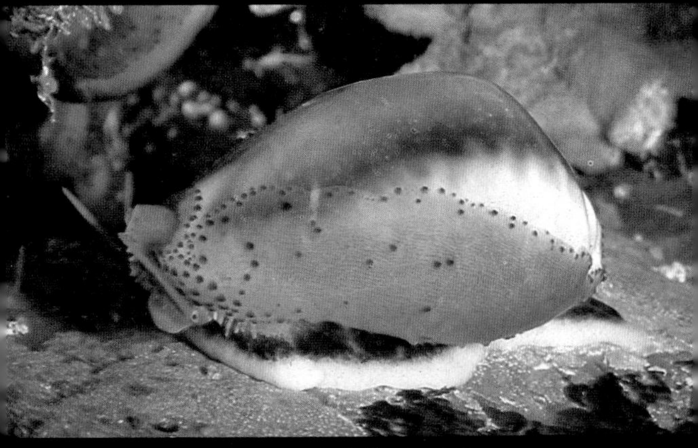

444 Chestnut Cowrie, *l.* 2⅝″, *p.* 489

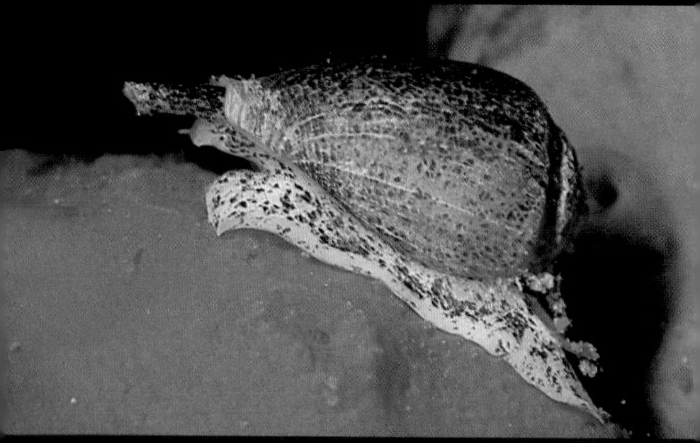

445 California Cone, *l.* 1⅝″, *p.* 513

446 Salt-marsh Snail, *l.* ½″, *p.* 532

447 Purple Dwarf Olive, *l.* 1¼″, *p.* 510

448 Salt-marsh Snail, *l.* ½", *p. 532*

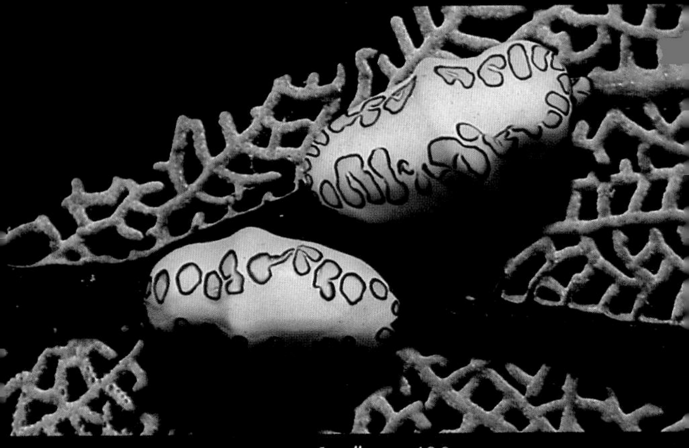

449 Flamingo Tongue, *l.* 1", *p. 490*

451 Unicorn, *l.* 1⅝", *p. 494*

452 Mud Dog Whelk, *l.* 1", *p. 505*

453 Lunar Dove Snail, *l.* ¼", *p. 501*

454 Scotch Bonnet, *l.* 4″, *p. 493*

455 Mottled Dove Snail, *l.* ¾″, *p. 500*

456 Atlantic Dogwinkle, *l.* 2″, *p. 499*

Periwinkles and Other Snails

Snails are known for the spiral shell into which they can withdraw when not moving about on a broad, muscular foot. The shape of a snail's shell ranges from high and conical in the top snails to low and compressed in the Common Slipper Snail. Many snail shells including those of the Northern Moon Snail, Common Baby's Ear and Shark Eye, are rounded with low spires. Still others have higher or pointed spires including those of some periwinkles and the Chink Snail. Peculiar to this group are the Common Purple Sea Snail, which hangs suspended from the water's surface by a bubble net, and the worm snails with their elongated spiral tubes. Also in this group is the Sinistral Spiral Tube Worm.

457 Purple-ringed Top Snail, *l.* 1½", *p.* 474

458 Pearly Top Snail, *l.* ½", *p.* 475

459 Red Top Snail, *w.* 3", *p.* 476

460 Black Turban Snail, *w.* 1¼″, *p. 476*

461 Northern Moon Snail, *l.* 4½″, *p. 491*

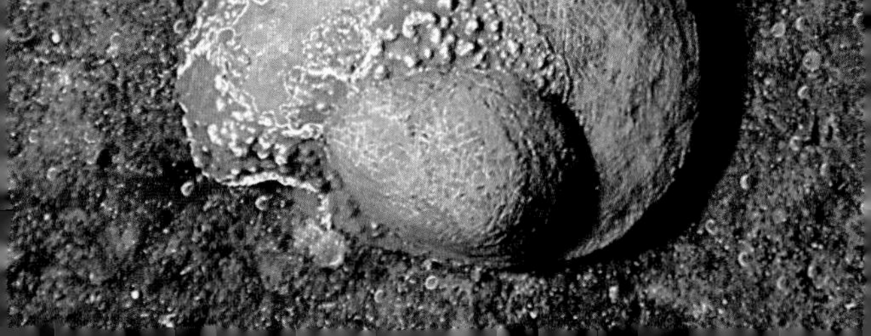

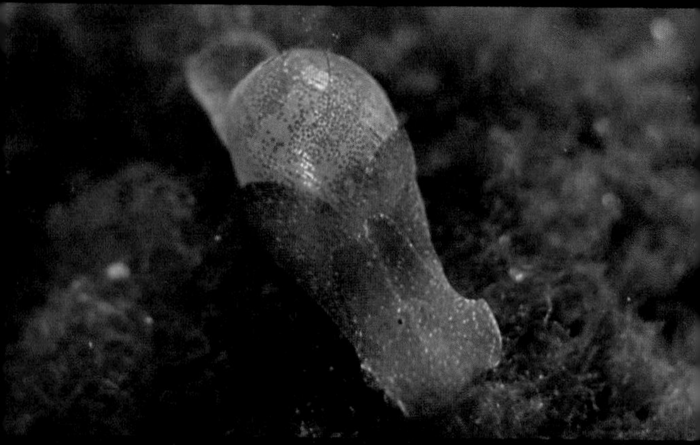

463 White Paper Bubble, *l.* 5/8", *p. 517*

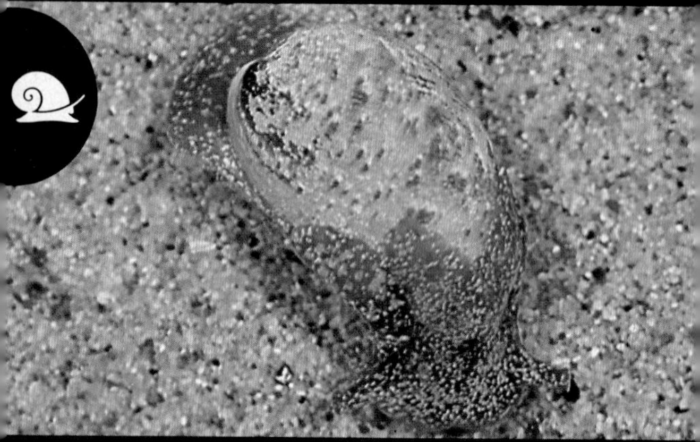

464 California Bubble, *l.* 2 1/4", *p. 517*

465 Common Baby's Ear, *l.* 2", *p. 492*

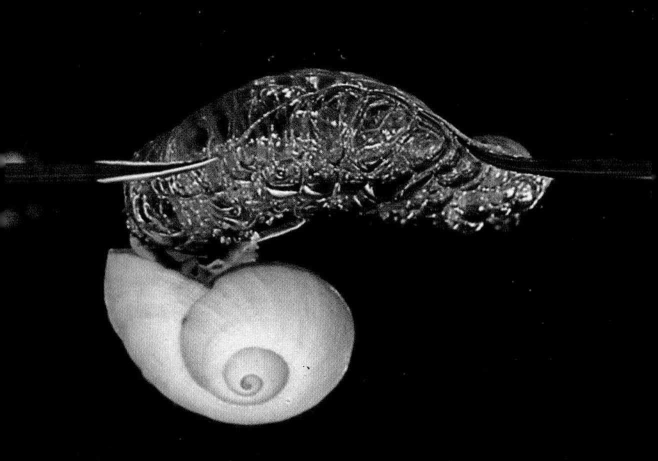

466 Common Purple Sea Snail, *w.* 1½″, *p. 486*

467 Shark Eye, *w.* 3¾″, *p. 490*

468 Greenland Top Snail, *w.* ½″, *p. 475*

469 Northern Yellow Periwinkle, *l.* ½", *p. 479*

470 Rough Periwinkle, *l.* ⅝", *p. 480*

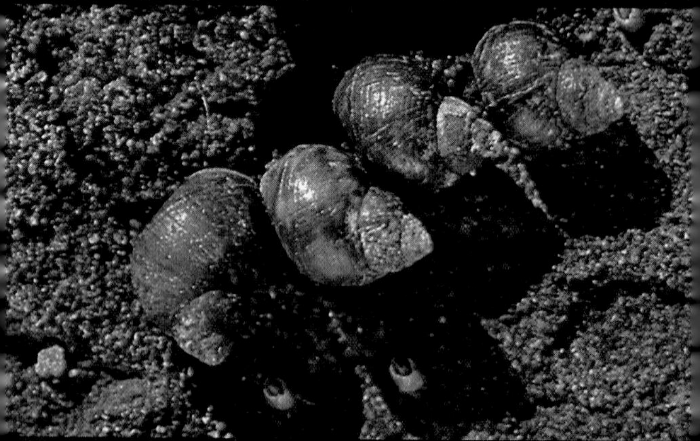

471 Checkered Periwinkle, *l.* ½", *p. 480*

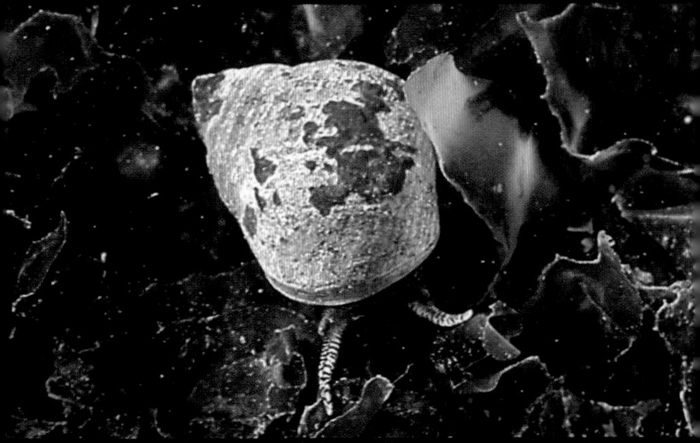

472 Common Periwinkle, *l.* 1", *p.* 478

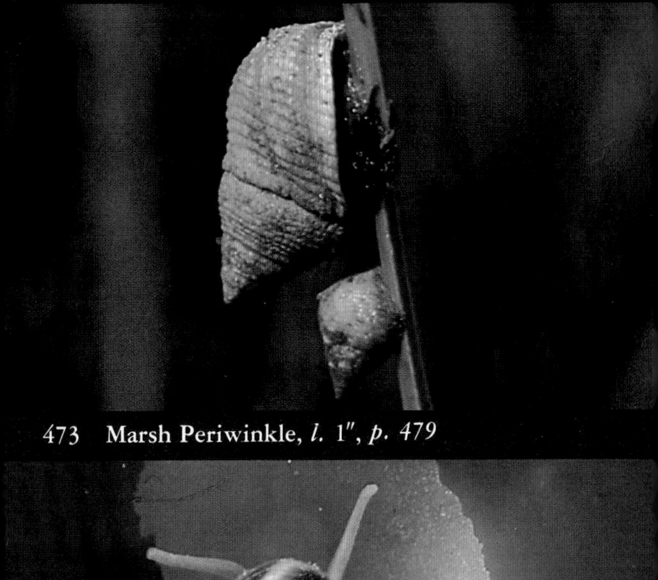

473 Marsh Periwinkle, *l.* 1", *p.* 479

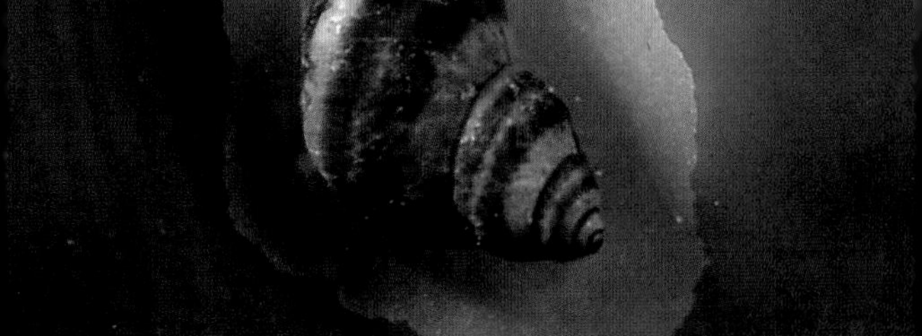

474 Chink Snail, *l.* ⅜", *p.* 477

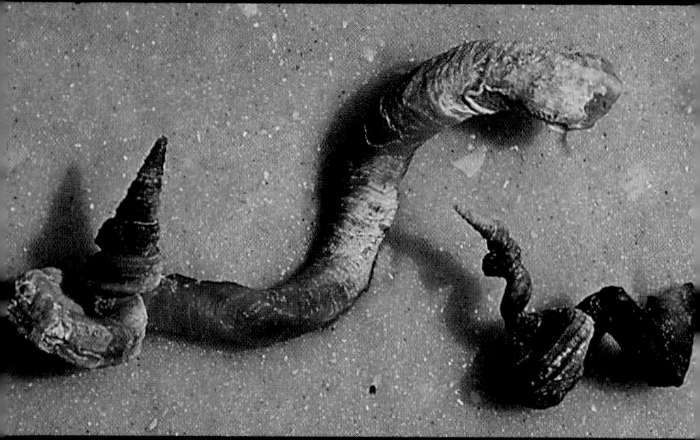

475 Common Worm Snail, *l.* 6″, *p.* *482*

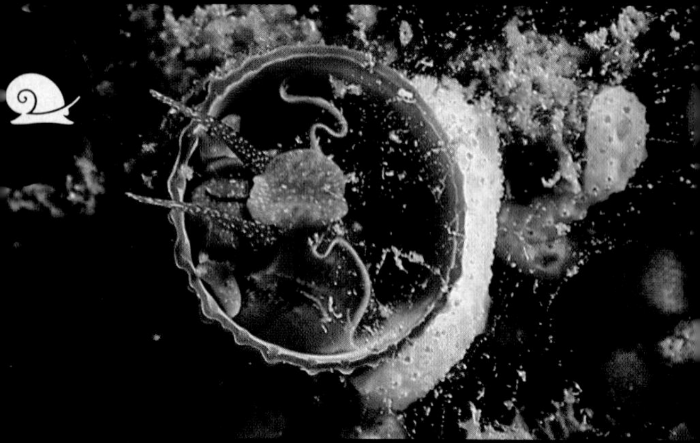

476 Scaled Worm Snail, *l.* 5″, *p.* *481*

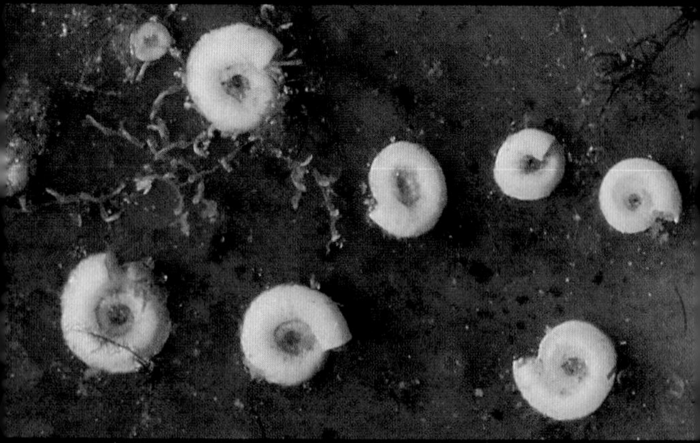

477 Sinistral Spiral Tube Worm, *l.* ⅛″, *p.* *442*

Octopods and Squids

The sinister reputation from which these mollusks suffer is generally undeserved, although the octopus beak can inflict a nasty bite. Octopods are basically secretive, hiding in crevices by day and foraging at night. Like squids, they have a round body with sucker-studded arms surrounding the mouth, and they move by jet propulsion, forcing water out through a siphon beneath the neck. Like squids, too, they can change their color in an instant.

478 Two-spotted Octopus, *l.* 30″, *p.* 581

479 Two-spotted Octopus, *l.* 30″, *p.* 581

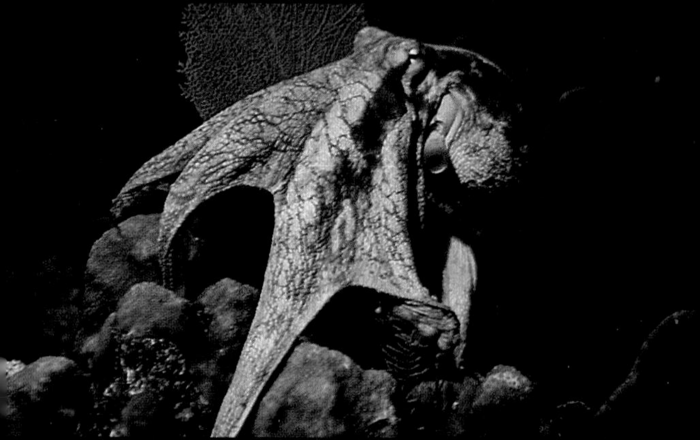

480 Common Atlantic Octopus, *l.* 120″, *p.* 579

481 Joubin's Octopus, *l.* 7″, *p. 580*

482 Long-armed Octopus, *l.* 36″, *p. 579*

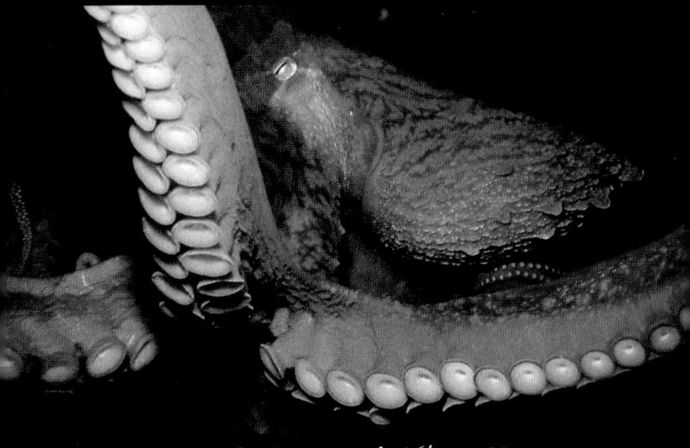

483 Giant Pacific Octopus, *l.* 16′, *p. 580*

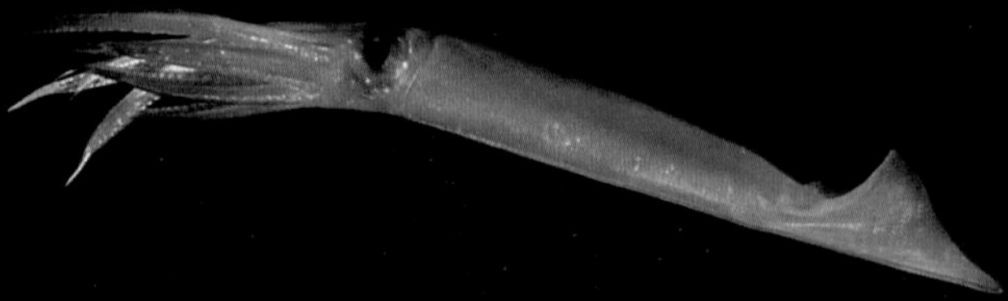

484 Short-fin Squid, *l.* 12″, *p.* 578

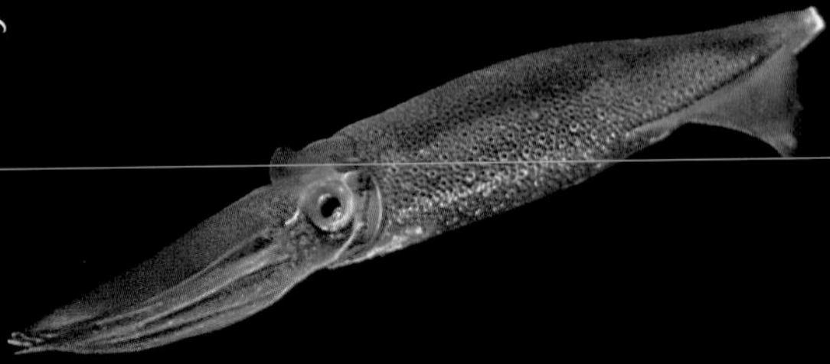

485 Opalescent Squid, *l.* 7⅝″, *p.* 577

486 Atlantic Long-fin Squid, *l.* 17″, *p.* 576

Jellylike Animals

Subject to the vagaries of tides, currents and winds, these diverse animals, pelagic for at least part of their lives, are generally either transparent or translucent. Salps and the Common Doliolid are barrel-shaped planktonic tunicates that are capable of swimming. Comb jellies are globular or compressed, with rows of cilia arranged along the long axis. Cnidarians in this group include hydromedusae, jellyfish and the Chain Siphonophore. Some, including the Sea Nettle and Portuguese man-of-war, can sting fiercely, even when beached.

487 Chain Siphonophore, *h.* 3″, *p.* 356

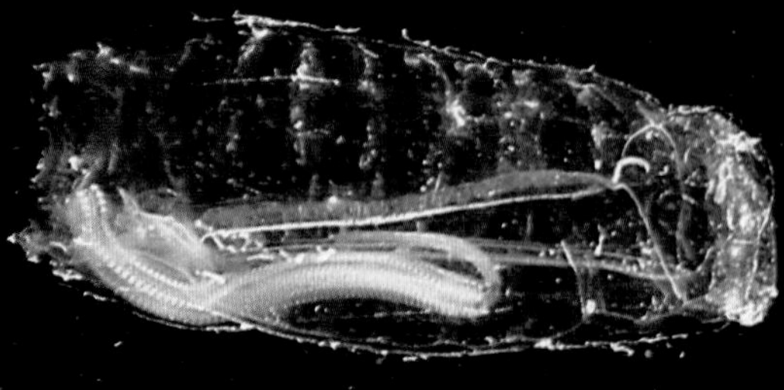

488 Common Salp, *l.* 3¼″, *p.* 744

489 Common Doliolid, *l.* ⅝″, *p.* 743

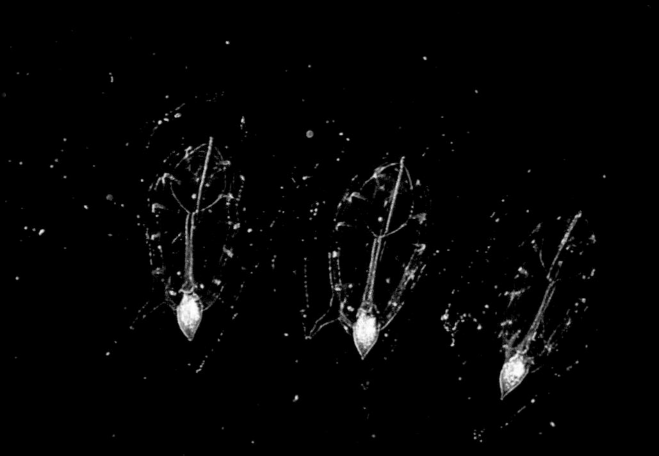

490 Horned Salp, *l.* 1″, *p.* 744

491 Common Northern Comb Jelly, *h.* 6″, *p.* 397

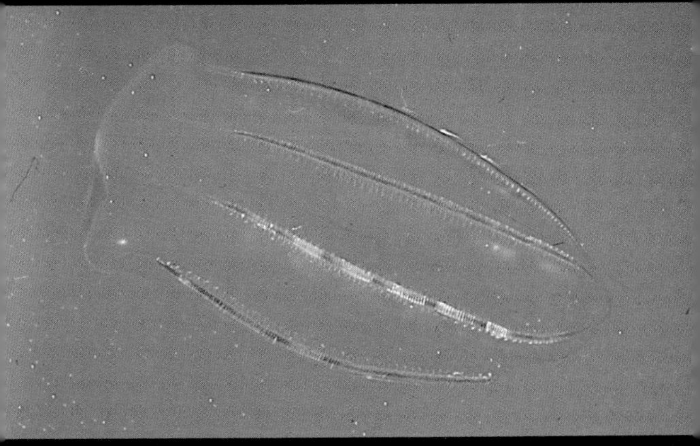

492 Beroë's Comb Jelly, *h.* 4½″, *p.* 398

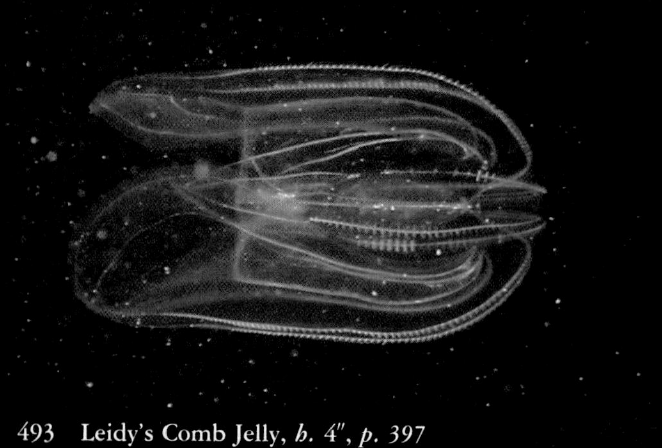

493 Leidy's Comb Jelly, *h.* 4″, *p.* 397

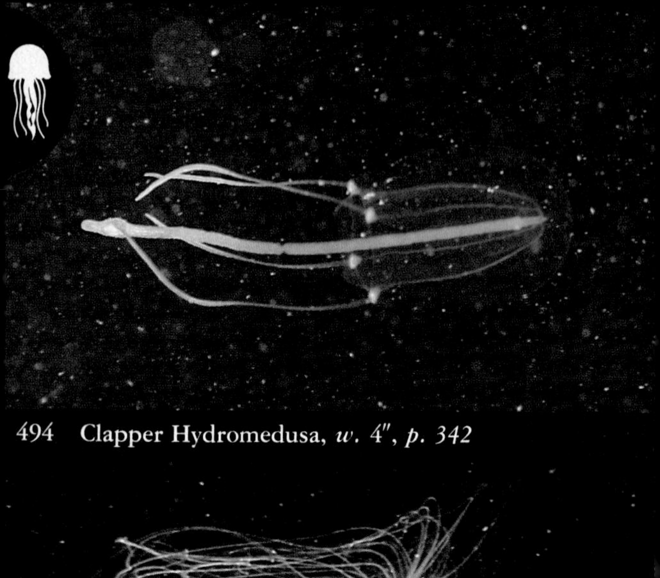

494 Clapper Hydromedusa, *w.* 4″, *p.* 342

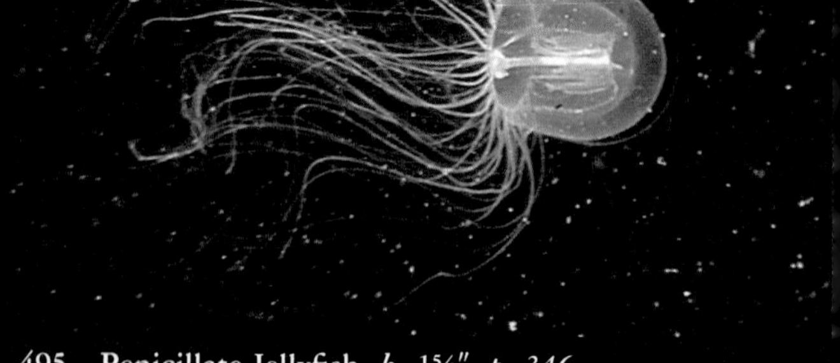

495 Penicillate Jellyfish, *h.* 1⅝″, *p.* 346

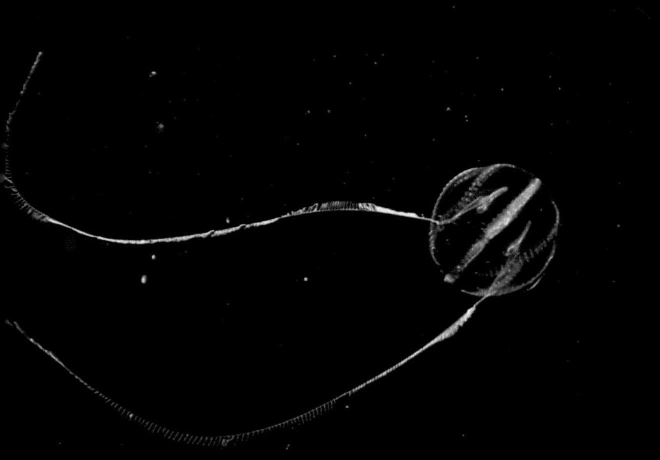

496 Sea Gooseberry, *h.* 1⅛", *p.* 396

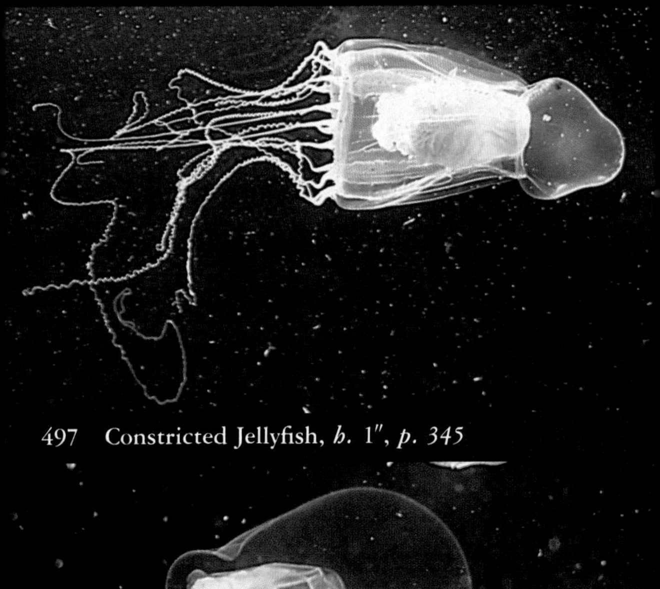

497 Constricted Jellyfish, *h.* 1", *p.* 345

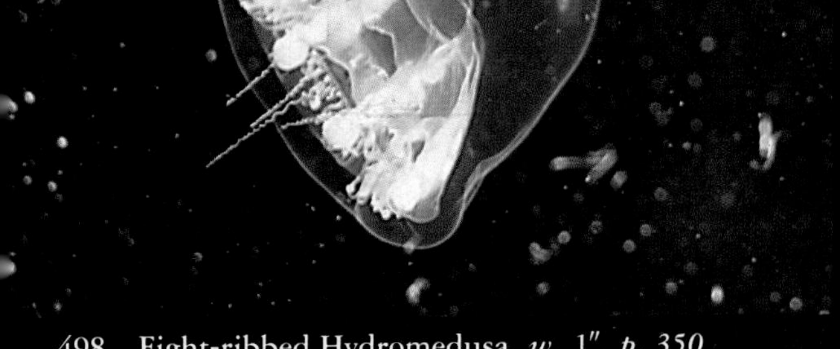

498 Eight-ribbed Hydromedusa, *w.* 1", *p.* 350

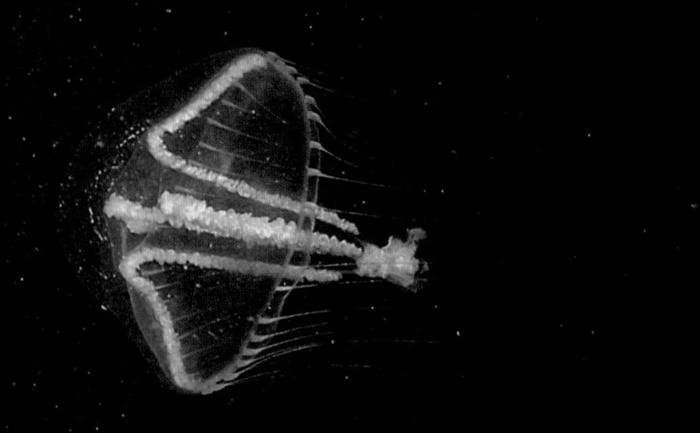

499 Elegant Hydromedusa, *w.* 4″, *p. 350*

500 Many-ribbed Hydromedusa, *w.* 7″, *p. 346*

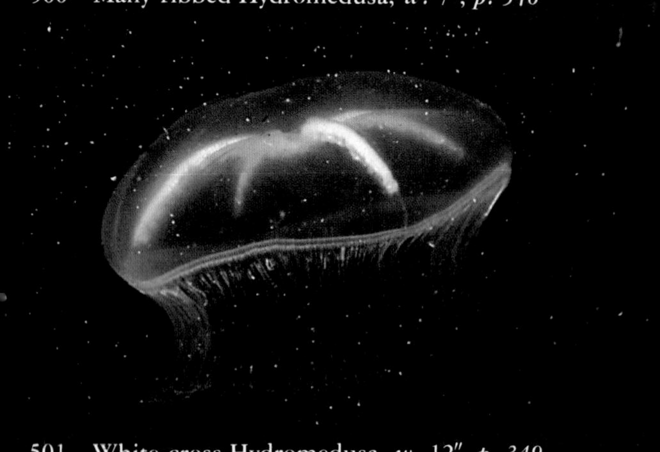

501 White-cross Hydromedusa, *w.* 12″, *p. 349*

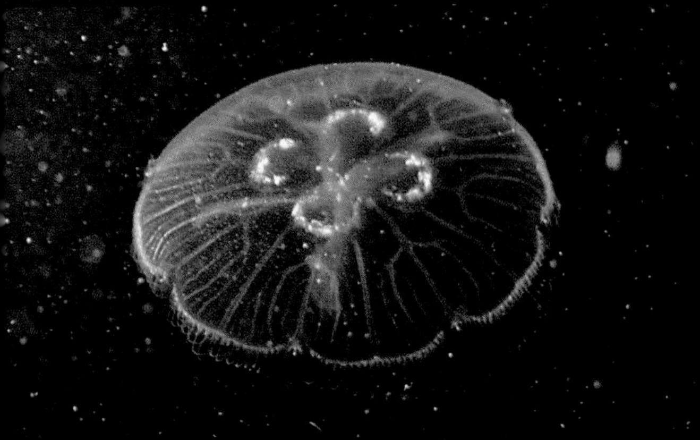

502 Moon Jellyfish, *w.* 16″, *p. 363*

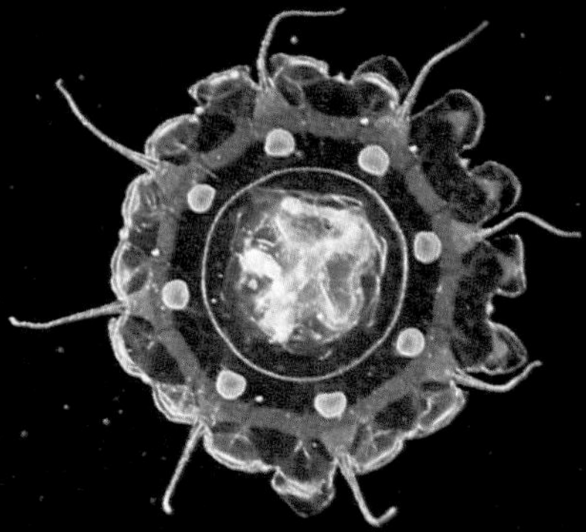

503 Crown Jellyfish, *w.* ⅝″, *p. 360*

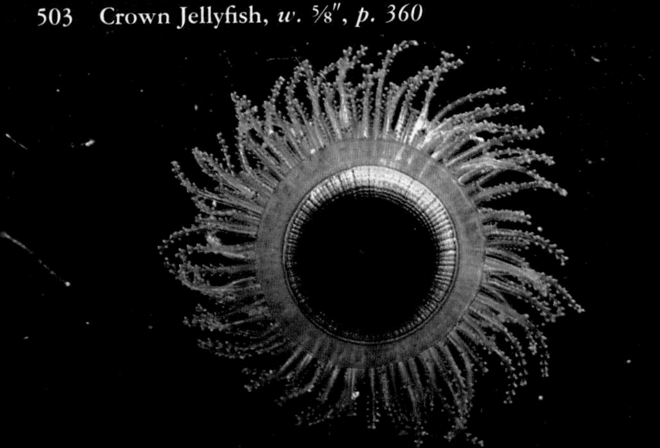

504 Blue Buttons, *w.* 1″, *p. 358*

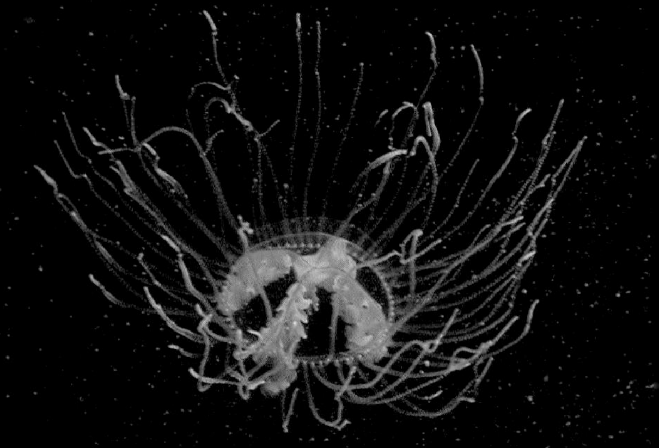

505 Angled Hydromedusa, *w.* ¾″, *p. 355*

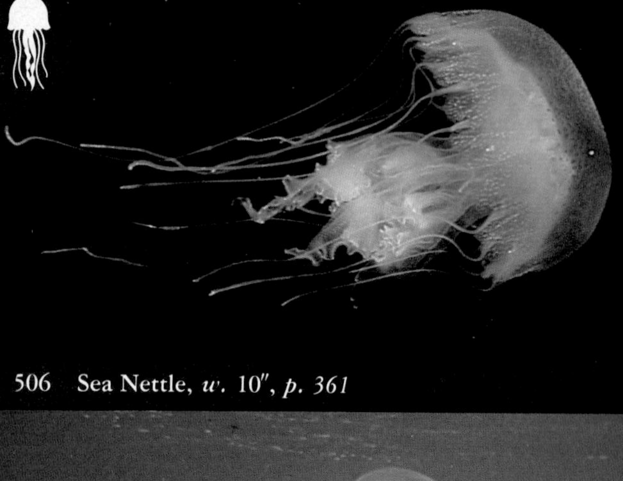

506 Sea Nettle, *w.* 10″, *p. 361*

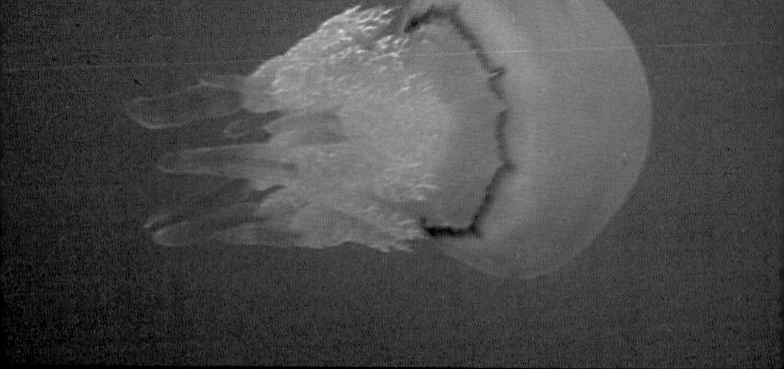

507 Cannonball Jellyfish, *w.* 7″, *p. 364*

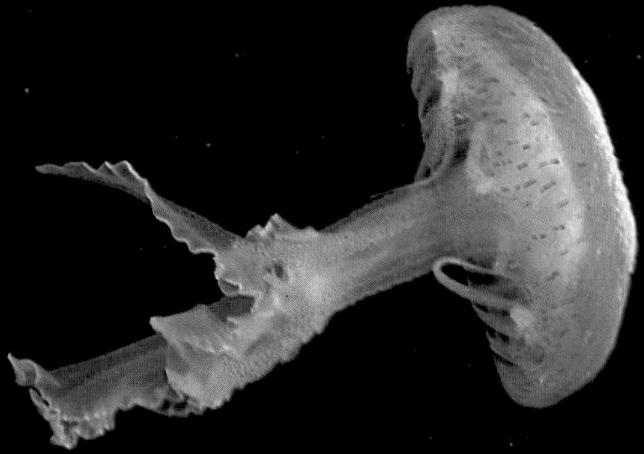

508 Purple Jellyfish, *w.* 4″, *p. 361*

509 Upside-down Jellyfish, *w.* 12″, *p. 363*

510 Sea Nettle, *w.* 10″, *p. 361*

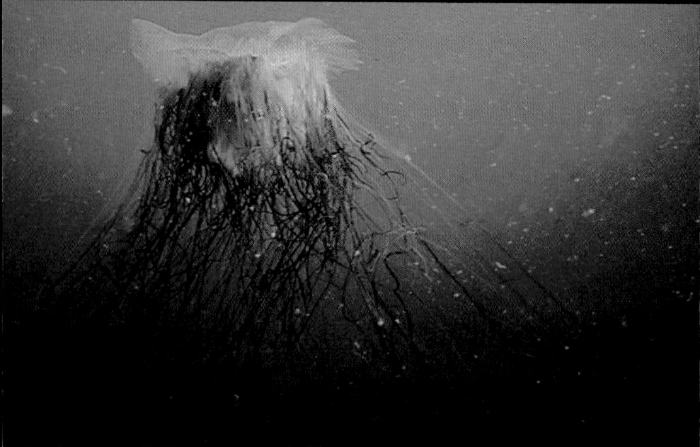

511 Lion's Mane, *w.* 96″, *p. 362*

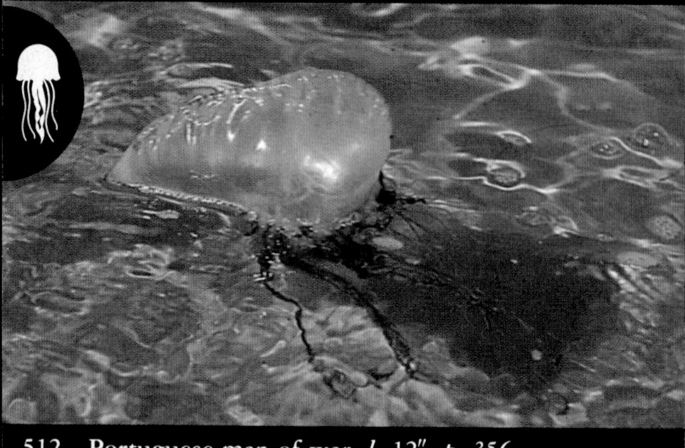

512 Portuguese man-of-war, *l.* 12″, *p. 356*

513 Portuguese man-of-war, *l.* 12″, *p. 356*

514 Cannonball Jellyfish, *w.* 7″, *p.* 364

515 By-the-wind Sailor, *l.* 4″, *p.* 357

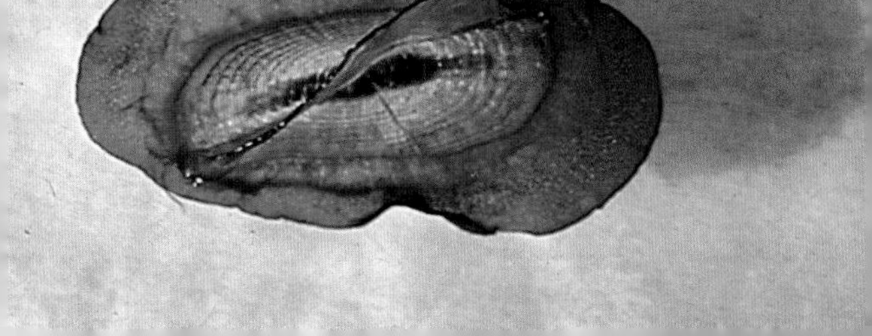

Urchins and Sand Dollars

The skeletons of sea urchins, which are shaped like doorknobs, and those of sand dollars, which are pancake-shaped, are frequently washed up on the beach. In life, sea urchins and heart urchins are covered with movable spines and vary widely in color from black and brown to white, green, red, and purple. The closely related sand dollars have shorter and more numerous spines and a distinctive, starlike pattern on the back. In life, their coloration varies from dark to light brown, but their dried skeletons, like those of the urchins, are chalky white.

517 Slate-pencil Urchin, *w.* 2¼″, *p. 688*

518 Atlantic Purple Sea Urchin, *w.* 2″, *p. 689*

519 Rock-boring Urchin, *w.* 2″, *p. 692*

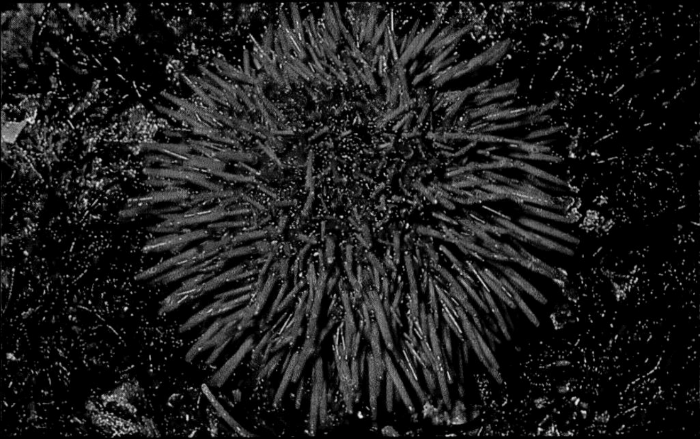

520 Red Sea Urchin, *w.* 5″, *p. 692*

521 Variegated Urchin, *w.* 3″, *p. 690*

522 Purple Sea Urchin, *w.* 4″, *p. 692*

523 Green Sea Urchin, *w.* 3¼″, *p. 691*

524 Long-spined Urchin, *w.* 4″, *p. 689*

525 Sea Egg, *w.* 6″, *p. 691*

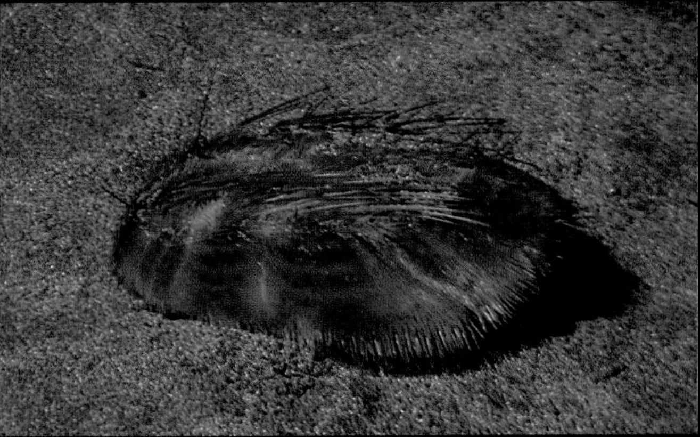

526 Long-spined Sea Biscuit, *l.* 10″, *p. 697*

527 Heart Urchin, *l.* 3″, *p. 697*

528 Variegated Urchin, *w.* 3″, *p. 690*

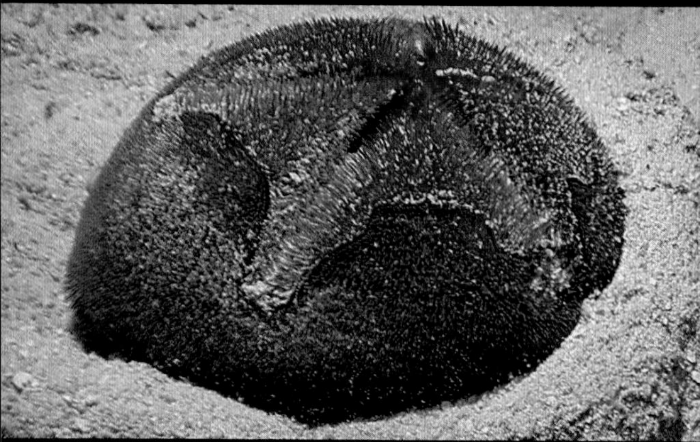

529 West Indian Sea Biscuit, *l.* 7″, *p. 698*

530 Common Sand Dollar, *w.* 3⅛″, *p. 694*

531 Eccentric Sand Dollar, *w.* 3″, *p. 694*

532 Six-hole Urchin, *w.* 4¼″, *p. 696*

533 Michelin's Sand Dollar, *l.* 6″, *p. 695*

534 Keyhole Urchin, *w.* 6″, *p. 695*

Sea Stars

Among the best known of all marine invertebrates, the sea stars are characterized by their five-armed shape. Their hues range from brown and green to red and orange. Sun stars, also often brightly colored, are frequently larger, with up to fourteen arms. Fragile, fast-moving brittle stars have long arms and a small central disk; the arms of the Caribbean Basket Star branch repeatedly. Also in this group are the Arrow Crab, and the sea spiders whose small bodies with eight legs resemble the many armed stars.

535 Mud Star, *w.* 4″, *p.* 670

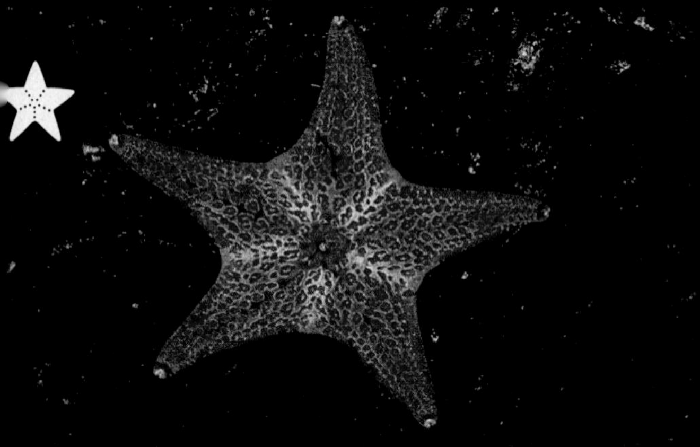

536 Leather Star, *w.* 9½″, *p.* 677

537 Bat Star, *w.* 8″, *p.* 678

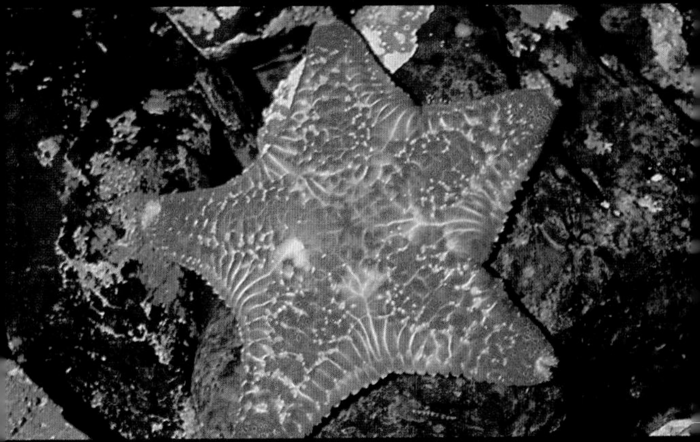

538 Badge Sea Star, *w.* 5½″, *p.* 674

539 Winged Sea Star, *w.* 5½″, *p.* 673

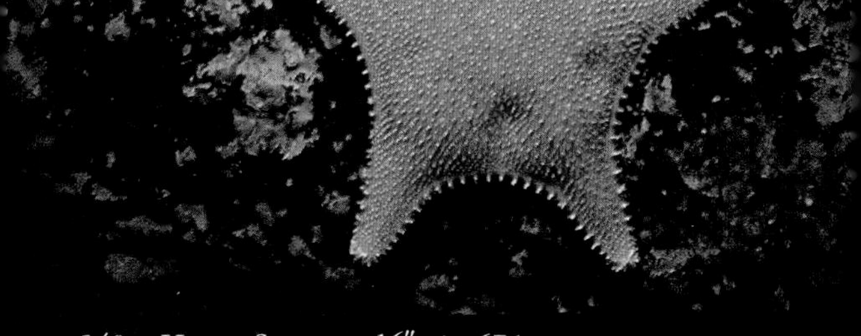

540 Horse Star, *w.* 16″, *p.* 671

541 Cushion Star, *w.* 20", *p. 671*

542 Smooth Sun Star, *w.* 16", *p. 672*

543 Smooth Sun Star, *w.* 16", *p. 672*

544 Sunflower Star, *w.* 52″, *p.* 682

545 Spiny Sun Star, *w.* 14″, *p.* 673

546 Dawson's Sun Star, *w.* 20″, *p.* 672

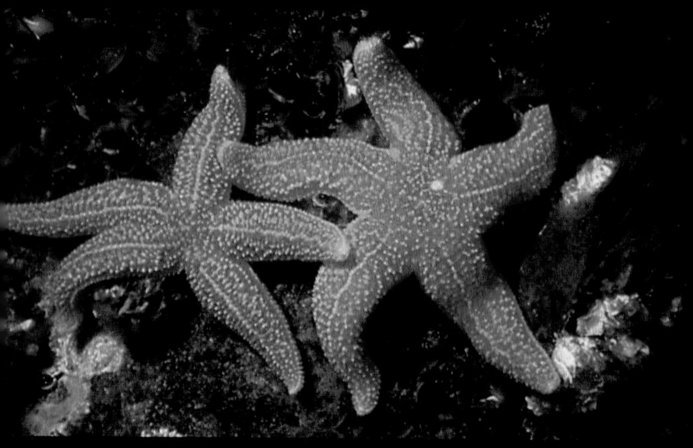

547 Northern Sea Star, *w.* 16″, *p. 678*

548 Ochre Sea Star, *w.* 20″, *p. 682*

549 Thorny Sea Star, *w.* 4½″, *p. 676*

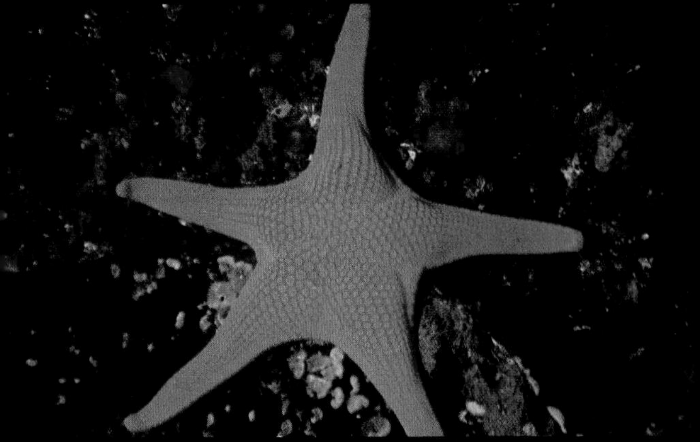

550 Equal Sea Star, *w.* 7″, *p.* *676*

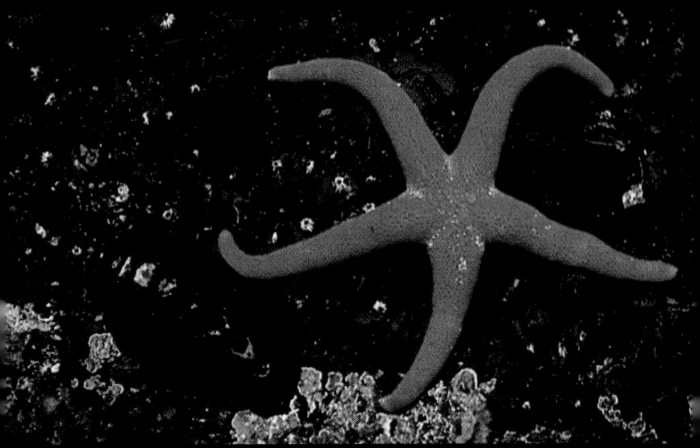

551 Pacific Henricia, *w.* 7¼″, *p.* *675*

552 Blood Star, *w.* 8″, *p.* *675*

553 Pacific Comet Star, *w.* 7¼″, *p. 677*

554 Troschel's Sea Star, *w.* 16″, *p. 679*

555 Broad Six-rayed Sea Star, *w.* 4″, *p. 681*

556 Giant Sea Star, *w.* 24″, *p. 681*

557 Forbes' Common Sea Star, *w.* 10¼″, *p. 679*

558 Forbes' Common Sea Star, *w.* 10¼″, *p. 679*

559 Northern Sea Star, *w.* 16", *p. 678*

560 Slender Sea Star, *w.* 3", *p. 680*

561 [illegible] Sea Star, *w.* 12", *p. 669*

562 Spiny Brittle Star, *w.* 12¾", *p. 687*

563 Spiny Sea Star, *w.* 10", *p. 670*

564 Banded Luidia, *w.* 12", *p. 668*

565 Reticulate Brittle Star, *w.* 6⅜″, *p. 687*

566 Panama Brittle Star, *w.* 21¾″, *p. 684*

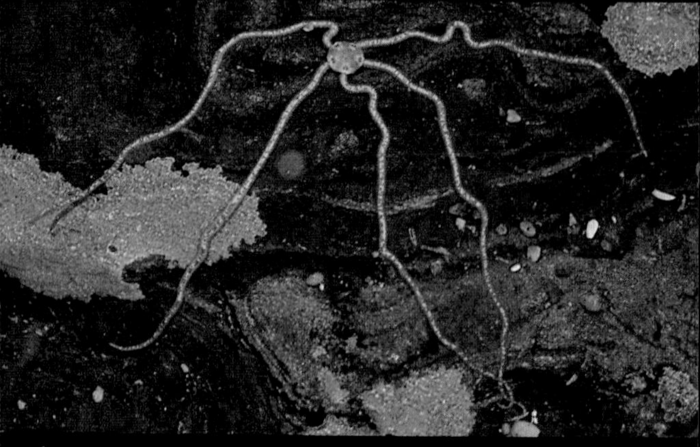

567 Burrowing Brittle Star, *w.* 13¾″, *p. 685*

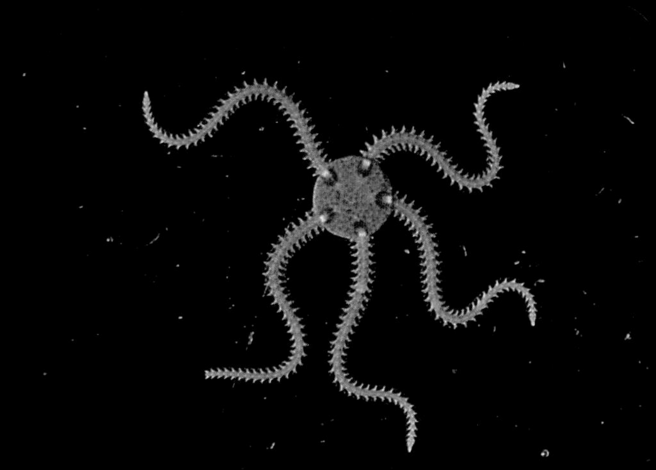

568 Dwarf Brittle Star, *w.* 2¼″, *p.* 686

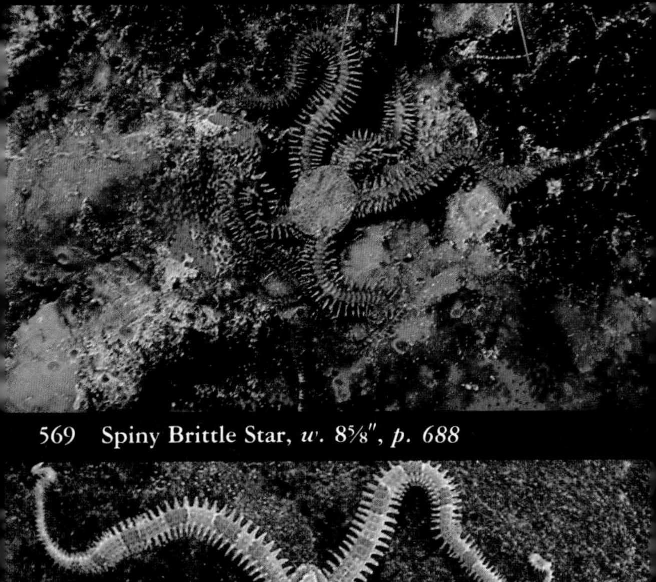

569 Spiny Brittle Star, *w.* 8⅝″, *p.* 688

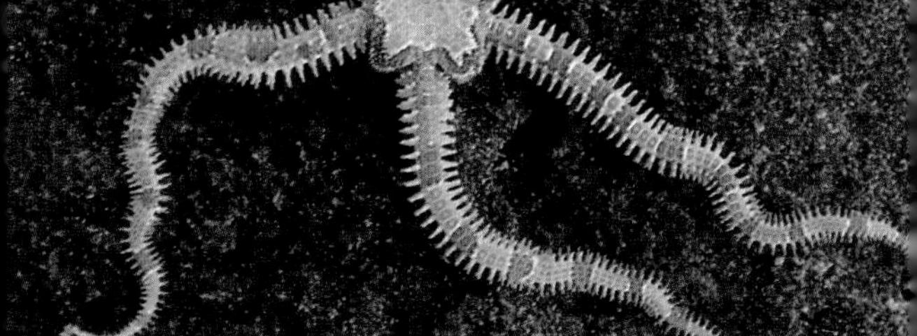

570 Daisy Brittle Star, *w.* 8″, *p.* 685

571 Esmark's Brittle Star, *w.* 8¾", *p. 686*

572 Northern Basket Star, *w.* 32", *p. 683*

573 Caribbean Basket Star, *w.* 32", *p. 683*

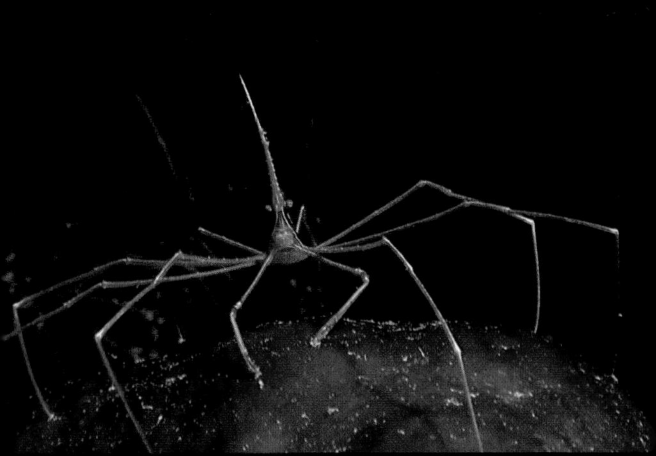

574 Arrow Crab, *l.* 2¼", *p.* 661

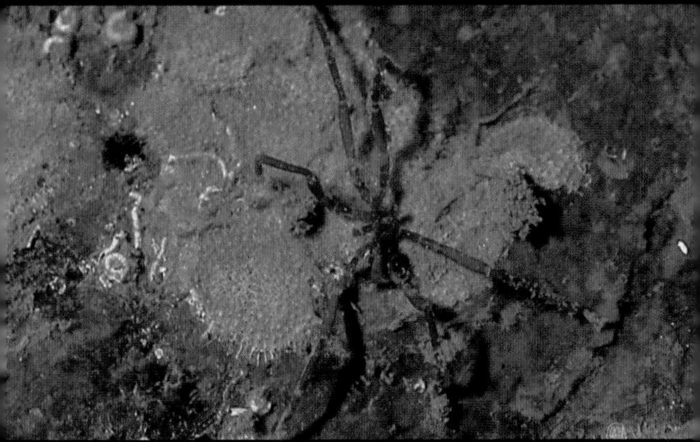

575 Clawed Sea Spider, *l.* ⅛", *p.* 590

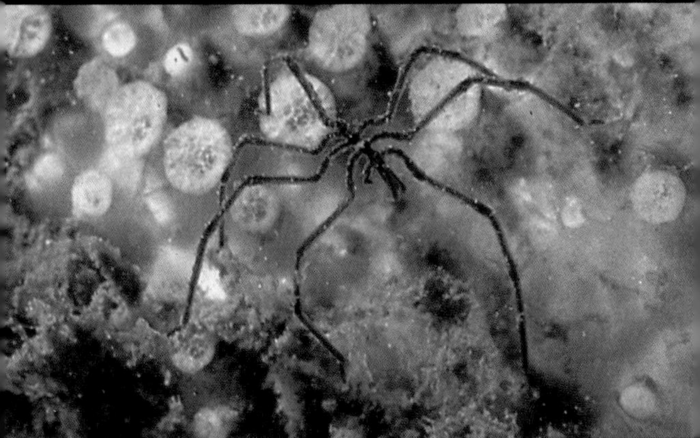

576 Lentil Sea Spider, *l.* ¼", *p.* 589

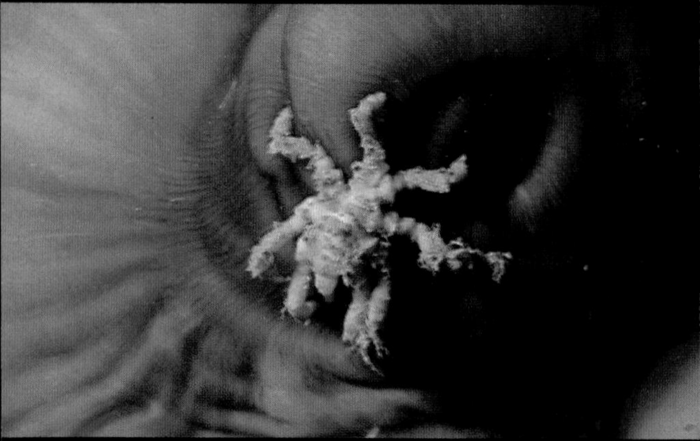

577 Anemone Sea Spider, *l.* ½", *p.* 587

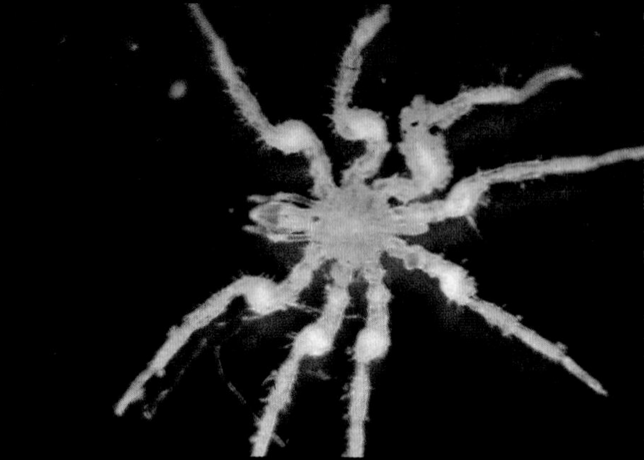

578 Ringed Sea Spider, *l.* ¹⁄₁₆", *p.* 588

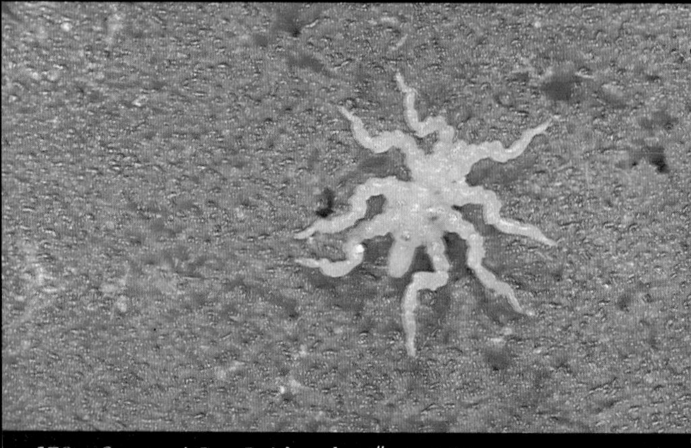

579 Stearns' Sea Spider, *l.* ½", *p.* 588

Lobsters and Shrimps

Some of these crustaceans are large enough to be prized as delicacies; others are inconspicuous enough to be almost unknown. Lobsters are readily distinguished by their substantial size, elongated body and large claws, but their color in life is considerably subtler than it becomes when the lobster is prepared for the dinner table. Shrimps' colors range from transparent silver or mottled brown to boldly patterned red and white, and the animals vary widely in size and form from the diminutive opossum shrimps, cleaning shrimps and Horned Krill to the ample Pink Shrimp and mantis shrimps. The smaller isopods, sea roaches, and amphipods share a compressed body that allows for very rapid movement.

580 Northern Sea Roach, *l.* 1″, *p. 603*

581 Western Sea Roach, *l.* 1″, *p. 603*

583 Vosnesensky's Isopod, *l.* 1⅜″, *p.* 602

584 Kirchansky's Isopod, *l.* ⅝″, *p.* 602

585 Baltic Isopod, *l.* 1″, *p.* 601

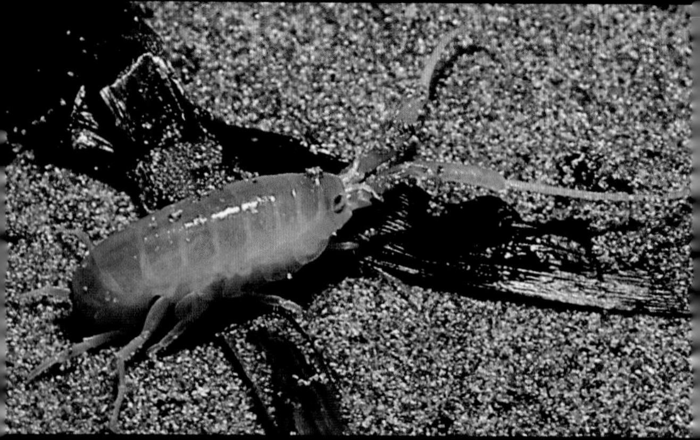

586 California Beach Flea, *l.* 1⅛", *p. 607*

587 Big-eyed Beach Flea, *l.* 1", *p. 606*

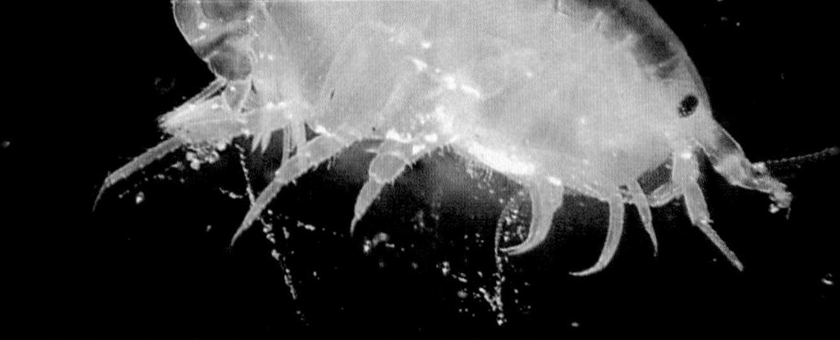

588 Noble Sand Amphipod, *l.* ¾", *p. 606*

589 Red-eyed Amphipod, *l.* ¾", *p. 604*

590 Mottled Tube-maker, *l.* ⅜", *p. 605*

591 Scud, *l.* 1", *p. 604*

592 Ciliated False Squilla, *l.* 4", *p.* 599

593 Sand Shrimp, *l.* 2¾", *p.* 618

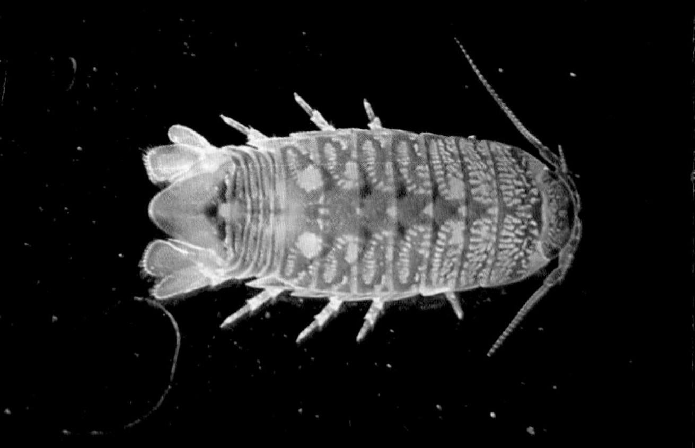

594 Harford's Greedy Isopod, *l.* ¾", *p.* 600

595 Common Mantis Shrimp, *l.* 10″, *p.* 599

596 Scaly-tailed Mantis Shrimp, *l.* 12″, *p.* 600

597 Swollen-clawed Squilla, *l.* 1⅝″, *p.* 598

598 Scud, *l.* 1", *p.* 604

599 Long-horn Skeleton Shrimp, *l.* 2⅛", *p.* 608

600 Smooth Skeleton Shrimp, *l.* 2", *p.* 608

601 Linear Skeleton Shrimp, *l.* ¾", *p. 607*

602 Bent Opossum Shrimp, *l.* 1", *p. 610*

603 Red Opossum Shrimp, *l.* ⅜", *p. 610*

604 Opossum Shrimp, *l.* 1¼", *p. 609*

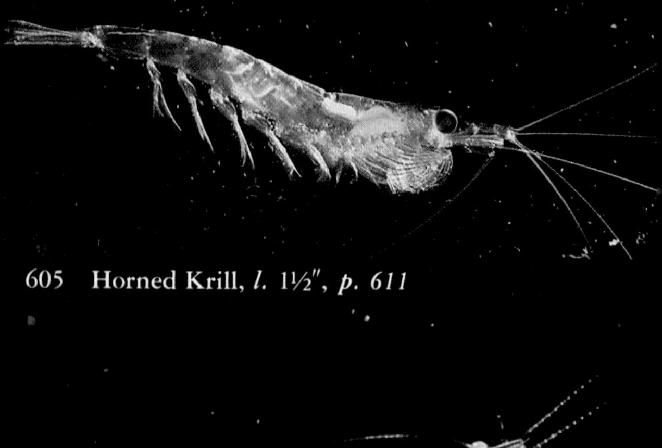

605 Horned Krill, *l.* 1½", *p. 611*

606 Opossum Shrimp, *l.* 1¼", *p. 609*

607 Common Shore Shrimp, *l.* 1¾", *p. 613*

608 Red Rock Shrimp, *l.* 2¾", *p. 615*

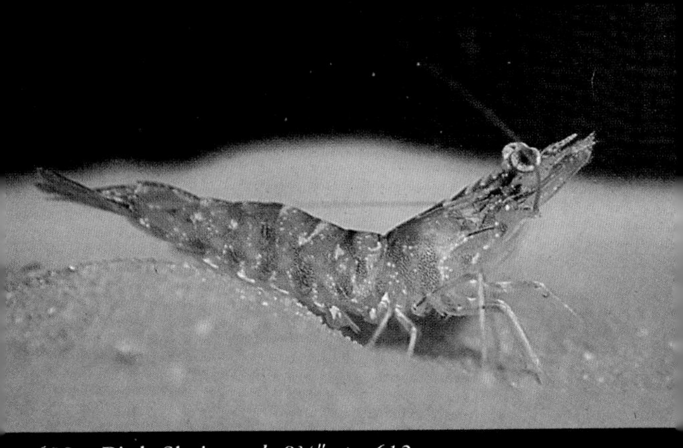

609 Pink Shrimp, *l.* 8⅜", *p. 612*

610 Coon-stripe Shrimp, *l.* 5¾", *p. 618*

611 Pink Shrimp, *l.* 8⅜", *p. 612*

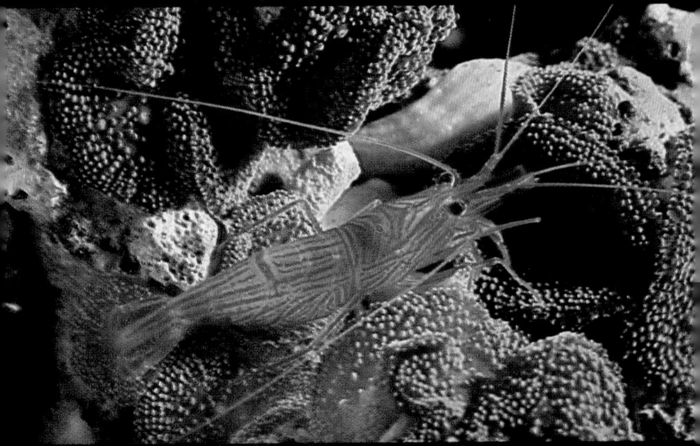

613 Red-lined Cleaning Shrimp, *l.* 2¾″, *p. 616*

614 Montague's Shrimp, *l.* 3¾″, *p. 617*

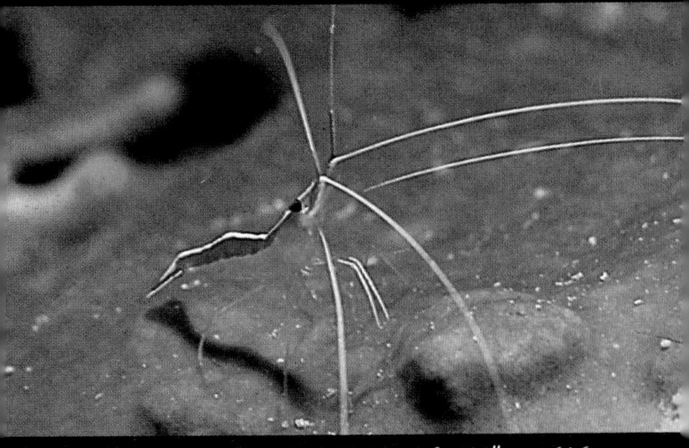

615 Grabham's Cleaning Shrimp, *l.* 2⅛″, *p. 616*

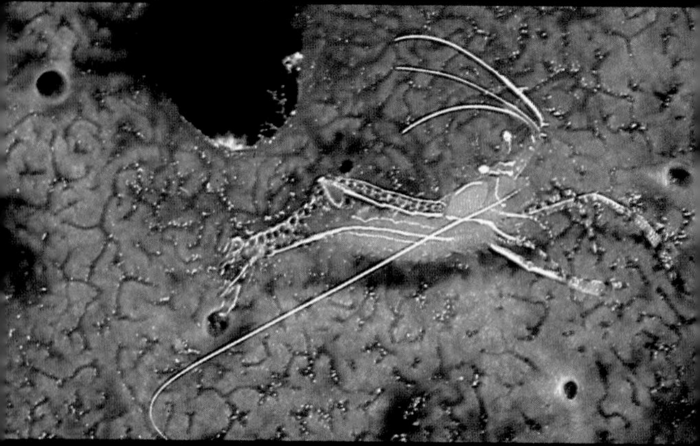

616 Pederson's Cleaning Shrimp, *l.* 1", *p. 613*

617 Spotted Cleaning Shrimp, *l.* 1", *p. 614*

618 Banded Coral Shrimp, *l.* 2", *p. 615*

619 Flat-browed Mud Shrimp, *l.* 2½", *p. 622*

620 Bay Ghost Shrimp, *l.* 4⅝", *p. 623*

621 Short-browed Mud Shrimp, *l.* 2½", *p. 622*

622 Brown Pistol Shrimp, *l.* 1½", *p. 614*

623 California Rock Lobster, *l.* 16", *p. 620*

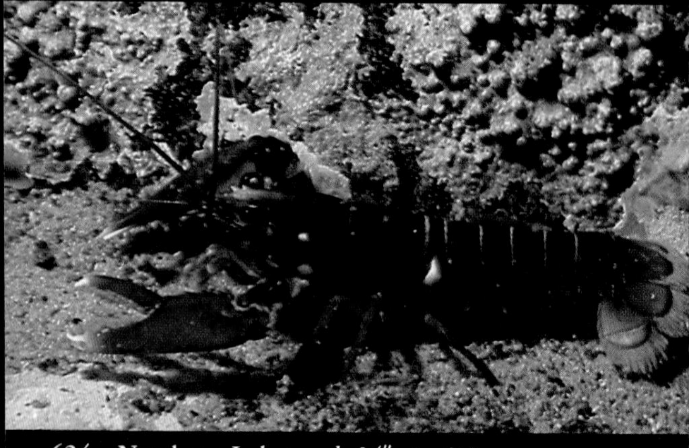

624 Northern Lobster, *l.* 34", *p. 619*

625 West Indies Spiny Lobster, *l.* 24″, *p. 620*

626 Spanish Lobster, *l.* 12″, *p. 621*

627 Spanish Lobster, *l.* 12″, *p. 621*

Crabs

The many forms and colors of crabs enable them to exploit the diverse habitats that exist along the sea coast. Most typical crabs are oval, rectangular, or triangular, with a front pair of legs ending in pincers, and four pairs of walking legs. Male fiddler crabs have a distinctive, large claw on the right front leg. Hermit crabs are readily identified by their habit of occupying snail shells to protect their vulnerable bodies. The small, egg-shaped mole crabs and the Pacific Sand Crab forage in the sand at the tide line. The Horseshoe Crab, not strictly a crab at all, is a unique relic, immediately recognized by its large horseshoe-shaped body and long, thin tail.

628 Sand Fiddler, *w.* 1½″, *p. 654*

629 Brackish-water Fiddler, *w.* 1½″, *p. 656*

630 California Fiddler, *w.* ¾″, *p. 654*

631 Ghost Crab, *w.* 2″, *p. 653*

632 Great Land Crab, *w.* 5″, *p. 652*

634 Commensal Crab, *w.* ⅝", *p. 648*

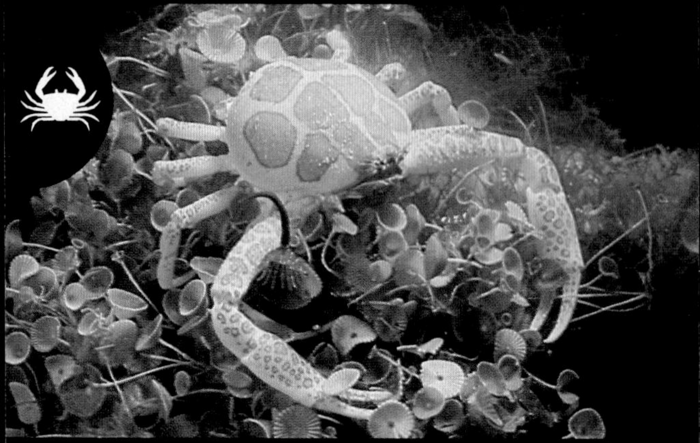

635 Purse Crab, *l.* 2¾", *p. 636*

636 Say's Mud Crab, *w.* ⅞", *p. 647*

637 Lady Crab, *w.* 3″, *p.* 638

638 Mountain Crab, *w.* 4″, *p.* 653

640 Spiny Spider Crab, *w.* 7⅜″, *p.* 658

641 Flat Porcelain Crab, *l.* 1″, *p.* 624

642 Stone Crab, *w.* 4⅝″, *p.* 647

643 Yellow Shore Crab, *w.* 1⅜″, *p. 650*

644 Pacific Rock Crab, *w.* 4⅝″, *p. 641*

645 Flat Mud Crab, *w.* ¾″, *p. 646*

646 Black-clawed Mud Crab, *w.* 1⅜", *p. 646*

647 Say's Porcelain Crab, *w.* ½", *p. 623*

648 Coral Crab, *w.* 6", *p. 645*

649 Sally Lightfoot, *w.* 3⅝", *p. 648*

650 Atlantic Rock Crab, *w.* 5¼", *p. 643*

651 Red Crab, *w.* 6¼", *p. 644*

652 Oregon Cancer Crab, *w.* 1⅞", *p. 642*

653 Jonah Crab, *w.* 6¼", *p. 642*

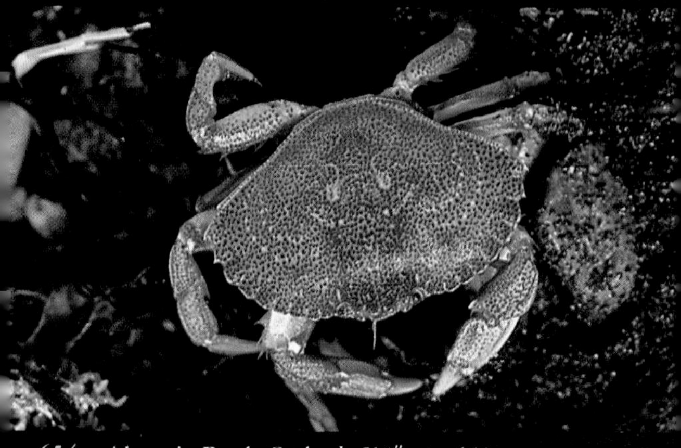

654 Altantic Rock Crab, *l.* 5¼", *p. 643*

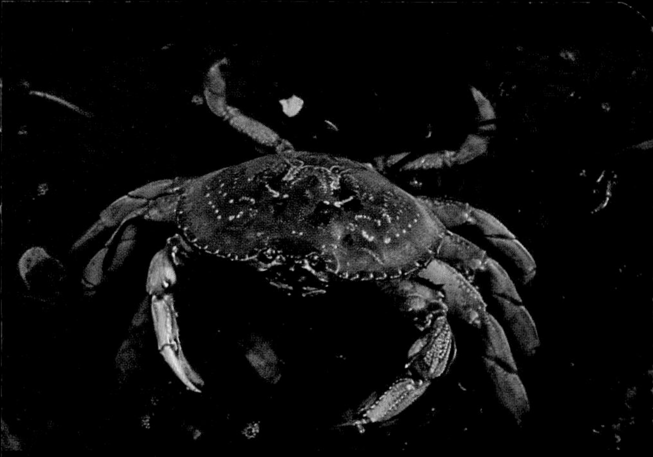

655 Dungeness Crab, *w.* 9¼″, *p. 643*

656 Common Spider Crab, *l.* 4″, *p. 657*

657 Blue Crab, *w.* 9¼″, *p. 639*

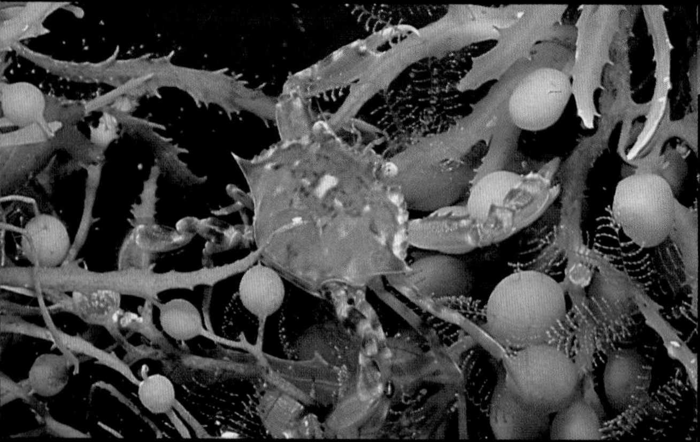

658 Sargassum Crab, *w.* 2⅛″, *p. 638*

659 Shield-backed Kelp Crab, *l.* 4¾″, *p. 659*

660 Toad Crab, *l.* 3¾″, *p. 656*

661 Toad Crab, *l.* 3¾″, *p. 656*

662 Striped Shore Crab, *w.* 1⅞″, *p. 650*

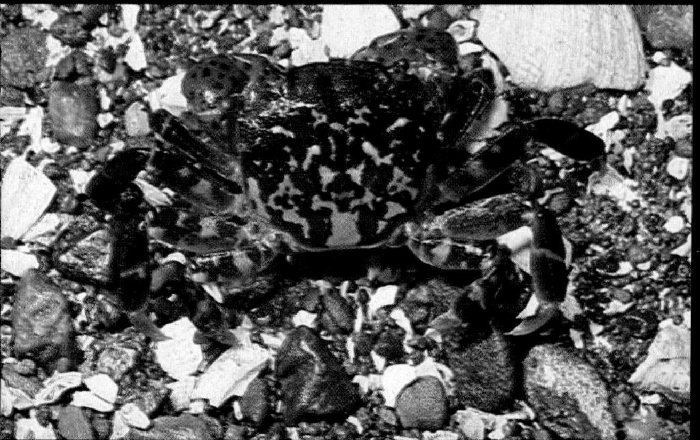

663 Purple Shore Crab, *w.* 2¼″, *p. 649*

664 Green Crab, *w.* 3⅛", *p. 641*

665 Wharf Crab, *w.* ⅞", *p. 651*

666 Horseshoe Crab, *l.* 24", *p. 586*

667 Turtle Crab, *w.* 2¾″, *p.* 626

668 Butterfly Crab, *w.* 2¾″, *p.* 626

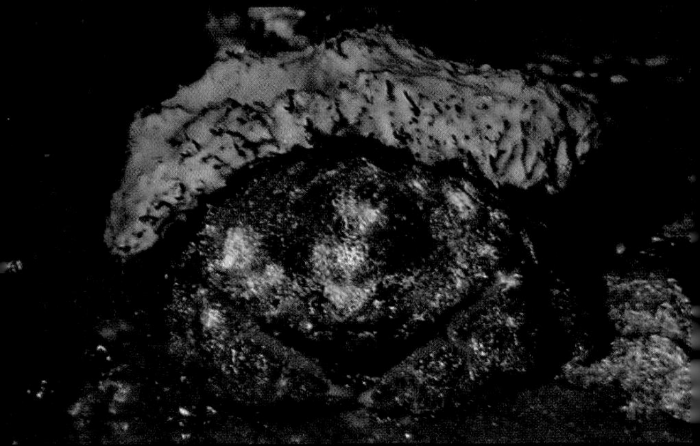

669 Lesser Sponge Crab, *l.* 1⅜″, *p.* 635

670 Sponge Crab, *w.* 5″, *p.* 634

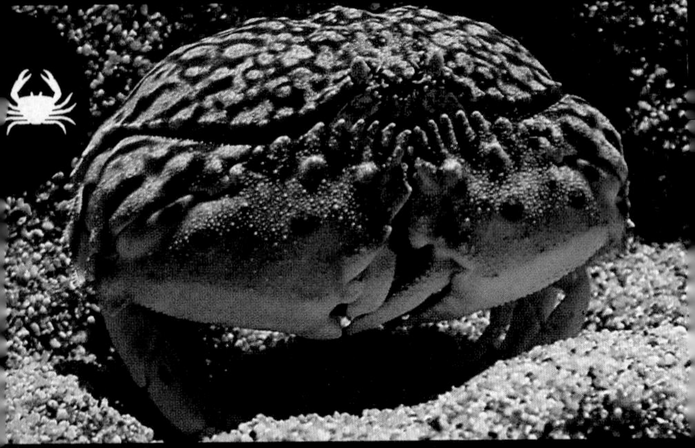

671 Flame-streaked Box Crab, *w.* 5½″, *p.* 636

673 Yellow Box Crab, *w.* 4″, *p. 637*

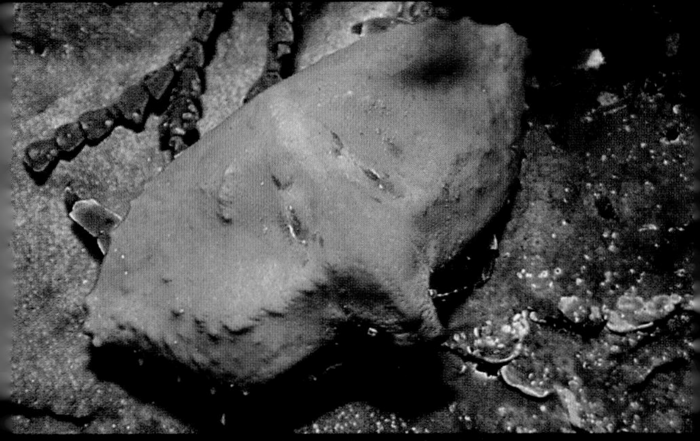

674 Butterfly Crab, *w.* 2¾″, *p. 626*

675 Masking Crab, *l.* 4″, *p. 658*

676 Flat-clawed Hermit Crab, *l.* 1¼", *p. 631*

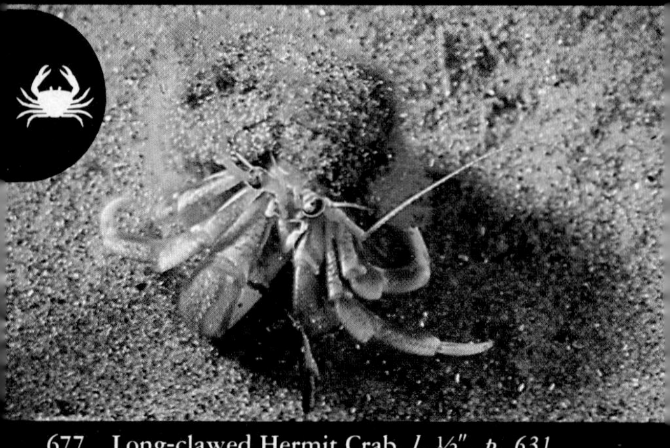

677 Long-clawed Hermit Crab, *l.* ½", *p. 631*

678 Sharp-nosed Crab, *l.* 1¾", *p. 659*

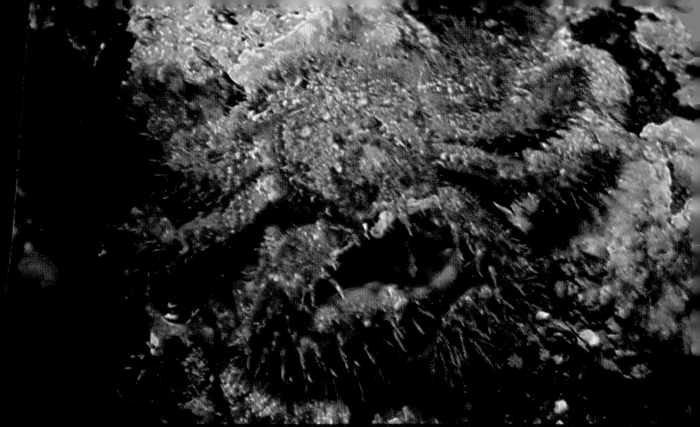

679 Fuzzy Crab, *w.* ¾″, *p. 625*

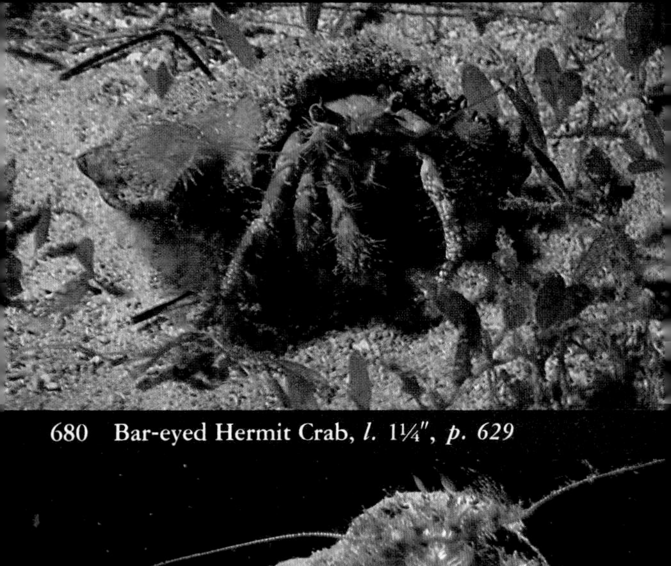

680 Bar-eyed Hermit Crab, *l.* 1¼″, *p. 629*

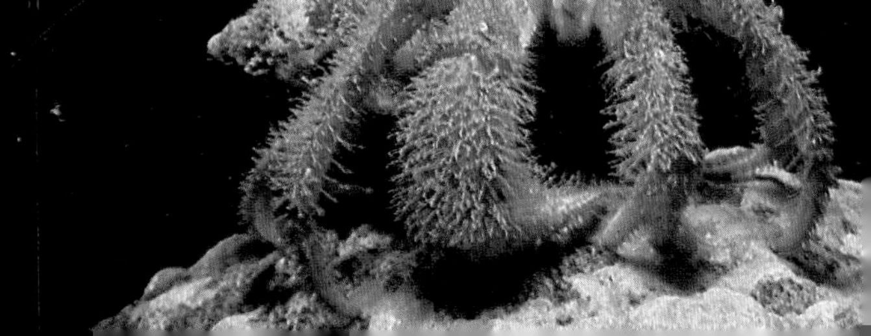

682 Giant Hermit Crab, *l.* 4¾″, *p.* 627

683 Grainy Hermit Crab, *l.* ¾″, *p.* 632

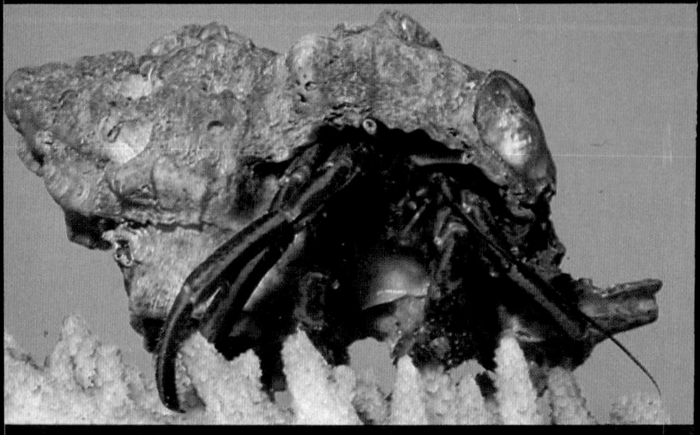

684 Striped Hermit Crab, *l.* 1¼″, *p.* 627

685 Land Hermit Crab, *l.* 1½″, *p. 626*

686 Acadian Hermit Crab, *l.* 1¼″, *p. 629*

688 Spiny Mole Crab, *l.* 2⅜", *p.* 634

689 Pacific Mole Crab, *l.* 1⅜", *p.* 633

Part II
Text

The numbers preceding the species descriptions in the following pages correspond to the plate numbers in the color section. If the description has no plate number, it is illustrated by a drawing that accompanies the text.

SPONGES
(Phylum Porifera)

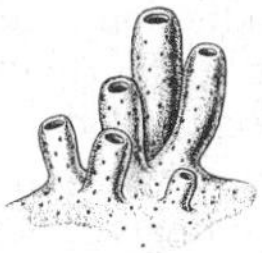

Sponges are the simplest many-celled animals. Their shapes vary from tiny cups, broad branches, or tall vases to encrustations and large, rounded masses. Sponges come in a variety of colors. Grays and browns predominate in deeper waters; brighter hues in the shallows. With differing growing conditions a species may vary greatly in size, shape, and color, and so can be difficult to identify.

A sponge consists of a cooperating community of individual cells, each performing a specific function. The cells surround a system of canals through which water is pumped, providing the basis for all the sponge's life functions. Water enters the canals through minute pores (*ostia*) that dot the surface of the sponge. It then passes into chambers lined with collar cells, each with a sticky, funnel-shaped collar. Out of each collar extends a *flagellum,* a hairlike structure whose beating creates a current. The combined action of all the collar cells drives water through the canals and out of the sponge through a larger pore, the *osculum.* The collar cells trap food particles brought in with the water, and either digest them or pass them to other cells to be digested. This flow of water through the animal also brings in oxygen and removes carbon dioxide and other waste products. A simple sponge has one chamber and one osculum; more complex sponges have many of each.

Most sponges have a skeleton that is a meshwork of tough protein (*spongin*), or of microscopic hard splinters (*spicules*), or a combination of both. Spicules are either limy or glasslike and appear in a variety of forms. Although too small to be helpful in field identification, spicules are examined in the laboratory to distinguish species that are superficially alike.

Sponges reproduce both asexually and sexually. Asexual reproduction occurs when a sponge constricts off a tip of one of its branches—a process called *budding*—or when parts are broken off by storms or by other animals. These fragments regenerate into complete sponges. Some sponges reproduce by means of *gemmules*—tiny clusters of cells, surrounded by a tough coating, that are released when the sponge body is broken up, as by winter storms. When spring comes, the gemmules germinate and differentiate into tiny sponges. Most sponges also have both male and female reproductive structures. Clouds of sperm are released into the water and are swept into another sponge, where they fertilize the eggs. These then develop into swimming larvae which are washed out of the parent's chamber by the water current. If a larva settles on a suitable surface, it becomes attached, changes its shape, and develops into a tiny sponge.

The Phylum Porifera is divided into three classes. The Calcispongiae, or Calcarea (with limy spicules), are represented here by four genera of small sponges: *Leucosolenia, Scypha, Leucandra,* and *Leucilla.* The Hyalospongiae, or Hexactinellida (glass sponges), are found only in deep waters, and so are not included in this book. The Class Demospongiae comprises all the remaining sponges. Members of this class may have skeletons of glasslike spicules, or of spongin, or both, or may lack a skeleton entirely.

54 **Organ-pipe Sponge**
(*Leucosolenia botryoides*)
Class Calcispongiae

Description: 1/16″ (2 mm) wide, 1/2″ (13 mm) high. *Branching, cylindrical tube rising from a creeping base.* White. Spicules limy, 3- or 4-pronged, visible with a hand lens. Grows in clusters of 2 or 3 to several hundred tubes.

Habitat: On the underside of overhanging rocks; on seaweeds and wharf piles, in protected areas; near low-tide line and below.

Range: Gulf of St. Lawrence to Cape Cod.

Comments: This small sponge spreads in patches whose surface resembles a web of starched lace. A similar species, Eleanor's Organ-pipe Sponge (*L. eleanor*), is found from British Columbia to California. As well as encrusting, *L. eleanor* also grows up in 3″ (76 mm) wide clumps of branching, cream-colored tubes 3/4″ (19 mm) high and 1/16″ (2 mm) wide.

---

45 **Little Vase Sponge**
(*Scypha ciliata*)
Class Calcispongiae

Description: 1/4″ (6 mm) wide, 1″ (25 mm) high. *Vaselike* tube, widest at middle. Creamy to pale tan. Opening at top surrounded by *circle of glistening, projecting spicules.* Fuzzy. Occurs in small clusters of 2–10.

Habitat: Attached to rocks, shells, and seaweeds; near low-tide line and below.

Range: Arctic to Rhode Island; British Columbia to c. California.

Comments: This is the classic simple sponge cited in most biology texts. Its generic name, *Scypha,* meaning "cup," indicates the sponge's resemblance to the deep drinking vessels of ancient Greece. Several species of *Scypha* occur from Washington to California.

109 **Heath's Sponge**
(*Leucandra heathi*)
Class Calcispongiae

Description: 3½" (89 mm) wide, 4⅜" (111 mm) high. *Globular to pear-shaped,* with *1 large pore at top,* giving volcanolike appearance. *Whitish to cream-colored.* Surface bristly with protruding limy spicules, those around pore ¼" (6 mm) long.

Habitat: Attached to rocks in crevices or protected places; from just above low-tide line to water more than 50′ (15 m) deep.

Range: British Columbia to s. California.

Comments: Despite their forbidding prickly appearance, the chalky spicules of Heath's Sponge crush easily when touched. The sponge sometimes takes the shape of the rock crevice in which it grows.

---

46 **Nutting's Sponge**
(*Leucilla nuttingi*)
Class Calcispongiae

Description: ⅜" (10 mm) wide, 2" (51 mm) high. *Urn-shaped, with narrow, stalked base* and 1 large pore at top. Creamy-white to tan. Smooth, usually with fringe of spicules around pore. Usually *occurring in clumps* of 12 or more.

Habitat: Suspended in rock crevices and caves, and from under shaded ledges; from low-tide line to water more than 100′ (30 m) deep.

Range: British Columbia to Baja California.

Comments: Large Nutting's Sponges may be found in Monterey Bay the year round, but north of San Francisco they are seasonal, and large ones, present in the spring, are scarce by August.

115 **Dujardin's Slime Sponge**
(*Halisarca dujardini*)
Class Demospongiae

Description: 4″ (102 mm) wide, ⅛″ (3 mm) high. Smooth, *slimy film. No skeleton.* Pale tan to light olive-brown. *Tiny pores* on surface.
Habitat: Underside of protected rocks and shells, and holdfasts of algae; from low-tide line to water more than 13′ (4 m) deep.
Range: Arctic to Cape Cod.
Comments: It is difficult even for the specialist to identify slime sponges because they lack the skeletal structures on which sponge classification is based. Dujardin's Slime Sponge is found mostly below low-tide line, and individuals have separate sexes. The Nahant Slime Sponge (*H. nahantensis*) is very similar, but it is hermaphroditic and is found between high- and low-tide line over much of the same geographic range.

52 **Finger Sponge**
"Eyed Sponge"
(*Haliclona oculata*)
Class Demospongiae

Description: More than 12″ (30 cm) wide, 18″ (46 cm) high. *Erect, branched.* Attached at base by short, *narrow stalk;* variable in number and shape of branches. Tan to gray-brown, sometimes rosy or red-orange. *Pores conspicuous, scattered* over surface.
Habitat: Attached to rocks; from low-tide line to water 400′ (124 m) deep.
Range: Labrador to North Carolina.
Comments: This species is called the "Eyed Sponge" because its pores are reminiscent of so many eyes. It is frequently broken free by storms and tossed up on the beach, where its skeleton bleaches to white.

126 **Purple Sponge**
(*Haliclona permollis*)
Class Demospongiae

Description: 36″ (91 cm) wide, 1⅝″ (41 mm) high. Encrusting, *with raised "volcanoes,"* each with a pore up to ¼″ (6 mm) wide. *Pink to lavender to purple.* Smooth; soft but not slimy.
Habitat: Encrusted on rocks in protected places; on floating docks and in tidepools; from midtidal zone to water 20′ (6 m) deep.
Range: New Brunswick to lower Chesapeake Bay; Washington to c. California.
Comments: The classification of this widespread species is still under investigation. On the West Coast it occurs higher up on rock faces than any other sponge. The nudibranch Ringed Doris feeds on the Purple Sponge, among others.

---

31 **Red Sponge**
(*Haliclona rubens*)
Class Demospongiae

Description: 16″ (41 cm) wide, 16″ (41 cm) high. Irregular globe with *blunt, rounded, upright branches.* Deep red or dark brown to nearly black. Surface smooth; *pores prominent.*
Habitat: Attached to coral rocks in back-reef lagoons and protected places, or attached to the bottom in beds of turtle grass; from shallow water to water 50′ (15 m) deep.
Range: Florida to Texas; West Indies.
Comments: Although the Red Sponge is harmless, it may be confused with the Fire Sponge and the Do-Not-Touch-Me Sponge, both of which are red and powerful stingers. It is a good policy to avoid contact with any red or orange sponge in subtropical waters.

29 **Red Beard Sponge**
(*Microciona prolifera*)
Class Demospongiae

Description: Varies from a *thin encrusting layer* less than ⅛″ (3 mm) high and covering a few square inches to 8″ (20 cm) wide and 8″ (20 cm) high with many *fanlike branches. Red to orange.* Pores inconspicuous.

Habitat: On rocks, pilings, oysters and other shells, and hard objects in protected bays and estuaries; below low-tide line.

Range: Nova Scotia to Florida and Texas; Washington to c. California.

Comments: The Red Beard Sponge was the first animal shown to reorganize its form from experimentally separated cells. Cells divided by squeezing the sponge through a fine mesh cloth into a bowl of sea water creep about on the bottom, stick to each other, and finally form a mass which reorganizes into a sponge. This sponge tolerates both the pollution and the reduced salinities of bays and estuaries.

128 **Velvety Red Sponge**
(*Ophlitaspongia pennata*)
Class Demospongiae

Description: More than 36″ (91 cm) wide, ¼″ (6 mm) high. *Flat,* encrusting. *Coral-red,* occasionally red-brown to mustard. *Velvety; pores starlike, close together.*

Habitat: Attached to overhanging rocks in shaded crevices, from between high- and low-tide lines to water 10′ (3 m) deep.

Range: Washington to California.

Comments: The red nudibranch Crimson Doris is almost invariably present on the Velvety Red Sponge on which it preys and lays coiled, red egg-masses that match the sponge in color.

28 **Common Palmate Sponge**
(*Isodictya palmata*)
Class Demospongiae

Description: 11″ (28 cm) wide, 12″ (30 cm) high. *Branches erect, flattened.* Yellow to light brown or reddish-brown. *Pores conspicuous,* distributed along flat sides of branches. Surface finely pebbled.
Habitat: Attached to rocks and other solid objects; from just below low-tide line to deeper water.
Range: Nova Scotia to North Carolina.
Comments: This sponge is known in the British Isles as "Mermaid's Gloves." It can be distinguished from Deichman's Palmate Sponge (*I. deichmani*) only by microscopic examination of the spicules.

---

30 **Smooth Red Sponge**
(*Plocamia karykina*)
Class Demospongiae

Description: More than 10″ (25 cm) wide, 1″ (25 mm) high. Encrusting. Bright red to salmon-orange. Smooth, firm; *pores large, variable in size, far apart.*
Habitat: Attached to rocks in protected places; from between high- and low-tide lines to just below low-tide line.
Range: British Columbia to Baja California.
Comments: This sponge is thicker and smoother than the Velvety Red Sponge, with fewer but larger pores. The red nudibranch Crimson Doris also feeds on this species.

---

123 **Crumb of Bread Sponge**
(*Halichondria panicea*)
Class Demospongiae

Description: More than 12″ (30 cm) wide, 2″ (51 mm) high. Thin crust. Yellow to greenish. *Pores prominent on low "volcanoes." Texture like bread crumbs.*

Habitat: Protected undersides of stones, wharf piles, and other solid substrata; from low-tide line to water more than 200′ (61 m) deep.

Range: Arctic to Cape Cod; Alaska to s. California.

Comments: Like most sponges, the Crumb of Bread Sponge has a strong odor. Bowerbank's Crumb of Bread Sponge (*H. bowerbanki*), which is distinguishable from *H. panicea* only by its spicules, is similarly distributed on both coasts.

---

53 Nipple Sponge
(*Polymastia robusta*)
Class Demospongiae

Description: 4″ (102 mm) wide, ⅛″ (3 mm) high. *Thin encrustation with upright ¾″ (19 mm) "nipples"* without obvious openings at tips. Creamy-white to orange-red. Pores abundant, tiny.

Habitat: Encrusting rocks and shells; below low-tide line and deeper.

Range: Gulf of St. Lawrence to Cape Cod.

Comments: The female sponge spawns eggs into a slimy layer that envelops it. The non-swimming larvae creep about until they find a suitable surface on which to develop. The Western Nipple Sponge (*P. pachymastia*), which is found on the California coast, is similar, but its nipples have pores at the tips.

---

17 Loggerhead Sponge
(*Spheciospongia vesparia*)
Class Demospongiae

Description: 36″ (91 cm) wide, 24″ (61 cm) high. *Roughly cylindrical,* narrower at base, with *depression on upper surface containing cluster of large pores.* Dark brown to black or purplish-black, but usually appearing grayish because of a coating of sediment. Texture woody when wet.

Habitat: Around coral reefs and on reef flats, firmly anchored to rocks on the sea bottom; below low-tide line to water 50′ (15 m) deep.

Range: North Carolina to Florida and Mexico; Bahamas; West Indies.

Comments: The Loggerhead Sponge is the largest sponge known. A biologist recorded 16,000 animals, most of them snapping shrimp, living in the canal system of one specimen. This species shares its common name with an unrelated sponge, *Ircinia strobilina.*

---

127 **Boring Sponge**
(*Cliona celata*)
Class Demospongiae

Description: One or more ⅛″ (3 mm) wide, 1/16″ (2 mm) high *yellowish pores* protruding from holes in mollusk shells or coral. Sponge may overgrow the host entirely.

Habitat: In living and dead mollusk shells and corals.

Range: Gulf of St. Lawrence to Gulf of Mexico; Washington to California.

Comments: The larvae of these sponges settle on shells and coral, develop into tiny sponges, and, secreting sulfuric acid, excavate pits and galleries, and loosen chips of the shell or coral which are then ejected. The host is weakened to the point of disintegration, and the sea bottom is thus kept free of accumulating shells.

Sheep's Wool Sponge
(*Hippiospongia lachne*)
Class Demospongiae

Description: 36″ (91 cm) wide, 24″ (61 cm) high. Almost round, *massive.* Dark olive-drab to black. Surface irregular; *pores large, raised.*

Habitat: On rocky, silt-free bottoms with good flow of water; from below low-tide line to water more than 150′ (46 m) deep.

Range: Florida to Mexico; Bahamas; West Indies.

Comments: This is the most important commercial sponge in the New World. Its unusual softness after cleaning suggests its common name. The Sheep's Wool Sponge once provided a thriving industry in the Caribbean and the Gulf of Mexico, but the industry could not compete with new synthetic cellulose sponges, and today natural sponges are sold chiefly to tourists.

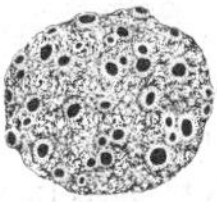

18 **Loggerhead Sponge**
"Cake Sponge"
(*Ircinia strobilina*)
Class Demospongiae

Description: 18″ (46 cm) wide, 24″ (61 cm) high. Cylindrical, *cake-shaped.* Dark gray to dull black. Surface with *irregular "peaks"; pores large.* Soft and resilient.

Habitat: Attached to rocky outcrops rising short distances from the bottom; from shallow water to water 50′ (15 m) deep.

Range: Florida to Mexico; Bahamas; West Indies.

Comments: The sponge *Spheciospongia vesparia,* a species totally different from *I. strobilina* but found in the same range, is known by the same common name—thus the alternate name, "Cake Sponge."

22 **Stinker Sponge**
(*Ircinia fasciculata*)
Class Demospongiae

Description: 8″ (20 cm) wide, 12″ (30 cm) high. Nodes, lobes, or thick branches.

Yellow to brown or purplish. Surface with *many short, rounded lobes. Pores large,* conspicuous.

Habitat: Attached to rocks rising a little above the substrate; in shallow water, on reefs, to water 50′ (15 m) deep.

Range: Florida to Mexico; Bahamas; West Indies.

Comments: While most sponges have an unpleasant smell, the Stinker Sponge, as its name implies, emits an especially repulsive odor. Though the sponge is resilient while alive, it becomes hard and brittle when dried. Stinker Sponges are commonly found washed ashore after storms.

19, 21 **Vase Sponge**
(*Ircinia campana*)
Class Demospongiae

Description: 24″ (61 cm) wide, 36″ (91 cm) high. *Vase-shaped or bell-shaped,* with a *deep central cavity.* Reddish to reddish-brown. Surface irregular, with small pores and coarse longitudinal ribs.

Habitat: Attached to rocks projecting a short distance above the bottom; from below low-tide line to water 50′ (15 m) deep.

Range: Florida to Mexico; Bahamas; West Indies.

Comments: An accumulation of coral sediment is usually present in this sponge's central cavity, apparently with no ill effects. This species, often washed ashore during storms, becomes tough and shrunken when dried.

---

20 **Do-not-touch-me Sponge**
(*Neofibularia nolitangere*)
Class Demospongiae

Description: 8″ (20 cm) wide, 12″ (30 cm) high. Massive. Brick-red to mahogany-brown. Dividing into *blunt, thick lobes,*

each with *large opening at top.*
Habitat: Often found at the base of Staghorn and Elkhorn corals and elsewhere around coral reefs; from shallow water to water 150′ (46 m) deep.
Range: Florida to Mexico; Bahamas; West Indies.
Comments: **HIGHLY TOXIC.** This sponge causes severe blisters if handled. Since another stinging sponge, the Fire Sponge, is also red, it is prudent to avoid touching any red sponge in subtropical waters.

23 Tube Sponge
(*Callyspongia vaginalis*)
Class Demospongiae

Description: 2″ (51 mm) wide, 36″ (91 cm) high. Cluster of 6 or more *erect, open-ended tubes.* Blue-green, gray-green, or gray-purple. Tough, rubbery; *outer surface pebbly,* inside surface velvety.
Habitat: On coral rocks; from below low-tide line to water more than 300′ (91 m) deep.
Range: Florida to Mexico; Bahamas; West Indies.
Comments: Brittle stars and basket stars commonly hide in these handsomely colored pipes, or perch atop them while feeding.

129 Fire Sponge
(*Tedania ignis*)
Class Demospongiae

Description: More than 12″ (30 cm) wide, 12″ (30 cm) high. Massive, *amorphous, usually lobed.* Bright red or red-orange. Smooth, with *few large pores.*
Habitat: In bays and lagoons, attached to rocks; from below low-tide line to water 50′ (15 m) deep.
Range: Florida to Mexico; Bahamas; West Indies.
Comments: **HIGHLY TOXIC.** This sponge causes

severe blistering and pain if touched with bare hands. Since this and the Do-not-touch-me Sponge, another stinger, are both red, it is wise to avoid touching any red sponge in subtropical waters. The Fire Sponge body is soft and easily torn, and its toxins may discourage damage or predation.

15 **White Sponge**
(*Geodia gibberosa*)
Class Demospongiae

Description: 10″ (25 cm) wide, 10″ (25 cm) high. *Rounded.* Whitish. Hard, smooth, with *clusters of tiny pores.*
Habitat: Commonly in turtle grass beds or around reefs, attached to rocks or plants; in shallow water.
Range: Florida to Mexico; Bahamas; West Indies.
Comments: The White Sponge's color and prevalence in turtle grass beds make it easy for the wader to find and identify.

24 **Chicken Liver Sponge**
(*Chondrilla nucula*)
Class Demospongiae

Description: 2″ (51 mm) wide, 2″ (51 mm) high. Mottled gray to black, with *several major lobes.* Surface *smooth and slippery,* with scattered small pores.
Habitat: Around coral reefs and in turtle grass beds, attached to rocks or plants; in shallow water.
Range: Florida to Mexico; Bahamas; West Indies.
Comments: This sponge's slick surface and its shape, size, and color are reminiscent of chicken livers, although its body is much firmer to the touch. It is common and easily located in turtle grass beds.

CNIDARIANS
(Phylum Cnidaria)

The Phylum Cnidaria includes the hydras, hydroids, jellyfishes, sea anemones, and corals. Members of this phylum are nearly all found in marine and brackish water.
The cnidarian body is radially symmetrical, and consists of a tube or sac with a single opening, the *mouth*, surrounded by *tentacles*. The body wall consists of an outer layer, the *epidermis*, which is separated from the *gastrodermis*, the layer lining the digestive cavity (*coelenteron*), by a middle layer, the *mesoglea*, which varies from a thin, non-cellular film to a thick layer which, in some forms, has so many cells that it resembles connective tissue. The epidermis includes cells specialized for production of nematocysts, a distinctive characteristic of cnidarians. Nematocysts provide an effective means of snaring prey animals and also offer protection against predators. The nematocyst is a capsule containing a long thread which is forcefully everted when triggered by contact with prey or other animals. It can be discharged only once. The cnidarian must constantly produce new nematocysts to replace used ones. There are three general types of nematocysts: those that penetrate and are toxic, those that entangle by spiraling around the bristles of prey, and those that entangle by being sticky. Humans coming into contact with certain cnidarians may sustain anything from a mild rash to severe blistering and, in extreme cases, congestive respiratory failure and death. Cnidarians nourish themselves chiefly by capturing and ingesting small animals. Exploded nematocysts both paralyze and entangle the prey, causing it to adhere to the tentacles, which then retract and bend in the direction of the mouth. A less common means of

feeding is the trapping of microscopic organisms in a thin sheet of mucus secreted onto the region surrounding the mouth (*hypostome*) and the tentacles. Some cnidarians, especially corals and sea anemones, harbor intracellular, symbiotic one-celled plants (*zoochlorellae, zooxanthellae*) which provide the host with nutrients and oxygen, the end-products of photosynthesis, and use up some of its carbon dioxide.

The cnidarian body may be a *polyp*—a tube with a mouth surrounded by tentacles, specialized for a sedentary (*sessile*) life attached to some solid object —or a saucer- or bowl-shaped *medusa* that floats free in the water and swims by pulsating contractions. The life cycle of a given cnidarian may include one or both of these forms. Polyps may be solitary or colonial, increasing colony size by forming outgrowths that develop into new individuals (*budding*). In a colony, polyps may be specialized for various functions, such as trapping food, defense, digestion, or reproduction.

The body or bell of a medusa is umbrella-shaped. An extension (*manubrium*) bearing the mouth is suspended from the underside. Tentacles are generally located on the margin of the bell and trail behind as the medusa swims.

Both asexual and sexual reproduction are found in this phylum. Asexual reproduction most commonly involves budding, but in some species is achieved by spontaneous splitting down the longitudinal axis (*binary fission*), or by regeneration of a whole organism or a colony from a fragment. Sexual reproduction also follows various patterns. Among species whose life cycle includes a medusa, it bears the reproductive organs (*gonads*). If the life cycle includes both medusa and polyp stages (*alternation of generations*), larvae produced sexually by medusae develop into polyps, which in turn bud off medusae asexually. Where the life cycle includes the polyp stage only, the polyp is sexual. In most cnidarians, the sexes are separate. Generally the egg is fertilized within the female's body and develops into a swimming *planula* larva which is released into the water. The planula is elongate, with a ciliated epidermis armed with nematocysts surrounding a solid mass of cells. In most species the planula attaches itself to its substrate by one end, mouth and tentacles develop at the other end, and the inner cell mass separates to form a coelenteron lined with gastrodermis. In those species whose life cycle includes only the medusa stage, the planula develops into a tiny medusa.

Classes of Cnidaria: The Phylum Cnidaria is divided into three classes:

Class Hydrozoa: Includes hydras, hydroids, hydromedusae, chondrophorans, siphonophorans, and hydrocorallines. These animals are characterized by having a non-cellular mesoglea, a gastrodermis lacking nematocysts, and, with a few exceptions, gonads in the epidermis. They may have either the polyp or the medusa body form, and a number of species pass through both stages in the life cycle.

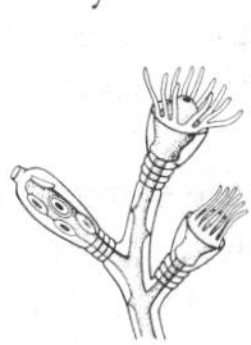

Class Scyphozoa: Includes those forms commonly known as jellyfish. The medusa is the dominant and in some cases the only stage. The small polyp stage (*scyphistoma*), when present, buds off small, 8-lobed medusae (*ephyrae*) by a series of transverse constrictions. Scyphozoan medusae have a thick, firm mesoglea. The coelenteron is subdivided into a number of chambers and canals. Its lining is equipped with nematocysts. The manubrium may be long or short, with 4 *oral arms*

surrounding the mouth. Gonads are located in pouches of the coelenteron.

Class Anthozoa: Includes the soft corals, the sea anemones, and the stony corals. All anthozoans are polyps with no known medusa stage. The coelenteron is subdivided into chambers by radial partitions, the *septa* or *mesenteries,* which extend from the body wall toward the center. The mouth is situated on an *oral disk,* surrounded by tentacles. It is generally slitlike and may have a *siphonoglyph,* a groove in the wall of the *pharynx,* at one or both ends; the pharynx opens into the coelenteron. The gastrodermis contains nematocysts and gonads. Sexes are usually separate. Some species are *oviparous,* shedding their eggs into the water. Others are *viviparous,* with fertilization taking place in the body of the female. After fertilization, the egg develops into a minute *planula* larva. Some sea anemones brood their young beyond the planula stage and release them as polyps. Except for sea anemones, nearly all anthozoans are colonial and undergo budding.

Subclass Octocorallia: These colonial animals are sometimes called "soft corals" because of the tough, elastic matrix they secrete, into which the polyps can retract. The sea whips and sea fans have a horny or woody core over which the softer tissues lie. Colonies may be bushy, whiplike, or fanlike. Their polyps have only 8 tentacles, each pinnately branched (like a feather).

Polyps of octocorals are usually small. Octocorals are common in tropical waters but rarer in cold waters.

Subclass Zoantharia: Stony corals, anemones, and their relatives have solitary or colonial polyps with over 8 rarely-branched tentacles.

Order Zoanthidea: The zoanthideans are small, anemonelike polyps without a skeleton and with one siphonoglyph. They are either solitary or colonial.

Order Actiniaria: Sea anemones are solitary, sessile cnidarians, generally cylindrical in form when fully extended. The anemone body consists of a *column,* at the bottom of which is a *pedal disk* that attaches the animal to the substrate and at the top of which is an *oral disk* bearing the slitlike mouth and surrounded by one or more rows of tentacles which vary greatly in size and number. The extended animal retracts by contracting its septal muscles, which lower the oral disk and tentacles, and by contracting its body wall to drive out water. The fully retracted animal is hemispherical in shape, with only an indentation at the site of the retracted oral disk. Some species have pores through which, when disturbed, they extrude filaments (*acontia*) loaded with nematocysts. While some anemones live with the entire body exposed, others bury themselves in a sandy or muddy bottom, exposing the oral disk only when fully extended, and retracting into the bottom when disturbed at low tide. Anemones are capable of creeping about on the pedal disk. A few are known to detach and "somersault" to a new position, while some can swim by flexing the body.

Order Ceriantharia: These cnidarians are anemonelike polyps with greatly elongate bodies adapted for living in sand or mud in secreted soft tubes. Their tentacles are arranged in two whorls, and they lack a basal disk.

Order Scleractinia: The stony corals are mostly warm-water creatures and are structurally similar to sea anemones. They deposit a skeleton of calcium carbonate at their base. The skeleton conforms to the configuration of the base of the *polyp,* including its pattern of internal *septa.* The deposited

skeleton extends partially up the column of the polyp and forms a *cup* in which the polyp sits. While some species of coral are solitary, consisting of only a single polyp, most are colonial, reproducing by budding and branching. The entire skeletal structure of the colony is called the *corallum*. The cup is commonly round in many species, with individual cups elongating into ovals as their polyp undergoes binary fission. In some forms this tendency to elongate is maintained as a growth pattern, and occasionally causes the initial cup to become so long as to form *valleys* and alternating *hills* (*e.g.* the Brain Coral, *Diploria*).
The corallum of different species varies in form from highly branched or bushy to solid, massive boulders.
Identification of corals is based largely on skeletal features, such as the form of the corallum, the size and form of the cup, and details of the septa. Species descriptions include information on skeletal details.

Order Corallimorpharia: Cnidarians in this order resemble true corals, but lack skeletons. They have a basal disk and radially arranged tentacles.

89 Solitary Hydroid
"One-armed Jellyfish"
(*Hybocodon pendula*)
Class Hydrozoa

Description: 4″ (102 mm) high, ¾″ (19 mm) wide. *Polyp* solitary, with slender, naked, bulbous-based stalk and rootlike holdfasts. Whitish-pink to scarlet. *Whorl of 30 long, threadlike tentacles near base of pear-shaped head and several rows of shorter ones encircling mouth.* Clusters of reproductive organs on branching stalks attached above basal whorl. *Medusa* ¼″ (6 mm) high, ¼″ (6 mm) wide. Dome-shaped, with 1 long tentacle. Feeding tube pink or pale purple, dotted with pink granules.

Habitat: Polyp rooted in sand, mud, or algal mat; below low-tide line. Medusa floats near surface.

Range: Arctic to Long Island Sound.

Comments: These hydroids feed by bending over and dragging their tentacles along the bottom. A related species, the Pacific Solitary Hydroid (*Corymorpha palma*), is found in similar habitats in southern California.

88 Tubularian Hydroid

(*Tubularia crocea*)
Class Hydrozoa

Description: *Polyp colony* more than 12″ (30 cm) wide. Single pink polyps 5″ (127 mm) high on sparsely branched stems rising from a creeping horizontal stem. Head pear-shaped, with *whorl of short, threadlike tentacles around mouth, longer ones around base;* 24 tentacles in each whorl. Grapelike clusters of reproductive organs, attached above basal whorl, hang down below it. *No medusa stage.*

Habitat: Attached to almost any solid object continuously submerged in shallow water.

Range: Nova Scotia to Cape Hatteras, possibly to Florida; Washington to California.

Comments: These hydroids commonly encrust boat hulls. Related species are the Ringed Tubularian (*T. larynx*), 2″ (51 mm) high, which has highly branched stems with circular constrictions, the Tall Tubularian (*T. indivisa*), with about 40 tentacles per whorl and polyps up to 12″ (30 cm) high, and the Sparsely-branched Tubularian (*T. spectabilis*), distinguished from *T. crocea* only by details of its reproductive organs.

84, 494 **Clapper Hydromedusa**
(*Sarsia tubulosa*)
Class Hydrozoa

Description: *Polyp colony* ¾″ (19 mm) high, 4″ (102 mm) wide. Polyp small. Colorless to pink. On sparsely branching stem. Head bulbous, with *12 or more knobbed tentacles scattered over surface.* Reproductive buds attached near base of head. *Medusa* ¾″ (19 mm) high, ⅝″ (16 mm) wide. Thimble-shaped. Feeding tube, canals, and tentacle bases yellowish, red, brown, or blue. *Feeding tube extends below rim of bell.* Mouth simple. 4 radial canals, 4 long, trailing tentacles, each with black eyespot at base.

Habitat: Polyps attached to rocks; below low-tide line. Medusa floats near surface.

Range: Arctic to Chesapeake Bay.

Comments: In this species' medusa stage, the mouth tube hangs below the body like the clapper of a bell.

86 **Feathered Hydroid**
(*Pennaria tiarella*)
Class Hydrozoa

Description: *Polyp colony* 6″ (15 cm) high, 6″ (15 cm) wide. Bushy, *branched alternately in one plane like a feather.* Stems covered with tough, horny, yellow to black sheath, and ringed above attachment of each branch. White to pink flask-shaped head has *5 irregular whorls of knobbed tentacles below mouth, basal whorl of 12 threadlike tentacles,* and a few white to rosy pink reproductive organs above basal tentacles. *Medusa* 1/16″ (2 mm) high, 1/25″ (1 mm) wide. Cup-shaped. Deep pink. 4 pink-spotted radial canals and 4 short, white tentacle bulbs along rim.

Habitat: Attached to solid objects; in shallow water below low-tide line. Medusa floats near surface.

Range: Maine to Florida and Texas; West Indies.

Comments: **Mildly toxic.** This hydroid is capable of delivering a mild sting when handled.

87 **Club Hydroid**
(*Clava leptostyla*)
Class Hydrozoa

Description: *Polyp colony* 3/8" (10 mm) high, 1" (25 mm) wide. *Unbranched,* rising from a network of creeping horizontal stems. Pink to reddish-orange. *30 threadlike tentacles scattered over top 1/4 of club-shaped heads.* Clusters of reproductive organs just below tentacles. *No medusa stage.*

Habitat: Usually attached to rockweeds and knotted wrack, occasionally on rocks; just above low-tide line and in bays and shallow water.

Range: Labrador to Long Island Sound; c. California.

Comments: These hydroids can be found growing in velvety clusters in tidepools.

38 **Snail Fur**
(*Hydractinia echinata*)
Class Hydrozoa

Description: *Polyp colony* 1/8" (3 mm) high. Dense, furry. Whitish, pale pink, or reddish-orange. Tough, *jagged-spined crust that coats entire snail shell occupied by hermit crab.* 3 kinds of polyps, connected by creeping stem: slender feeding polyps, with up to 30 threadlike tentacles in irregular whorl below mouth; long, flexible protective polyps with no mouth or tentacles, ending in large knob of stinging cells; reproductive polyps with stalked, saclike organs, either male or female. *No medusa stage.*

Habitat: On snail shells occupied by hermit crabs; occasionally on stones.

Range: Labrador to Florida and Texas; British Columbia to c. California.

Comments: As the hermit crab moves, the Snail Fur polyps reach new water from which to capture food. The stinging cells of the hydroid may in turn protect the crab from being eaten. Miller's Hydractinia (*H. milleri*), ⅛″ (3 mm) high, a related species, occurs as pink patches on rocks and pilings on the West Coast, from British Columbia to California. The similar Smooth-spined Snail Fur (*Podocoryne carnea*), ⅛″ (3 mm) high, differs in the texture of its spines, and buds off tiny medusae which bear the sex organs.

---

80 Bougainvillia Hydroids
(*Bougainvillia* spp.)
Class Hydrozoa

Description: *Polyp colony* 12″ (30 cm) high, 8″ (20 cm) wide. Dense, bushy, irregularly branched. Whitish. *Sheathed stems* single, or several fused together, *branches usually have ringlike constrictions* at base. *Cone-shaped heads,* partly enclosed in stem sheath, have whorl of 8–20 basal tentacles. Reproductive bodies roundish, with short stalks, attached to branches singly or in clusters. *Medusa* ½″ (13 mm) high and equally wide. Egg-shaped. Transparent. 4 radial canals, *4 clusters of 3–9 threadlike tentacles* with yellow or reddish bases. 4-sided feeding tube short, with *4 branching tentacles.*

Habitat: Polyps on rocks and other solid objects; near low-tide line and below in shallow water. Medusa floats near surface.

Range: Arctic to Florida and Texas; Bahamas; West Indies; Mexico; Alaska to s. California.

Comments: This genus includes many species. Because of the small size of the polyp, laboratory examination is necessary to determine species.

497 **Constricted Jellyfish**
(*Catablema vesicarium*)
Class Hydrozoa

Description: *Medusa* 1″ (25 mm) high, ⅜″ (10 mm) wide. Transparent. *Constriction at the level of attachment of feeding tube demarking a separate region.* Four radial canals. *32 threadlike tentacles* of various lengths, with dark red eyespot on base of each long tentacle. Feeding tube almost as long as inside of bell, with 4 prominent, frilly reproductive organs attached. Mouth with 4 frilled lips. Manubrium and tentacle bases golden-brown. *Polyp stage unknown.*

Habitat: Floating in open water, occasionally near shore.

Range: Arctic to Cape Cod.

Comments: When contracted, this medusa clearly shows its constriction as a kind of waistline.

82 **Stick Hydroid**
(*Eudendrium ramosum*)
Class Hydrozoa

Description: *Polyp colony* 6″ (15 cm) high, 6″ (15 cm) wide. *Highly and irregularly branched.* White to pinkish-green. Stems fused together in bundles; circular constrictions at base of each branch and stalk. Heads have *flared, projecting mouths,* each surrounded by *24 threadlike tentacles. Reproductive organs at base of tentacles,* or below. *No medusa stage.*

Habitat: Attached to rocks and other hard objects and the bases of seaweeds; near low-tide line and below.

Range: Arctic to Florida and Texas; Bermuda; Bahamas; West Indies; s. California.

Comments: These hydroids resemble a miniature forest. Related Atlantic species include the Red Stick Hydroid (*E. carneum*), 4¾″ (121 mm) high and 4″ (102 mm) wide, and bright red; the White Stick

Hydroid (*E. album*), with colonies only ⅜" (10 mm) high and ¾" (19 mm) wide; and the Slender Stick Hydroid (*E. tenue*), ⅝" (16 mm) high and 1" (25 mm) wide, with simple stems and branches without ringlike constrictions. The California Stick Hydroid (*E. californicum*), 6" (15 cm) high and 3" (76 mm) wide, found from British Columbia to central California, has stiff brown stems and branches with abundant ringlike constrictions.

495 **Penicillate Jellyfish**
(*Polyorchis penicillatus*)
Class Hydrozoa

Description: *Medusa* 1⅝" (41 mm) high, 1⅜" (35 mm) wide. Bell globular, transparent, with 90 tentacles on margin, each with a dark red base bearing an eyespot. 4 radial canals with *numerous short side branches,* and numerous long, slender gonads suspended below each canal; feeding tube short, with 4-cornered mouth at top. *Polyp stage unknown.*
Habitat: Floating; in coastal and bay waters.
Range: British Columbia to s. California.
Comments: This is among the largest of hydromedusae on the West Coast.

500 **Many-ribbed Hydromedusa**
(*Aequorea aequorea*)
Class Hydrozoa

Description: *Medusa* 1½" (38 mm) high, 7" (18 cm) wide. Saucer-shaped. Glassy-transparent with thick jelly. *Radial canals narrow, 80 or more.* Marginal tentacles long, numbers vary. Feeding tube short, wide; mouth with ruffled lips as numerous as radial canals. Gonads slender, extending most of length of radial canals, bluish in male, rosy in female. *Polyp stage unknown.*

Habitat: Floating in open water, occasionally near shore.
Range: Maine to Texas; Alaska to California.
Comments: This worldwide species of large, familiar jellyfish is frequently washed up on beaches. It is luminescent, and at night one can see the outline of its parts in "living light."

75, 77 **Wine-glass Hydroids**
(*Campanularia* spp.)
Class Hydrozoa

Description: *Polyp colony* 10″ (25 cm) high, 6″ (15 cm) wide. Sheathed stems regularly or irregularly branched, or unbranched; single, or several fused together. Whitish. Upright stems arise from creeping base; stems and branches with or without circular constrictions. *Feeding head sheath flaring, wine-glass-shaped, stalked;* margin smooth or toothed. Polyp has flaring mouth, *single whorl of 20 threadlike tentacles.* Sheath surrounding reproductive buds more than twice as long as feeding head sheath; cylindrical; slightly flared, *stalkless, without neck. No medusa stage.*
Habitat: On rocks, pilings, and other hard objects, and on seaweeds; from low-tide line to water 1380′ (420 m) deep.
Range: Labrador to Florida; Bermuda; Bahamas; West Indies to Venezuela; Alaska to s. California.
Comments: These hydroids belong to a large family characterized by the stalked, wine-glass-shaped sheath which surrounds and protects the feeding head. Unlike its close relative *Obelia, Campanularia* does not release free-swimming medusae. Instead, the medusae remain within the sheath without developing mouth or tentacles, mature sexually, and produce ciliated larvae which escape into the water and disperse. Laboratory examination is necessary to determine species.

78 Zig-zag Wine-glass Hydroid
(*Obelia geniculata*)
Class Hydrozoa

Description: *Polyp colony* 1″ (25 mm) high, more than 12″ (30 cm) wide. *Stems simple, unbranched, zig-zagged, rising in a row* from creeping horizontal base. Whitish. Ringed stalks of polyps rising from each "knee" of zig-zag stem. Head sheath conical, margin smooth, feeding head with flared mouth. Single whorl of up to 20 basal tentacles. *Sheath around reproductive buds urn-shaped, with collar,* attached by short, ringed stalk in angle between stem and stalk of head sheath. *Medusa* 1/16″ (2 mm) high, 1/4″ (6 mm) wide. Almost flat. 100 short tentacles at rim; 8 marginal sense organs; 4 radial canals with yellow gonad under each. Feeding tube yellow, short. Mouth has 4 simple lips.

Habitat: Polyps attached to large kelp, Sargasso weed, rocks; in shallow water. Medusa floats near surface.

Range: Arctic to Florida and Texas; West Indies; British Columbia to s. California.

Comments: The medusae of *O. geniculata* often swim with the bell turned inside out like wind-blown umbrellas. The Two-branched Wine-glass Hydroid (*O. dichotoma*) ranges from Alaska to Baja California. The slender colony rises from a creeping base, and can grow to 4¾″ (121 mm) high in 2–3 months. It is sparsely branched, with 2 forks at each branching.

---

79 Bushy Wine-glass Hydroids
(*Obelia* spp.)
Class Hydrozoa

Description: Colony 8″ (20 cm) high, 4″ (102 mm) wide. Bushy. Whitish. Colony highly branched, each branch with *rings just above attachment to stem.* Feeding heads, head sheath, and sheath around reproductive buds like those of the Zigzag Wine-glass Hydroid.

Habitat: Attached to rocks, shells, pilings, floats, and other solid objects; from low-tide line to water 165′ (50 m) deep.

Range: Both coasts of the United States.

Comments: Recent work on the classification of *Obelia* suggests that there are only two species of Bushy Wine-glass Hydroids, *O. bidentata* and *O. dichotoma.* They cannot be distinguished in the field. The Double-toothed Bushy Wine-glass Hydroid (*O. bidentata*) ranges from Cape Cod to Florida and the West Indies, and from Puget Sound to s. California, and the Two-branched Wine-glass Hydroid (*O. dichotoma*) is found from Alaska to Baja California.

---

501 **White-cross Hydromedusa**
(*Staurophora mertensi*)
Class Hydrozoa

Description: *Medusa* 2″ (51 mm) high, 12″ (30 cm) wide. Flattened hemisphere. Transparent blue. Numerous short, coiling, rosy or yellowish marginal tentacles, each with a brown eyespot at base. Digestive cavity, mouth, and 4 radial canals combine to form a *large, white cross reaching nearly across the bell.* Mouth has 4 slitlike arms with much-folded lips. White gonads attached to 4 radial canals. *Polyp stage unknown.*

Habitat: Found floating near shore, close to surface at night, deeper during the day.

Range: Arctic to Rhode Island.

Comments: The life cycle of this species is not known. The medusae feed on small crustaceans and other medusae.

498 Eight-ribbed Hydromedusa

(*Melicertum octocostatum*)
Class Hydrozoa

Description: *Polyp colony* ⅛″ (3 mm) high, 1″ (25 mm) wide. Transparent, whitish. Unbranched, club-shaped feeding heads rising from horizontal creeping stem. Mouth conical, with whorl of threadlike tentacles. Slender stalk bears globular reproductive buds, each producing 3 medusae at a time, ½″ (13 mm) high and equally wide. *Medusa* bell-shaped. Transparent, white to yellow or yellowish-brown. *8 radial canals,* each with a gonad reaching nearly to margin. Feeding tube short, 8-sided, wider at base. Mouth with 8 slightly folded lips. *72 long marginal tentacles, coiled at free end,* alternate with equal number of shorter ones.

Habitat: Polyps attached to rocks and shells; well below low-tide line. Medusa floats near surface.

Range: Arctic to Cape Cod.

Comments: This medusa swims by remarkably rapid contractions of its bell. The species name, *octocostatum,* means "8 ribs" and describes the 8 radial canals and gonads.

499 Elegant Hydromedusa

(*Tima formosa*)
Class Hydrozoa

Description: *Medusa* 2½″ (64 mm) high, 4″ (102 mm) wide. Rounded to slightly conical. Transparent, with white or pink organs. *8 long tentacles, and 8 intermediate-length tentacles alternate with 16 short tentacles.* Feeding tube extends well below margin of bell, ending in mouth with *4 long lips with ruffled margins.* 4 radial canals, with convoluted gonads suspended along most of their length. *Polyp stage unknown.*

Habitat: Floating in open water, occasionally near shore.
Range: Arctic to Cape Hatteras.
Comments: This large medusa is found north of Cape Cod in late summer and fall, and year-round south to Rhode Island.

72 **Halecium Hydroid**
(*Halecium halecinum*)
Class Hydrozoa

Description: *Polyp colony* 3″ (76 mm) high, 1″ (25 mm) wide. Erect, rigid. Grayish-white. The few primary branches cemented together. *Head sheath reduced to a slight flare.* Feeding head small, with 12 short, rather thick tentacles. Sheath of reproductive organ pouchlike, attached directly to stem. Female has opening at one side of end bearing two small feeding heads. Reproductive organs reduced to saclike structures bearing gonads. *No medusa stage.*
Habitat: Attached to rocks, shells, and seaweed; from low-tide line to water 40′ (12 m) deep.
Range: Arctic to Cape Hatteras; British Columbia to c. California.
Comments: Among related species on the Atlantic Coast are the Flared Halecium Hydroid (*H. tenellum*), which is sparsely branched, ⅝″ (16 mm) tall and ¼″ (6 mm) wide, and Bean's Halecium Hydroid (*H. beani*) and the Graceful Halecium Hydroid (*H. gracile,*) which are taller, 2″ (51 mm) high and ½″ (13 mm) wide.

71 **Fern Garland Hydroids**
(*Abietinaria* spp.)
Class Hydrozoa

Description: *Polyp colony* 12″ (30 cm) high, 6″ (15 cm) wide. Bushy. Whitish, yellowish, orange, or reddish-brown. Stems stout, *branches alternate, all in one plane. Polyp encased in flask-shaped tube* with narrow neck, *tubes alternate on sides of branch;* polyp small, with 12 slender tentacles. Reproductive structures are oval cases with smooth, ringed, or ridged walls. *No medusa stage.*

Habitat: On ledges, rocks, and shells; from low-tide line to water 1440′ (439 m) deep.

Range: Labrador to Cape Cod; Alaska to s. California.

Comments: The arrangement of branches in one plane gives these hydroids a fernlike appearance. The largest specimens occur well below low-tide line, and are collected by dredging or by scuba diving. The small size of structures used for identification makes laboratory examination necessary for accurate determination of species.

---

81 Garland Hydroid
(*Sertularia pumila*)
Class Hydrozoa

Description: *Polyp colony* 2″ (51 mm) high, 1″ (25 mm) wide. Clusters of stiff stems with opposite branching. Creamy to pale tan. *Head sheaths in opposite pairs,* tubular, lower ½ attached to stem, upper ½ curved outward. *Margins have 2 teeth.* Aperture closed by lid with 2 flaps. Feeding heads small, slender, with a single whorl of 12 threadlike tentacles. Gonad sheath large, egg-shaped, smooth, with short collar and large opening. *No medusa stage.*

Habitat: Attached in tufts to rockweed, knotted wrack, rocks, and other solid objects; near low-tide line and below.

Range: Arctic to Cape Hatteras.

Comments: This species is representative of a large family of widely distributed hydroids of similar structure. *Sertularella* differs in having the head sheaths alternate instead of opposite on the stem. *Abietinaria* also has alternate head sheaths, and is highly branched in 1 plane. All 3 of these genera have West Coast representatives ranging from British Columbia to southern California. The Forked Garland Hydroid (*Sertularia furcata*), 2" (51 mm) high and 1" (25 mm) wide, is common within that range. It is golden or tan, with numerous upright stalks rising from a mesh of creeping bases on surfgrass and various algae, and is found on rocky shores from near low-tide line to water 165′ (50 m) deep.

74 Silvery Hydroid
(*Thuiaria argentea*)
Class Hydrozoa

Description: *Polyp colony* 12" (30 cm) high, 4" (102 mm) wide. Branches rise from all sides of each slender stem. Silvery. *Head sheaths alternate up 2 sides of stem;* tubular, tapered, slanting outward with about ⅓ free of stem. Margin has 2 teeth, one often longer. Feeding head small, with 12 tentacles. Gonad sheath triangular, with two shoulder spines, one on either side of raised central collar containing the wide aperture. *No medusa stage.*

Habitat: Attached to rocks, shells, and other solid objects; below low-tide line.

Range: Arctic to Cape Hatteras; Alaska to s. California.

Comments: This graceful hydroid is called "White Weed" in England, where it is collected, dyed green, dried, and shipped to the United States for sale in flower shops as "Sea Fern."

70 **Tropical Garland Hydroid**
(*Sertularella speciosa*)
Class Hydrozoa

Description: *Polyp colony* 8″ (20 cm) high, 6″ (15 cm) wide. Flat, branching. Yellowish to tan. *Main stem stout,* sometimes several stems fused, with *branches almost at right angles and alternate.* Polyps small, with 12 slender basal tentacles encased in slender tubes with 4 teeth at opening, placed alternately on sides of branch or stem; *reproductive organs enclosed in cone-shaped sheath 5 times as long as polyp tube,* with longitudinal ridges. *No medusa stage.*

Habitat: On rocks and coral boulders; from below low-tide line to water 528′ (161 m) deep.

Range: Bermuda; s. Florida and West Indies.

Comments: This genus resembles the genus *Sertularia,* but differs in having its polyps arranged alternately instead of oppositely on the stem. Among the many species is the Turgid Garland Hydroid (*Sertularella turgida*), ranging from British Columbia to southern California. *S. turgida* is yellowish, 2″ (51 mm) high, 1½″ (38 mm) wide, and highly branched, with conspicuous, plump reproductive bodies.

67, 69 **Feathery Hydroids**
(*Aglaophenia* spp.)
Class Hydrozoa

Description: *Polyp colony* ⅝–24″ (16 mm–61 cm) high, ⅛–4″ (3–102 mm) wide. Featherlike. Whitish, yellowish, tannish, or reddish. Stem simple or branching, usually in clusters rising from creeping base; *upright stems jointed, with parallel jointed branches in one plane* like a feather. *Head sheaths on upper side of branch only,* tubular, flared, one side fused to branch. *Row of reproductive buds* on branches, *covered by pairs of leaflike structures* bordered by knobs of stinging cells. *No medusa stage.*

Habitat: Attached to rocks, shells, or seaweeds; from low-tide line to water more than 8000′ (2438 m) deep.

Range: Cape Cod to Florida and Texas; Bermuda; Bahamas; West Indies to Brazil; Alaska to s. California.

Comments: Most of the many species described for this genus occur in warm, subtropical Atlantic waters. It is possible that upon further study some of the hydroids described will no longer be considered separate species, since the differences cited in some cases are very small.

505 Angled Hydromedusa
(*Gonionemus vertens*)
Class Hydrozoa

Description: *Medusa* ½″ (13 mm) high, ¾″ (19 mm) wide. Dome-shaped. Transparent. Feeding tube not quite reaching bell margin, thickest where attached. Mouth has 4 slightly frilled lips. Ruffled sex organs extend along most of length of 4 radial canals, creating cross-shaped marking. 60 long marginal tentacles, each with *spiral or ringlike clusters of stinging cells and an adhesive sucker near end.* Feeding tube, gonads, and tentacle bases yellowish-tan to reddish-brown. *Polyp stage unknown.*

Habitat: Floating in shallow water; sometimes clinging to eelgrass.

Range: Arctic to Cape Cod; Alaska to c. California; common in Puget Sound.

Comments: When the medusa attaches itself to a rock or seaweed, the tentacles form an angle at the sucker, hence the species' name.

487 **Chain Siphonophore**
(*Stephanomia cara*)
Class Hydrozoa

Description: Colony 3″ (76 mm) high, ¼″ (6 mm) wide. *Parts attached in a line* to a stemlike axis. Transparent, iridescent. Oil-filled float at the front end, 6 *pairs of vase-shaped swimming bells* that contract rhythmically like a medusa; long, tubular, white or pink-scarlet feeding polyps, partially concealed by flat, leaflike covers; long tentacles with numerous branches, each tipped with tuft of stinging cells; reproductive polyps among feeding polyps on latter half of stemlike axis.

Habitat: Floats in shallow water.

Range: Nova Scotia to Chesapeake Bay.

Comments: Like all siphonophores, this species is a floating, free-swimming colony made up of individual polyps. The colony moves slowly forward by the action of the pulsating bells, trailing its tentacles. This creature is delicate and breaks readily when handled.

512, 513 **Portuguese man-of-war**
(*Physalia physalis*)
Class Hydrozoa

Description: Float 12″ (30 cm) long, 6″ (15 cm) high, 5″ (127 mm) wide. *Float gas-filled. Iridescent pale blue* and pink, with large, *deflatable, pink-ridged crest* above. Dense cluster of 3 kinds of polyps suspended underneath. Tentacles of different lengths, some more than 60′ (18 m) long, containing blue, beadlike stinging cells; blue tubular feeding parts with terminal mouths and no tentacles; treelike, branching gonads, salmon-pink when mature.

Habitat: Surface of the sea.

Range: Florida to Texas and Mexico; Bahamas; West Indies. Driven ashore by storms from Gulf Stream to Cape Cod.

Comments: **HIGHLY TOXIC.** This siphonophore floats on the surface by means of a gas-filled, balloonlike float that changes shape to catch the prevailing wind. Its tentacles contain one of the most powerful poisons known in marine animals and can inflict severe burns and blisters even when the animal is dead on the beach. The Horse Mackerel (*Nomeus gronovii*) lives among the tentacles with impunity, acting as a lure to other fish while receiving protection and fragments of food from the man-of-war.

515, 516 By-the-wind Sailor
(*Velella velella*)
Class Hydrozoa

Description: Float 4″ (102 mm) long, 3″ (76 mm) wide, 2″ (51 mm) high. Float consisting of flat, oval, *cartilagelike skeleton* full of gas-filled pockets, with *vertical triangular crest set diagonally across the top,* serving as a sail. *Blue, transparent.* Single large-mouthed feeding tube, surrounded by rows of reproductive bodies. Numerous blue tentacles around the rim.

Habitat: Surface of the sea.

Range: Warm waters. Driven ashore from the Gulf Stream by storms, as far north as Cape Hatteras, or occasionally farther, and as far north as c. California from the tropical Pacific.

Comments: Although they contain stinging cells, the tentacles of the Sailor are harmless to man. The By-the-wind Sailor can "tack" in the manner of a sailboat.

504 **Blue Buttons**
(*Porpita linneana*)
Class Hydrozoa

Description: Float ¼" (6 mm) high, 1" (25 mm) wide. *Round, almost flat.* Bright blue. Single mouth underneath, surrounded by blue gonads and *tentacles with numerous branchlets ending in knobs* of stinging cells.
Habitat: Surface of the sea.
Range: Driven ashore from Gulf Stream by storms, as far north as Cape Hatteras or occasionally farther, and as far north as c. California from the tropical Pacific.
Comments: Since Blue Buttons does not have a sail, it is blown ashore less frequently than the Portuguese man-of-war or the Sailor. In tropical waters it can be seen by the thousands, dotting the water with blue for miles.

124 **Purple Stylasterine**
(*Allopora porphyra*)
Class Hydrozoa

Description: 6" (15 cm) wide, more than ⅛" (3 mm) high. Colony encrusting, limy. *Vivid purple. Covered with scalloped pits,* each containing 12 feeding polyps surrounded by mouthless stinging polyps.
Habitat: On protected, shaded faces of exposed rocks; near low-tide line and below.
Range: British Columbia to c. California.
Comments: Members of this genus seem to thrive best on vertical surfaces where they are not silted or overgrown by algae.

25 **Fire Coral**
(*Millepora alcicornis*)
Class Hydrozoa

Description: More than 24" (61 cm) high, more than 18" (46 cm) wide. Upright, *branching or platelike,* on coral rock or encrusting mollusk shells or skeletons of horny corals. Brown to creamy-yellow. Covered with *tiny pores occupied by whitish polyps.*

Habitat: Coral reefs, cement pilings, and other submerged structures.

Range: Florida to Mexico; Bahamas; West Indies.

Comments: **HIGHLY TOXIC.** People touching the Fire Coral suffer a severe burning sensation and blistery rash.

---

90 **Horned Stalked Jellyfish**
(*Lucernaria quadricornis*)
Class Scyphozoa

Description: 3″ (76 mm) high, 2″ (51 mm) wide. Olive, brown, or white. *4-sided cone* with each of its *8 arms ending in pompom of 100 knobbed tentacles.* Alternate notches between arms shallow so that arms are paired together like horns. Mouth has 4 lips. Reproductive organs along length of arms. Stalk at end opposite mouth ½ total height.

Habitat: Attached to large, brown kelp; near and below low-tide line.

Range: Greenland to Cape Cod.

Comments: This species and the Goblet Stalked Jellyfish (*Craterolophus convolvulus*) closely match the color of the seaweeds on which they sit. *Craterolophus* is smaller, 1½″ (38 mm) high and ¾″ (19 mm) wide, with a much shorter stalk and evenly spaced pompoms of tentacles.

---

40 **Trumpet Stalked Jellyfish**
(*Haliclystus salpinx*)
Class Scyphozoa

Description: 1″ (25 mm) high, ½″ (13 mm) wide. Translucent, with red, orange, yellow, or tan. Widely flared when expanded, with each of 8 *short arms ending in a pompom of 100 knobbed tentacles.* Notches between arms spaced equally, each notch with *trumpet-shaped anchor* that has small tentacle in middle and ridged ring around base. Mouth has 4 lips. Reproductive organs situated along length of arms.

Habitat: Attached to eelgrass, kelp, rockweed, other seaweeds, and occasionally rocks; near low-tide line and below in shallow water.

Range: New Brunswick to Cape Cod.

Comments: The Trumpet Stalked Jellyfish is a trap for small crustaceans, and when one of these comes into contact with a tentacle, it is immediately put into the mouth and swallowed. The Eared Stalked Jellyfish (*H. auricula*) is similar in appearance, but with a somewhat shorter stalk and holdfasts that are plump, oval, and shaped like an earlobe with a short tentacle in the center. It ranges farther south, to the south side of Cape Cod. *Haliclystus* species are also found from British Columbia to northern California.

---

503 Crown Jellyfish
(*Nausithoe punctata*)
Class Scyphozoa

Description: ⅜" (10 mm) high, ⅝" (16 mm) wide. Almost flat. Pale green to pale brown, with reddish spots. *Circular groove* separates center from ring of *16 marginal lobes.* Alternate lobes bear 8 tentacles and 8 marginal sense organs. *16 flaps* underneath lobes. Mouth simple, with 4 lips. Eight, *large, yellow, red, or brown gonads.*

Habitat: Floats near the surface.

Range: Cape Hatteras (and occasionally north) to Florida; Gulf Stream; West Indies.

Comments: The circular groove on the top surface gives the Crown Jellyfish its name. It is frequently found in warmer waters.

508 **Purple Jellyfish**
(*Pelagia noctiluca*)
Class Scyphozoa

Description: 3″ (76 mm) high, 4″ (102 mm) wide. Bell hemispherical. Rose-pink to purple or yellow. Warty. Margin has *16 rectangular lobes* which alternate with *8 long, rosy pink tentacles* and 8 marginal sense organs. Feeding tube long, thick, extending far below the bell as *4 long, frilly, pink-edged lips* surrounding the mouth. Four pink, ribbonlike gonads, placed horizontally. Luminescent at night.

Habitat: Surface of open ocean; sometimes washed ashore by storms.

Range: Warm waters off N. and S. America.

Comments: **Mildly toxic.** *Pelagia noctiluca* occurs in large swarms, which appear as glowing white balls at night. The Purple Banded Jellyfish (*P. colorata*), a much larger form, 32″ (80 cm) wide and 24″ (61 cm) high, is a Pacific species that is sometimes seen in the tidepools and along the West Coast of the United States. It is a handsome creature, with deep purple radial bands on a silvery white background. Although it is **HIGHLY TOXIC,** it is eaten by the Ocean Sunfish and the Blue Rockfish.

506, 510 **Sea Nettle**
(*Chrysaora quinquecirrha*)
Class Scyphozoa

Description: 5″ (127 mm) high, 10″ (25 cm) wide. Pink with radiating red stripes. 40 tentacles. Bay form 2″ (51 mm) high, 4″ (102 mm) wide. White. 24 tentacles. *Bell covered with fine warts. Margin divided into scalloped, shallow lobes. Long, yellow tentacles* alternate with marginal sense organs between lobes. Feeding tube extends well below bell margin as *4 long, ruffled, lacy lips.* 4 gonads, each a convoluted loop.

Habitat: Floats near surface.
Range: Cape Cod to Florida and Texas. Abundant in Chesapeake Bay.
Comments: **Mildly toxic.** Contact with a Sea Nettle usually results in a mild itchy irritation, but a person stung severely may require hospitalization. A related species, the Lined Sea Nettle (*C. melanaster*), is often washed ashore from Alaska to southern California. It is larger, 12″ (30 cm) wide and 8″ (20 cm) high, with yellow to brown radial lines on the bell. It may also sting severely.

511 **Lion's Mane**
(*Cyanea capillata*)
Class Scyphozoa

Description: 24″ (61 cm) high, 96″ (244 cm) wide. Bell saucer-shaped, upper surface smooth. Color varies with age and, thus, size: pink and yellowish to 5″ (127 mm), reddish to yellow-brown to 18″ (46 cm), darker red-brown when larger. *16 marginal lobes.* Shaggy clusters of *more than 150 tentacles* attached beneath 8 deep clefts between lobes, marginal sense organs in 8 shallower clefts. Feeding tube stout, extending as 4 much-folded, membranous lips around mouth. 4 highly folded, ribbonlike gonads suspended under bell alternate with lips.
Habitat: Floats near surface.
Range: Arctic to Florida and Mexico; Alaska to s. California.
Comments: **HIGHLY TOXIC.** This is the largest jellyfish in the world. Specimens 8 feet wide have been found. Contact with *Cyanea*'s tentacles produces severe burning and blistering. Prolonged exposure may cause muscle cramps and breathing difficulties. In Sir Arthur Conan Doyle's story, "The Adventure of the Lion's Mane," Sherlock Holmes

solves a homicide caused by contact between the victim and this medusa in a tidepool.

502 **Moon Jellyfish**
(*Aurelia aurita*)
Class Scyphozoa

Description: 3″ (76 mm) high, 16″ (41 cm) wide. Saucer-shaped. Whitish, translucent. *8 shallow marginal lobes,* sense organs in 8 clefts between lobes. Numerous short, *fringelike tentacles.* Feeding tube short, stout, expanding as 4 long oral arms with frilly margins. Numerous branching radial canals. *Reproductive organs horseshoe-shaped or round.* Ripe female organs: yellowish, pink, or violet; males': yellow, yellow-brown, or rose; immatures': whitish.
Habitat: Floats near surface; just offshore.
Range: Arctic to Florida and Mexico; Alaska to s. California.
Comments: **Mildly toxic.** This is the jellyfish most commonly washed up on beaches during high tide or after a storm. Its sting causes a slight rash that may itch for several hours.

509 **Upside-down Jellyfish**
(*Cassiopeia xamachana*)
Class Scyphozoa

Description: 2″ (51 mm) high, 12″ (30 cm) wide. Bell gray-green to brownish-yellow. Rather flat, with rounded edges. *80 marginal lobes,* with *16 marginal sense organs,* no marginal tentacles. Feeding tube stout, with 8 long, green or brown, fleshy *oral arms with grapelike clusters* on 15 primary branches and several large, ribbon-shaped filaments suspended beneath. Mouth subdivided into *many tiny pores* on oral arms.
Habitat: In warmer shallow waters, back-reef

lagoons, and mangrove bays.

Range: Florida to Texas and Mexico; Bahamas; West Indies.

Comments: **Mildly toxic.** Contact with the Upside-down Jellyfish causes itching followed by a rash. Thousands of these creatures lie side by side on their backs in shallow water, basking in the sun and exposing their oral arms to the currents.

507, 514 **Cannonball Jellyfish**
(*Stomolophus meleagris*)
Class Scyphozoa

Description: 5″ (127 mm) high, 7″ (18 cm) wide. Hemispherical, thick, tough. Milky-bluish or yellowish, with pale-spotted brown band around margin. Margin has 128 small lobes with *8 deep notches* containing prominent sense organs; *no marginal tentacles.* Feeding tube stout, with *16 short, forked oral arms.* Primary mouth present.

Habitat: Floats near shore.

Range: Chesapeake Bay to Florida and Texas; Bahamas; West Indies.

Comments: This jellyfish occurs in huge swarms along shores of the Gulf of Mexico. One swarm observed at Port Aransas, Texas, was estimated drifting through the channel at a rate of 2 million per hour.

37, 39 **Red Soft Coral**
(*Gersemia rubiformis*)
Class Anthozoa
Subclass Octocorallia

Description: 6″ (15 cm) high, 3″ (76 mm) wide. Soft, fleshy, with stout, *club-shaped branches* or cluster of pear-shaped lobes rising from main stem. Red to orange. Branches terminate in *clusters of polyps* set close together, each with *8 short, featherlike tentacles.*

Habitat: Attached to rocks, pilings, and other solid objects; below low-tide line.
Range: Arctic to Gulf of Maine; Alaska to n. California.
Comments: The needlelike limestone spicules imbedded in the Red Soft Coral's stem lend support to its structure.

68 **Corky Sea Fingers**
(*Briareum asbestinum*)
Class Anthozoa
Subclass Octocorallia

Description: 12″ (30 cm) high, 3″ (76 mm) wide. Purplish-gray. Thick, *corky, fingerlike branches.* Surface smooth or slightly bumpy, evenly covered with *numerous small pores.*
Habitat: On coral reefs; below low-tide line.
Range: Florida to Texas; Bahamas; West Indies.
Comments: When dried, the corky, spongy body of this species becomes tough and woody, and turns yellowish-brown. The Corky Sea Fingers is often washed ashore.

63 **Sea Whip**
(*Leptogorgia virgulata*)
Class Anthozoa
Subclass Octocorallia

Description: 36″ (91 cm) high, ½″ (13 mm) wide. Stems and branches *long, whiplike, tough,* pitted with small, *evenly distributed pores.* Purple, red, yellow-orange, or tan.
Habitat: Attached to rocks and other hard objects in shallow water; near low-tide line and below.
Range: New Jersey to n. Florida.
Comments: This horny coral is so named because its branches have a horny central core. The closely related Straight Sea Whip (*L. setacea*), 72″ (183 cm) high and ½″ (13 mm) wide, is unbranched. It ranges

up into Chesapeake Bay as far north as the Patuxent River in Maryland.

60, 66 **Sea Plumes**
(*Pseudopterogorgia* spp.)
Class Anthozoa
Subclass Octocorallia

Description: 36″ (91 cm) or more high, 12″ (30 cm) wide. Plumy, branching. Bluish-gray, violet, purple, yellow, or whitish. Branches alternate or opposite on stem, long or short, with *round or slitlike openings along edges of branches* into which polyps retract.
Habitat: Attached to rocks around reefs; below low-tide line in shallow water.
Range: Florida; Bahamas; West Indies.
Comments: Sea plumes, along with their relatives the sea fans and sea whips, are sensitive to bright light and expand their polyps only at night or on overcast days.

64 **Sea Fans**
(*Gorgonia* spp.)
Class Anthozoa
Subclass Octocorallia

Description: 36″ (91 cm) high, 36″ (91 cm) wide. Flattened; outline egg-shaped, with *small branches fusing* to make a continuous latticework. Pinkish-purple, yellow, or, rarely, white. *Smaller branches wider than thick,* compressed in same plane as stem. Numerous fine pores on branches.
Habitat: Attached to rocks around reef and on reef flat; below low-tide line.
Range: S. Florida; Bermuda; Bahamas; West Indies.
Comments: Sea fans evoke idyllic tropical seas, and are often collected, dried, and sometimes spray-painted for souvenirs.

59, 62 **Yellow Sea Whip**
(*Pterogorgia citrina*)
Class Anthozoa
Subclass Octocorallia

Description: 24″ (61 cm) high, 12″ (30 cm) wide; stem ½″ (13 mm) wide. Low; *branches flattened,* few, with *slitlike openings along edges.* Yellow, occasionally olive-gray.

Habitat: Attached to hard bottom; in shallow water.

Range: Florida; West Indies; c. and s. American Atlantic.

Comments: This sea whip occurs on reefs, but is also found in shallow water great distances away from them. The name *citrina* describes its color. The Angular Sea Whip (*P. anceps*) is about the same size, but is more highly branched, with branches flattened and folded to form 3 or 4 ridgelike edges. It ranges in color from purplish-brown to yellowish, and is usually found in water deeper than 10′ (3 m).

61 **Spiny Muricea**
(*Muricea muricata*)
Class Anthozoa
Subclass Octocorallia

Description: 24″ (61 cm) high, 12″ (30 cm) wide. *Branches flattened at point of origin* all in one plane, ¾″ (19 mm) wide, few, broad. Pale tan or yellowish. *Spiny appearance* caused by elevated, *tubular, reinforced pores.*

Habitat: Attached to rocks or hard bottom, around reefs and on reef flat; in shallow water.

Range: Florida; Gulf of Mexico; West Indies.

Comments: Both this animal's genus and species names are derived from a Latin word meaning "spiny." A closely related species, the Drooping Muricea (*M. pendula*), so named because of its drooping posture, is found over the same range and north to North

Carolina. Another related species, the Bushy Sea Fan (*M. appressa*), is common in southern California.

57 **Black Sea Rod**
(*Plexaura homomalla*)
Class Anthozoa
Subclass Octocorallia

Description: 24″ (61 cm) high, 12″ (30 cm) wide. Bushy, flattened, *two branches at each fork,* mostly in one plane. Yellowish-brown. *Polyp pores on slightly raised projections* on otherwise smooth surface.
Habitat: Attached to rocks and hard bottom on reef tract; in moderately shallow water.
Range: Florida; West Indies.
Comments: This soft coral turns black when dried. It has been found to be an important source of prostaglandins, hormones whose functions in human beings include inducing labor and preventing blood from clotting. The Bent Sea Rod (*P. flexuosa*) is a bushier colony, 24″ (61 cm) high, 12″ (30 cm) wide, which branches in one plane and occurs in the same habitat and range as *P. homomalla.*

56 **Eunicea Sea Rods**
(*Eunicea* spp.)
Class Anthozoa
Subclass Octocorallia

Description: 36″ (91 cm) high, 18″ (46 cm) wide. Colony of slender, flexible branches. Yellowish-brown, grayish-brown, or purplish-gray. Branches ⅛–½″ (3–13 mm) thick, almost parallel to each other; *surface knobby,* each projection with a *pore.*
Habitat: Attached to rocks around reef; below low-tide line in shallow water.
Range: Florida; Bahamas; West Indies.
Comments: Eunicea Sea Rods are common components of the reef fauna. The

Warty Sea Rod (*E. calyculata*) and the Mammillated Sea Rod (*E. mammosa*) are widely distributed throughout the Bahamas and the West Indies. The Warty Sea Rod is covered with warty projections, each with a large, gaping pore. The Mammillated Sea Rod has rounded projections, each with a smaller pore directed upward. Palmer's Sea Rod (*E. palmeri*) has conical projections, each with a pore. This latter is found only on reefs around Florida.

58 Double-forked Plexaurella

(*Plexaurella dichotoma*)
Class Anthozoa
Subclass Octocorallia

Description: More than 24″ (61 cm) high, 8″ (20 cm) wide. Bushy, with *2 stout, straight or crooked branches at each fork.* Yellow to light brown. Smooth, covered with *slitlike pores* into which polyps retract.

Habitat: Attached to rocks on reefs, and to solid bottom of reef flats; in shallow water.

Range: Florida; Gulf of Mexico; West Indies.

Comments: The Double-forked Plexaurella forms a colony whose shape depends on environmental conditions. The slitlike pores are good characteristics for field identification. The closely related Gray Plexaurella (*P. grisea*), 24″ (61 cm) high and 6″ (15 cm) wide, not only differs in color, but has fewer, more slender branches, which stand stiffly erect. Both species have the same range and habitat.

Common Sea Pansy
(*Renilla reniformis*)
Class Anthozoa
Subclass Octocorallia

Description: 2″ (51 mm) high, colony 2½″ (64 mm) wide. Fleshy, flattened, *kidney-shaped; stalk fleshy,* anchored in sand, longer than radius of colony. Rosy to light purple, sometimes white or yellowish; polyps on upper surface only; pale. Stem purple.

Habitat: On sandy bottoms in shallow water.

Range: Cape Hatteras to Florida; West Indies; uncommon in Gulf of Mexico.

Comments: The Sea Pansy is held above the soft bottom by its stalk. It can reestablish its position if dislodged and, if covered by shifting sands, can free itself by extending and contracting its muscular stalk. Sea pansies are bioluminescent, emitting light in the dark. Müller's Sea Pansy (*R. muelleri*), 2″ (51 mm) high and 4″ (102 mm) wide, which ranges from the western coast of Florida to Texas and Mexico, differs in having a width twice the length, and a slender stalk shorter than the colony radius. The Western Sea Pansy (*R. koellikeri*) has a more heart-shaped colony 3″ (76 mm) high and 3¼″ (83 mm) wide. It ranges from southern California to Baja California.

44 Gurney's Sea Pen
(*Ptilosarcus gurneyi*)
Class Anthozoa
Subclass Octocorallia

Description: 18″ (46 cm) high, 4″ (102 mm) wide. Stout, *plumelike.* Tannish-yellow to orange, translucent. Midrib bearing 20 pairs of *flat, wide side branches* with rows of polyps along both edges, and *swollen base* for anchorage.

Habitat: Anchored in soft bottoms; from below low-tide line to water more than

100′ (30 m) deep.

Range: British Columbia to c. California.

Comments: When disturbed, this plumy colony can contract and completely withdraw into the muddy bottom by releasing large amounts of water from a system of internal canals.

---

12 **Mat Anemone**
(*Zoanthus pulchellus*)
Class Anthozoa
Order Zoanthidea

Description: 24″ (61 cm) wide, up to several square feet in area. Irregularly shaped, *dense mat of anemonelike polyps,* each 1″ (25 mm) long, ¼″ (6 mm) wide when fully extended; mouth at top surrounded by *single ring of short tentacles.* Greenish or bluish.

Habitat: Attached to rocks and in reef-flat areas; just below low-tide line.

Range: Florida; West Indies.

Comments: The greenish color of the Mat Anemone comes from the tiny green plant cells, zoochlorellae, in its tissues.

---

8 **Knobbed Zoanthidean**
(*Palythoa mammillosa*)
Class Anthozoa
Order Zoanthidea

Description: 24″ (61 cm) wide, 1″ (25 mm) high. Irregularly shaped. Tan. Hard, *woody,* with *large, round openings,* each with a low, rounded edge, giving a knobby appearance. Polyps with *short, thick tentacles* continuous beneath skeleton tissue.

Habitat: On reefs, covering dead coral skeletons.

Range: Florida; Bahamas; West Indies.

Comments: The living colony forms a tough mat over dead coral and, when expanded, resembles the Large Star Coral.

191 Elegant Burrowing Anemone
(*Edwardsia elegans*)
Class Anthozoa
Order Actiniaria

Description: 2″ (51 mm) high, ¼″ (6 mm) wide. Slender, wormlike. Usually enclosed in thin, parchmentlike tube to which sand grains adhere. Pale red or purplish-brown, with pale lemon-yellow spots. Collar bearing mouth smooth, surrounded by ring of *16 slender tentacles; middle section grooved and warty; basal section inflatable,* whitish or pinkish. Tentacles yellowish, with orange to red stripe.

Habitat: Under stones, in sandy mud; near low-tide line and below.

Range: New Brunswick to Chesapeake Bay.

Comments: Unlike most sea anemones, which have flat basal disks, the Elegant Burrowing Anemone has an inflatable basal section with which it can probe and dig in the sandy mud.

166 Lined Anemone
(*Fagesia lineata*)
Class Anthozoa
Order Actiniaria

Description: 1⅜″ (35 mm) high, ¼″ (6 mm) wide. Cylindrical, slender. Cream-colored to tan or brownish, with pale longitudinal lines. Oral disk surrounded by *40 slender tentacles in 3 rings, inner ones longest. Base in thin mucus tube,* flattened or rounded. Column smooth.

Habitat: Among worm tubes and other growth on and under rocks; from below low-tide line to water more than 76′ (23 m) deep.

Range: Cape Cod to Cape Hatteras.

Comments: This little anemone occurs locally in numbers great enough to carpet the bottom.

Warty Burrowing Anemone
(*Haloclava producta*)
Class Anthozoa
Order Actiniaria

Description: 6″ (15 cm) high, ¾″ (19 mm) wide. Slender, contractile. Whitish to salmon. *20 rows of wartlike papillae* on upper part of column. *20 short tentacles, knobbed at tips,* arranged in 2 rings around mouth.

Habitat: On sandy flats and bottoms; in shallow water near low-tide line.

Range: Cape Cod to South Carolina.

Comments: The Warty Burrowing Anemone buries itself in the sand with only its crown of tentacles extended above the bottom. When the tide ebbs it retreats into the sand, leaving a hole surrounded by radiating lines where the tentacles lay.

182 **Northern Red Anemone**
(*Tealia crassicornis*)
Class Anthozoa
Order Actiniaria

Description: 5″ (127 mm) high, 3″ (76 mm) wide. Columnar, smooth. Red, sometimes mottled, occasionally with dull green spots. *100 thick, blunt tentacles arranged in several rings* around the mouth; white, frequently ringed with red, white, or dark pigment. Particles of shell and gravel often attached to column.

Habitat: Firmly attached to rocks, usually in protected places shaded by seaweeds or overhanging rocks; near low-tide line and below in shallow water.

Range: Arctic to Cape Cod; Alaska to c. California.

Comments: Specimens found above low-tide line are usually small; larger ones are subtidal. This handsome anemone is listed in some references as *T. felina*.

183 Strawberry Anemone
(*Tealia lofotensis*)
Class Anthozoa
Order Actiniaria

Description: 6″ (15 cm) high, 4″ (102 mm) wide. Similar to Northern Red Anemone, but *tentacles lacking bands or markings.*
Habitat: On rocks, walls of channels, pilings, and floats; from low-tide line to water 50′ (15 m) deep.
Range: Washington to s. California.
Comments: These beautiful anemones survive well in marine aquaria. One specimen accepted clam meat, small crustaceans, snails, and fish meat as food, and tripled its size in 8 months.

184, 195 Leathery Anemone
(*Tealia coriacea*)
Class Anthozoa
Order Actiniaria

Description: 5⅝″ (143 mm) high, 4″ (102 mm) wide. Columnar. Brownish-red to bright reddish-orange. Short, stout, green, blue or pink *tentacles banded with white or pink, arranged in 4 rings* about mouth. Body covered with *thick, rounded, suckerlike projections* to which sand, gravel, and bits of shell adhere.
Habitat: In sand, gravel, and rocky crevices; near low-tide line and below to water 50′ (15 m) deep.
Range: Alaska to c. California.
Comments: The Leathery Anemone usually lies buried in sand or gravel with only its crown of tentacles visible. It is preyed upon by the Leather Star.

196 **Silver-spotted Anemone**
(*Bunodactis stella*)
Class Anthozoa
Order Actiniaria

Description: 1½″ (38 mm) high, 2″ (51 mm) wide. Columnar. Olive, gray-green, or bluish-green, occasionally reddish. *6 glistening white lines radiating from mouth,* surrounded by *120 long, smooth tentacles* with whitish ring in middle and white spot at base; basal disk firmly attached to rock.

Habitat: Attached to rocks near sandy substrate; in shallow water near low-tide line.

Range: Nova Scotia to the Gulf of Maine.

Comments: When seen out of water at low tide, this species will have a layer of gravel and sand covering its surface.

---

179 **Buried Sea Anemone**
(*Anthopleura artemisia*)
Class Anthozoa
Order Actiniaria

Description: 10″ (25 cm) high, 2″ (51 mm) wide. Columnar. Upper part black, gray, or brown, becoming whitish or pinkish at base. *Tentacles arranged in 5 rings about mouth:* red, orange, white, blue, or black, solid or patterned; slender, tapering, numerous. Upper part of column covered with *rounded, suckerlike projections.*

Habitat: Attached to rocks buried in sand, in holes of rock borers, sometimes on pilings and floats; near low-tide line and below in shallow water.

Range: Alaska to s. California.

Comments: The Buried Sea Anemone usually lies covered by sand or hidden in a hole or crevice in a rock, with only its crown of tentacles exposed.

185 Aggregating Anemone
(*Anthopleura elegantissima*)
Class Anthozoa
Order Actiniaria

Description: Aggregating individuals 6″ (15 cm) high, 3⅛″ (79 mm) wide; solitary individuals 20″ (51 cm) high, 10″ (25 cm) wide. Cylindrical. Column pale gray-green to white; pale, variously colored *tentacles with pink, lavender, or blue tips, in 5 rings around oral disk,* numerous, thick, pointed; ring of knobs with stinging cells just under tentacles. Column covered with *vertical rows of adhesive projections.*

Habitat: Either in dense populations or solitary, on rock walls, boulders, or pilings; from between high- and low-tide lines to low-tide line.

Range: Alaska to Baja California.

Comments: Aggregation in this species is not a matter of many individuals coming together, but the result of many asexual longitudinal divisions of one founding anemone.

---

186 Giant Green Anemone
(*Anthopleura xanthogrammica*)
Class Anthozoa
Order Actiniaria

Description: 12″ (30 cm) high, 10″ (25 cm) wide. Cylindrical. *Column greenish-brown;* tentacles green, bluish, or white; oral disk green, grayish, or bluish-green. Numerous short, thick, tapered *tentacles in 6 or more rings* around flat oral disk; basal disk adhesive, flat. Column covered with *scattered adhesive projections.*

Habitat: On exposed coastline and in bays and harbors, on rocks, seawalls, and pilings; in tidepools; from above low-tide line to water more than 50′ (15 m) deep.

Range: Alaska to Panama.

Comments: Each Giant Green Anemone is solitary,

but is often in tentacle-tip contact with another in favorable tidepools and channels.

181, 197 **Proliferating Anemone**
(*Epiactis prolifera*)
Class Anthozoa
Order Actiniaria

Description: 1½″ (38 mm) high, 2″ (51 mm) wide. Cylindrical, squat. Solid, spotted, or striped, gray, green, blue, brown, orange, or reddish; base usually with white vertical lines. *96 short, tapered tentacles* around oral disk; *column smooth, with pits near base where young develop.* Basal disk flat, adhesive, spread wider than trunk.

Habitat: Attached to rocks, large algae, and eelgrass, on unprotected coasts and in bays; from between high- and low-tide lines to water 30′ (9 m) deep.

Range: Alaska to s. California.

Comments: This anemone has unusual breeding habits. At any time of the year, half the animals observed in nature will be brooding and may have 30 or more young developing around the base of the column.

187, 188 **Pink-tipped Anemone**
(*Condylactis gigantea*)
Class Anthozoa
Order Actiniaria

Description: 6″ (15 cm) high, 12″ (30 cm) wide. Large, showy, columnar. White, light blue, pink, orange, pale red, or light brown. Mouth surrounded by *100 or more long, tapering tentacles tipped with pink, scarlet, blue, or green* in several rings, usually paler than body. Basal disk firmly attached.

Habitat: Attached to hard objects in shallow water; common around reefs in both forereef and lagoonal areas and in turtle grass beds.

Range: From s. Florida through the Florida Keys; West Indies.

Comments: This impressive anemone is the largest in American Atlantic tropical waters. It is sometimes called the "passion flower" of the Caribbean.

193 **Warty Sea Anemone**
(*Bunodosoma cavernata*)
Class Anthozoa
Order Actiniaria

Description: 4″ (102 mm) high, 2″ (51 mm) wide. Columnar. Olive-green to dull brown, frequently with rounded reddish projections. Mouth usually red, surrounded by *100 sticky tentacles* mottled yellow to olive, in 5 rings, inner ring usually with red stripe. Body covered with *40 longitudinal rows of closely set, non-adhesive vesicles,* giving a warty appearance. Basal disk flat, adhesive.

Habitat: Jetties and sandy or gravelly bottoms which offer hard substrate for attachment; in shallow water.

Range: North Carolina to the Florida Keys and Texas; West Indies.

Comments: This anemone is usually fully expanded only at night, contracting during the day.

178 **Red Stomphia**
(*Stomphia coccinea*)
Class Anthozoa
Order Actiniaria

Description: 2″ (51 mm) high, 1½″ (38 mm) wide. Cylindrical. Scarlet, pink, orange, or greenish, sometimes mottled. Oral disk surrounded by 96 *pale, red-tinged tentacles in 3 rings.* Basal disk adhesive, usually wider than column. *Column smooth.*

Habitat: On rocks; from low-tide line to water more than 150′ (46 m) deep.

Range: Bay of Fundy to Cape Cod.

Comments: The Red Stomphia cannot always be distinguished from the Northern Red Anemone on the basis of color and general form alone. Laboratory study is necessary for sure identification.

189, 192 **Speckled Anemone**
(*Phymanthus crucifer*)
Class Anthozoa
Order Actiniaria

Description: 2″ (51 mm) high, ¾″ (19 mm) wide. Columnar, smooth. *Speckled brown. Oral disk greenish-brown with white speckles or radiating white lines,* surrounded by more than *200 short tentacles* arranged in several rings and occasionally speckled white. Basal disk attached to rock or gravelly bottom.

Habitat: On gravelly bottoms or base of rocks in shallow water, and back-reef areas.

Range: Florida Keys; Bahamas; West Indies.

Comments: This little anemone is frequently attached to a buried pebble and, when disturbed, contracts and completely disappears into the bottom.

---

194 **Tricolor Anemone**
(*Calliactis tricolor*)
Class Anthozoa
Order Actiniaria

Description: 3″ (76 mm) high, 2″ (51 mm) wide. Columnar. Dull brown to olive, tan, cream, orange, red, and purple, sometimes with longitudinal stripes, and usually with *dark spots around the base.* Mouth surrounded by *500 slender, graceful tentacles,* usually pink, rose, whitish, or pale gray-green. Column smooth, but with pores through which the anemone, when disturbed, extrudes *bright orange threads* loaded with stinging cells. Strong pedal disk firmly attached to substrate.

Habitat: Commonly attached to shells occupied by hermit crabs, less commonly to living snails or large crabs; in shallow water below low-tide line.

Range: North Carolina to Florida and Texas; Mexico; West Indies.

Comments: This anemone has the unusual habit of attaching to crabs. The anemone probably protects the crab, and gets more food than it would if fixed to one spot.

167 **Pale Anemone**
(*Aiptasia pallida*)
Class Anthozoa
Order Actiniaria

Description: 2″ (51 mm) high, ½″ (13 mm) wide at narrowest diameter, *flaring at oral disk.* Slender, smooth. Brown, gray, or whitish; translucent and difficult to see when contracted. Oral disk surrounded by a *few long, slender tentacles and a large number of shorter ones.* White threads of stinging cells discharged through mouth when animal is disturbed.

Habitat: Attached to rocks, oyster shells, jetties; in shallow water.

Range: North Carolina to Florida and Texas; West Indies.

Comments: These anemones usually occur in large numbers in their habitat. Their brownish color comes from zooxanthellae—intracellular, symbiotic, one-celled plants.

190 **Ringed Anemone**
(*Bartholomea annulata*)
Class Anthozoa
Order Actiniaria

Description: More than 2″ (51 mm) high, 1½″ (38 mm) wide. Columnar, smooth. Brownish; whitish at base. *200 tentacles with pale rings* varying in size, light brown.
Habitat: On or beneath rocks and other solid objects; in shallow water.
Range: Florida to Texas; Bahamas; West Indies.
Comments: The rings on this anemone's tentacles are actually batteries of stinging cells.

172 **Smooth Burrowing Anemone**
(*Actinothoe modesta*)
Class Anthozoa
Order Actiniaria

Description: 2½″ (64 mm) high, ½″ (13 mm) wide. Long, wormlike. Grayish-cream. Mouth surrounded by *60 delicate, slender, grayish tentacles, each with dark band near base.* Strong basal disk attached to a buried stone. Smooth, with threads of stinging cells present.
Habitat: On sand or gravel bottoms; near low-tide line in shallow water.
Range: From Cape Cod south, probably as far as Cape Hatteras.
Comments: Like its relatives, this burrowing anemone lies in the sand with only its flowerlike oral disk and tentacles showing. When disturbed, it quickly withdraws.

171 Frilled Anemone
(*Metridium senile*)
Class Anthozoa
Order Actiniaria

Description: 18″ (46 cm) high, 9″ (23 cm) wide. Smooth. Reddish- to olive-brown or lighter, to cream-colored and white; paler forms may be mottled. *Oral disk lobed;* tentacles slender and very abundant, 1000 in large specimens, producing a *frilled appearance.* Long, white threads of stinging cells discharged when animal is disturbed.

Habitat: Attached to rocks, wharf piles, and other solid objects; near low-tide line and below in shallow water.

Range: Arctic to Delaware; Alaska to s. California.

Comments: These anemones reproduce either sexually or, asexually, by dividing lengthwise or by leaving behind, as they creep over a surface, bits of tissue from the pedal disk that regenerate into complete anemones.

168 Striped Anemone
(*Haliplanella luciae*)
Class Anthozoa
Order Actiniaria

Description: ¾″ (19 mm) high, ¼″ (6 mm) wide. Brown to olive-green. Column usually with *longitudinal stripes of orange, yellow, or cream;* some specimens lack stripes. Tentacles surrounding oral disk long, slender, numbering up to 50 in larger specimens. White threads of stinging cells discharged through mouth when animal is disturbed.

Habitat: On solid objects in shallow water; also found in brackish water and in salt marshes.

Range: Maine to Chesapeake Bay; reported from Texas; Washington to s. California.

Comments: The Striped Anemone was apparently

introduced from Japan in the late 19th century. Its species name, *luciae,* was given by the American naturalist A. E. Verrill in honor of his daughter Lucy.

169 **Ghost Anemone**
(*Diadumene leucolena*)
Class Anthozoa
Order Actiniaria

Description: 1½″ (38 mm) high, ½″ (13 mm) wide. Translucent, whitish, pink, or olive. Columnar. Mostly smooth, but with *low, scattered projections.* 60 slender, pale, ½″ (13 mm) long tentacles surrounding mouth. White threads of stinging cells discharged when animal is disturbed.
Habitat: On or under rocks, among marine growth on pilings and jetties; in shallow water of bays and other protected areas.
Range: Maine to North Carolina; California.
Comments: This little anemone is easily confused with immature forms of the Frilled Anemone, but it lacks the threads of stinging cells found in that species.

170 **Northern Cerianthid**
(*Cerianthus borealis*)
Class Anthozoa
Order Ceriantharia

Description: 18″ (46 cm) high, 1½″ (38 mm) wide. Wormlike, smooth, slimy. Various tones of brown. *Mouth expanded,* trumpetlike, with *two rings of tentacles,* some in outer row about twice as long as others. *Inner tentacles short.* Completely *lacking pedal disk;* basal end tapered to blunt point.
Habitat: Burrowing among rocks on bottom; from shallows to water 1310′ (400 m) deep.
Range: Arctic to Cape Cod.

Comments: This species secretes a tube of many layers of mucus, mud, and sand to line its burrow.

180 **Club-tipped Anemone**
(*Corynactis californica*)
Class Anthozoa
Order Corallimorpharia

Description: 1¼″ (32 mm) high, 1″ (25 mm) wide. Cylindrical, short, with flared crown of tentacles. Red, pink, orange, buff, purple, brown, or nearly white. Oral disk surrounded by many *club-tipped tentacles,* usually white, in radial rows. Column smooth. Flat, adhesive *basal disk with tissue usually continuous* with that of its neighbors.
Habitat: On rocks, ledges, and pilings, on open shores and in bays; from low-tide line to water 95′ (29 m) deep.
Range: N. California to Baja California.
Comments: These are not true anemones, but are more closely related to the stony corals. Their knobby tentacles contain the largest stinging cells known, but they are not toxic to humans.

176, 177 **Orange Cup Coral**
(*Balanophyllia elegans*)
Class Anthozoa
Order Scleractinia

Description: ⅜″ (10 mm) high, ⅜″ (10 mm) wide. Cylindrical, solitary. Orange or yellow. Oral disk surrounded by *36 long, tapered, translucent tentacles dotted with masses of stinging cells.* Base set in *stony, cup-shaped skeleton* into which the polyp can retract. Skeleton has radial plates showing pattern of internal septa.
Habitat: In shaded places on and under ledges and boulders along open coast and in bays; from above low-tide line to water 70′ (21 m) deep.

Range: British Columbia to Baja California.
Comments: This is the only stony coral to occur between the high- and low-tide lines on the Pacific Coast. Its orange color is a fluorescent pigment and even at depths of 30′ (9 m) or more, where red light is lacking, it shows the same bright hue.

55 **Staghorn Coral**
(*Acropora cervicornis*)
Class Anthozoa
Order Scleractinia

Description: 10′ (3 m) high, more than 60″ (152 cm) wide. Loosely branched, with 1″ (25 mm) wide, *cylindrical branches of variable length which do not fuse together.* Yellowish or purplish-brown, paler at the tips. Surface covered with small, *protruding, round cups oriented toward branch tip.*
Habitat: In protected areas of reefs; on windward side of reefs in water more than 10′ (3 m) deep.
Range: Florida Keys; Bahamas; West Indies.
Comments: This coral is heavily collected for tourists, to the great detriment of many reefs.

32 **Elkhorn Coral**
(*Acropora palmata*)
Class Anthozoa
Order Scleractinia

Description: 10′ (3 m) high, more than 60″ (152 cm) wide. *Treelike, with flattened, fanlike branches* of extremely variable length and width extending outward from a short, thick stalk. Brownish-yellow to cream-colored, tips of branches white. Surface of branches covered with small, *protruding, round cups oriented toward growing edge.*
Habitat: On windward side of reefs; below low-tide line.

Range: Florida Keys; Bahamas; West Indies to Brazil.

Comments: In more protected areas of the reef, this coral forms fingerlike margins on its branches which resemble the flat, branching antlers of a moose or European elk. Cuts and scratches resulting from contact with this coral can be slow to heal.

---

33 **Lettuce Coral**
(*Agaricia agaricites*)
Class Anthozoa
Order Scleractinia

Description: 6″ (15 cm) long, colony 6″ (15 cm) wide. *Leafy or frondlike,* forming semicircular, *flattened sheets.* Brown or purplish-brown. Tentacles of polyps short, white, in cups on both sides of sheet, in *rows paralleling edge of colony;* rows separated by projecting ridge.

Habitat: Encrusting rocks and dead corals, sometimes attached to base of Elkhorn Coral or to mangrove roots; common on all reefs below low-tide line.

Range: Florida Keys; Bahamas; West Indies to Brazil.

Comments: The body form of the Lettuce Coral ranges from delicate to massive, depending on growth conditions.

---

16 **Starlet Coral**
(*Siderastrea radians*)
Class Anthozoa
Order Scleractinia

Description: 12″ (30 cm) wide, 12″ (30 cm) high. *Rounded mass.* Brown, salmon, yellow, or gray. Young forms encrust and eventually cover small stones. Surface covered with *deep, starlike cups* ⅛″ *(3 mm) wide, with 40 septa.*

Habitat: On coral reefs and reef flats in shallows.

Range: Florida to Texas and Mexico; Bermuda;

Bahamas; West Indies to South America.

Comments: This very common species tolerates the great temperature fluctuations of a shallow reef flat as well as considerable variations in salinity.

14 **Reef Starlet Coral**
(*Siderastrea siderea*)
Class Anthozoa
Order Scleractinia

Description: 36″ (91 cm) wide, 36″ (91 cm) high. Boulderlike. Brown or grayish. Surface covered with *round cups* ¼″ (6 mm) wide and equally high, shallow, with *60 septa*.

Habitat: Associated with reefs, usually in protected areas and shallow water; sometimes on reef front in water more than 15′ (5 m) deep.

Range: Florida; Bermuda; Bahamas; West Indies to Brazil.

Comments: A common inhabitant of the reef and surrounding areas.

9 **Porous Coral**
(*Porites astreoides*)
Class Anthozoa
Order Scleractinia

Description: Encrusting colony 24″ (61 cm) high, 24″ (61 cm) wide. *Flattened or rounded, lump-covered masses.* Yellow or greenish-brown to neon-green. Surface covered with *closely set, small cups,* 1/16″ (2 mm) wide, 1/16″ (2 mm) high, each with *12 septa*.

Habitat: Back-reef areas, sand bottoms, and turtle grass beds; in shallow water.

Range: Florida to Texas and Mexico; Bermuda; Bahamas; West Indies to Brazil.

Comments: The massive coral boulders of this species are among the most common features of American reefs. The Spiral-gilled Tube Worm is a frequent inhabitant of the Porous Coral.

13 **Clubbed Finger Coral**
(*Porites porites*)
Class Anthozoa
Order Scleractinia

Description: Colony 12" (30 cm) or more long, 12" (30 cm) wide, forming thick clumps of *irregular branches swollen at their ends,* 1" (25 mm) wide. Pale beige, yellowish-brown to purplish. Branches covered with *closely set cups* about 1/16" (2 mm) wide and equally high.

Habitat: Throughout the reef; in shallow water.

Range: Florida to Texas and Mexico; Bermuda; Bahamas; West Indies.

Comments: This species is a common inhabitant of the back reef, although in that habitat it does not achieve the large size of specimens growing on the fore reef.

4 **Knobbed Brain Coral**
(*Diploria clivosa*)
Class Anthozoa
Order Scleractinia

Description: 24" (61 cm) high, colony 48" (122 cm) wide. Large, heavy, low, with *irregular knobs on surface.* Cups with polyps greatly elongated, forming "valleys" or grooves winding over the surface. *Valleys not all connected with each other,* 1/4" (6 mm) wide, shallow. Greenish-brown in the valleys, brown on the hills; tentacles of polyps bright green with white tips. *40 septa per cm.*

Habitat: On seaward and lagoon sides of reefs, in turtle grass beds, or on mangrove roots; in shallow water.

Range: Florida; West Indies.

Comments: This is one of the several types of corals

known as brain corals because their convolutions resemble those of the human brain.

3 **Labyrinthine Brain Coral**
(*Diploria labyrinthiformis*)
Class Anthozoa
Order Scleractinia

Description: 96″ (244 cm) high, 96″ (244 cm) wide. Convex, heavy boulders with *winding, interconnected valleys* ¼″ (6 mm) deep, and ⅜″ (10 mm) wide. Bright orange-yellow to brownish-yellow. *Walls between valleys thick, with longitudinal groove.*
Habitat: Abundant on reefs; in shallow water.
Range: Florida to Texas and Mexico; Bermuda; Bahamas; West Indies.
Comments: The young brain coral newly settled on a solid substrate has one polyp in a round cup. The cup begins to elongate as it grows, folding and twisting into "valleys." The polyp is thus stretched into a long and contorted mass that appears brainlike.

2 **Smooth Brain Coral**
(*Diploria strigosa*)
Class Anthozoa
Order Scleractinia

Description: 24″ (61 cm) high, 48″ (122 cm) wide. *Smoothly convex* boulders. Greenish or yellowish-brown. Winding *valleys not all connected with each other,* ¼″ (6 mm) deep, ⅜″ (10 mm) wide; *ridge between valleys low,* not sharp. *20 septa per cm.*
Habitat: On lagoon side of reefs, in turtle grass beds; in shallow water.
Range: Florida; Bahamas; West Indies.
Comments: The surface of this coral, like that of its relatives, resembles that of the human brain. Its colony form is smoother and more rounded than other brain corals.

175 **Rose Coral**
(*Manicina areolata*)
Class Anthozoa
Order Scleractinia

Description: 4″ (102 mm) high, 2″ (51 mm) wide. Solitary, *attached to a solid when young, unattached when older.* Base conical, stalked. Young with cup oval, becoming elongate, border folding into broad lobes. Brown to yellow or green, valleys usually green, tentacles white-tipped. Valley ⅜″ (10 mm) deep, 1″ (25 mm) wide, branching into each lobe. *20 septa per cm.*

Habitat: Quiet areas behind reefs and in turtle grass beds; in shallow water.

Range: Florida; Bahamas; West Indies.

Comments: This coral is relatively hardy in marine aquaria, so free specimens can be conveniently taken, without damage, into laboratories for physiological studies.

11, 174 **Large Star Coral**
(*Montastrea cavernosa*)
Class Anthozoa
Order Scleractinia

Description: 36″ (91 cm) high, 60″ (152 cm) wide. Massive, boulderlike. Greenish or yellowish-brown. *Cups crowded, large, elevated above the surface,* ½″ (13 mm) wide, ⅛″ (3 mm) deep. 48 septa extending into the space around cups.

Habitat: On coral rock bottom; in shallow water.

Range: Florida to Texas and Mexico; Bahamas; West Indies; Bermuda.

Comments: An important reef species, but also occurs by itself on flats behind the reef. It feeds actively on plankton.

10 **Common Star Coral**
(*Montastrea annularis*)
Class Anthozoa
Order Scleractinia

Description: 48″ (122 cm) wide, 24″ (61 cm) high. Boulderlike. Greenish or yellowish-brown. *Cups crowded, small,* ¼″ (6 mm) wide, ⅛″ (3 mm) high, elevated above surface, their *septa in 3 rings of 12,* the largest saw-toothed and extending into the space around cups.
Habitat: On front and back sides of reef; from shallows to water 25′ (8 m) deep.
Range: Florida Keys; Bahamas; West Indies.
Comments: This species is a common inhabitant of the back reef, although in that habitat it does not achieve the large size of specimens growing on the fore reef.

Northern Stony Coral
(*Astrangia danae*)
Class Anthozoa
Order Scleractinia

Description: 5″ (127 mm) long, 5″ (127 mm) wide. Forms *thin crust over rocks or shells,* sometimes with short, thick branches. Pinkish to white. Cups to ¼″ (6 mm) wide, closely set, 30 in a colony 4″ (102 mm) wide, with *30 septa.*
Habitat: Attached to rocks and shells; from shallow water to water 135′ (40 m) deep.
Range: Cape Cod to Florida.
Comments: This is the only shallow-water species of stony coral found north of Cape Hatteras. The tropical species, the Dwarf Cup Coral (*A. solitaria*), may occur either as a solitary polyp or form thin crusts of several polyps at the base of a larger coral. It is ¼″ (6 mm) wide and about as high, and has 48 septa of 4 sizes. It ranges from Bermuda to Brazil.

35 **Ivory Bush Coral**
(*Oculina diffusa*)
Class Anthozoa
Order Scleractinia

Description: Colony 12″ (30 cm) high, 12″ (30 cm) wide. *Densely branching* and bushy. Ivory or faintly greenish. *Branches irregular, knobby,* ½″ (13 mm) thick. Cups ⅛″ (3 mm) wide, shallow, widely spaced, with 24 septa.
Habitat: Lagoons, protected regions of reefs, and open stretches of coral sand bottom; in shallow water.
Range: Florida; Bermuda; Bahamas; West Indies.
Comments: In some shallows this coral may cover acres of the bottom, forming the basis of a large community of organisms including brittle stars, worms, snails, anemones, hydroids, and algae.

5 **Meandrine Brain Coral**
(*Meandrina meandrites*)
Class Anthozoa
Order Scleractinia

Description: 12″ (30 cm) high, 12″ (30 cm) wide. Heavy, convex boulders. Yellowish-brown; polyps with white tentacles. *Valleys twisted, long, not all continuous,* ⅜″ (10 mm) deep. 7 large septa per cm; neighboring valleys with *septa meeting in zig-zag line* at crest of wall, which may be grooved in places.
Habitat: Reefs, especially near the seaward edge; in water 5′ (2 m) or more deep, with good circulation.
Range: Florida; West Indies.
Comments: Like those in the genus *Diploria,* this coral resembles a human brain. It does not get as big as *Diploria,* and differs from it by the zig-zag line formed by adjacent septa at the crest of the wall.

7 **Stokes' Star Coral**
(*Dichocoenia stokesii*)
Class Anthozoa
Order Scleractinia

Description: 12″ (30 cm) high, 12″ (30 cm) wide. Heavy, convex boulders. Yellowish-tan; tentacles of polyps white. *Cups of various shapes*—round, oval, elongate, or Y-shaped. *Cups form short valleys* ¼″ (6 mm) wide, ¼″ (6 mm) deep. *10 thick septa per cm,* alternately thick and thin.

Habitat: Back reef and forereef areas; in water 5′ (2 m) or more deep, with good circulation.

Range: Florida; Bahamas; West Indies.

Comments: The retracted polyps of this coral show the starlike pattern of its septa against a darker yellow or brown background.

---

6 **Pillar Coral**
(*Dendrogyra cylindrus*)
Class Anthozoa
Order Scleractinia

Description: 48″ (122 cm) high, 72″ (183 cm) wide. Massive, with *erect, cylindrical branches* 8″ (20 cm) wide at the base and 24″ (61 cm) long. Brownish-yellow. *Valleys short, contorted, discontinuous,* ¼″ (6 mm) wide and ⅛″ (3 mm) deep. Between valleys narrow *walls with shallow grooves. Septa alternately thick and thin,* upper part arching over wall to the groove, 10 per cm.

Habitat: Throughout reefs in water 20′ (6 m) or more deep.

Range: Florida; Bahamas; West Indies.

Comments: This is one of the most massive and obvious corals on the reef, with heavy branches towering above its base. It is commonly inhabited by the Spiral-gilled Tube Worm.

1 **Large Flower Coral**
(*Mussa angulosa*)
Class Anthozoa
Order Scleractinia

Description: 24″ (61 cm) high, 36″ (91 cm) wide. Convex. Yellowish- or greenish-brown, pinkish or purplish. Branches 2″ (51 mm) wide and 5″ (127 mm) high, each ending in a *cup which is round, oval, or elongate,* and Y-shaped when budding. *Margins of septa spiny,* with *rows of spines* corresponding with the septa extending down side of branch.
Habitat: Among other corals throughout reefs.
Range: Bahamas; West Indies; rare in Florida.
Comments: Branches in this coral may be so dense as to give the colony a solid, boulderlike appearance when polyps are expanded. The polyps are among the largest of the stony corals.

173, 198 **Flower Coral**
(*Eusmilia fastigiata*)
Class Anthozoa
Order Scleractinia

Description: 10″ (25 cm) high, 12″ (30 cm) wide. Colony with small, thick branches, each ending in a cup. Dark brown, polyps green, tentacles white. *Cups round, oval, or dividing,* 1½″ (38 mm) long and ½″ (13 mm) wide, with 1–3 mouths. *Rim sharp.* 18 septa per cm. *Sides of branches spiny.*
Habitat: Common on both sides of reefs, or under larger overhanging corals; in shallow water.
Range: Florida; Bahamas; West Indies.
Comments: This coral is equally beautiful in its living state and as a skeleton. It is a prized item in shell shops, a fact which has led to the decimation of its population on some reefs.

COMB JELLIES
(Phylum Ctenophora)

Comb jellies share with cnidarians a modified radial symmetry and the presence of differentiated tissues, without true organ systems. Unlike cnidarians, comb jellies do not sting. Body form varies, but is commonly globular or somewhat compressed. Water represents more than 95% of the comb jelly's body weight. Some species have a pair of tentacles equipped with adhesive cells, some have oral lobes (2 large flaps around the mouth), and some have neither. The comb jelly's mouth leads into a gullet, or *pharynx,* which in turn opens into a digestive cavity, the *stomach.* From the stomach, numerous canals extend throughout the animal, and digested food can thus reach all parts of the body. Ctenophores are carnivorous, feeding on a large variety of prey.

The comb jelly's *comb plates* (ctenae) consist of transverse rows of cilia fused together by a thin membrane and arranged in 8 lines down the long axis of the animal. The beat of the comb plates moves the animal through the water, mouth-end forward, coordinated by a sensory structure, the *apical organ* —a dome-shaped cyst containing a heavy granule, the *statolith.* As the animal changes position, gravity pulls the granule downward, causing it to stimulate the nervelike tissue beneath it. Branches of this tissue extend beneath each line of comb plates and regulate their beat. All species found in American waters swim feebly, and are thus at the mercy of ocean currents.

Comb jellies are hermaphroditic, with both male and female sex organs. Most species shed both eggs and sperm into the sea, where fertilization and development take place. A few (e.g. *Mnemiopsis*) hold their fertilized eggs within the canals where they develop

into tiny, motile larvae.
Refraction imparts a jewel-like quality to ctenophores seen in the sunlight. In the dark, comb jellies are bioluminescent, and their whole form, including the rows of comb plates, can be seen outlined in flashes of light.

496 Sea Gooseberry
(Pleurobrachia pileus)

Description: 1⅛″ (28 mm) high, 1″ (25 mm) wide. Round to egg-shaped. Transparent, iridescent. *2 tentacles,* each fringed on 1 side, can extend over 20 times body length or retract completely. *8 rows of comb plates, equally spaced,* extend nearly full length of body. Pharynx, stomach and its branches, and tentacles and sheaths white, pink, yellow, or orange-brown.

Habitat: Near shore; usually in large swarms.

Range: Maine to Florida and Texas.

Comments: Unlike many jellyfish, Sea Gooseberries do not sting. The sticky filaments of the trailing tentacles capture small crustaceans, fish eggs and larvae, and other planktonic animals. The tentacles then contract and wipe the prey off on the mouth, which immediately swallows it. *P. bachei,* called "Cats Eyes" on the Pacific Coast, ranges from Alaska to Baja California and is a voracious carnivore; swarms can decimate schools of young herring and cod. *P. bachei* is indistinguishable from *P. pileus* except in the laboratory. The Arctic Sea Gooseberry (*Mertensia ovum*) is more egg-shaped, flatter, and larger, 2″ (51 mm) high and 1″ (25 mm) wide. It occurs from the Arctic to the Gulf of Maine, and sometimes in winter to Cape Cod, and in central California.

491 Common Northern Comb Jelly
(*Bolinopsis infundibulum*)

Description: 6″ (15 cm) high, 2″ (51 mm) wide. Elongate, oval, somewhat flattened, narrow at top. Transparent, iridescent. *2 large lobes less than half the total body length.* 4 fingerlike structures around mouth; 8 comb plate rows, 2 extending down each lobe nearly to end, 1 down each structure. Comb plate rows, pharynx, stomach and its branches faintly white.

Habitat: In shallow water.

Range: Arctic to Gulf of Maine; sometimes in Massachusetts Bay.

Comments: This species is the most common comb jelly north of Cape Cod in the summer. The corresponding species, the Short-lobed Comb Jelly (*B. microptera*), comparable in size, occurs from Washington to California.

493 Leidy's Comb Jelly
(*Mnemiopsis leidyi*)

Description: 4″ (102 mm) high, 2″ (51 mm) wide. Oval, somewhat flattened, broad at top end. Milky-transparent; iridescent. 2 large *lobes longer than half the total body length.* 1 pair of short tentacles in sheaths between lobes, 4 ribbonlike structures around mouth, between lobes. 8 rows of comb plates, 2 extending down each lobe nearly to end, 1 down each structure.

Habitat: In shallow water; penetrates into brackish waters.

Range: South of Cape Cod to the Carolinas; common in Chesapeake Bay as far north as Baltimore.

Comments: Leidy's Comb Jelly has been used in marine laboratories to study bioluminescence and problems of regeneration. Its relative, McCrady's Comb Jelly (*M. mccradyi*), ranges from Florida through the West Indies. It is about the same size and shape as *M. leidyi,* but is greenish-tan and less transparent, and sometimes has 2 brown spots on each side.

492 **Beroë's Comb Jelly**
(*Beroe cucumis*)

Description: 4½″ (114 mm) high, 2″ (51 mm) wide. *Flattened, purselike, lacking tentacles and lobes.* Translucent; pink to rust. Mouth broad, occupying entire oral end. 8 rows of comb plates extend over ½ length of body. Under each row lies a canal with abundant branches which end blindly.

Habitat: In nearshore waters and bays.

Range: New Brunswick to Cape Cod; Alaska to s. California.

Comments: Beroë's Comb Jelly feeds chiefly on cnidarian medusae and other comb jellies. A related Atlantic species, the Ovate Comb Jelly (*B. ovata*), ranges from Cape Cod south and into the Gulf of Mexico. It differs in having the branches of neighboring canals united. A West Coast species, Forskal's Comb Jelly (*B. forskali*), likewise has united branches, but is roughly triangular. Both are the same size as *B. cucumis.*

FLATWORMS
(Phylum Platyhelminthes)

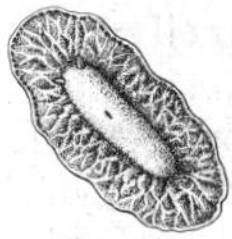

Flatworms have the simplest body plan of all bilaterally symmetrical animals. As their name suggests, flatworms' bodies are compressed, their thickness small compared with their length and breadth. As with cnidarians, the mouth is the only opening into the digestive cavity; through it food is taken in and wastes discharged. Unlike cnidarians, flatworms have well-defined nervous, muscular, excretory, and reproductive systems which lie within a solid matrix of tissue (*parenchyma*). Distribution of digested food is achieved by a digestive cavity that branches into all parts of the body. Because of the worms' flatness, all cells are close enough to the surface for exchange of oxygen and carbon dioxide with the environment. The flat body also enables these creatures to hide in narrow crevices or enter the body openings of other animals.

The phylum includes three classes, the Turbellaria, Trematoda, and Cestoidea. The latter two parasitize various vertebrate animals and will not be considered in this book. The Class Turbellaria includes mostly free-living forms, of which only the orders Tricladida and Polycladida have members large and obvious enough to catch the eye of the beachcomber. Triclads have a digestive cavity with three major branches, one toward the head and two toward the rear. Their eggs are laid in tough capsules which they attach to some solid object. Development is direct: the young are tiny replicas of the adult. Polyclads have numerous branches radiating from the central digestive cavity. Their eggs are not enclosed in a shell. In some species development is direct, but others have a specialized planktonic larva which later undergoes metamorphosis to the adult form.

Members of both orders have an epidermis covered with cilia, simple eyespots (*ocelli*), and sensory structures for taste and perhaps smell on the head end, as well as a mouth situated somewhat posteriorly on the underside, and a muscular, sucking pharynx. Locomotion is achieved by rippling contractions of body muscles aided by the action of cilia, allowing the worm to glide smoothly over a surface. Both groups include predators and scavengers that feed on the bodies of dead animals. Some species are *commensals,* living in close relation with another kind of animal.

---

258 Limulus Leech
(*Bdelloura candida*)
Order Tricladida

Description: ⅝" (16 mm) long, ¼" (6 mm) wide. Elongate, with narrow head end and squarish *rear with "sucker."* White to pale yellowish. Brown intestine has 1 front and 2 rear *branches that do not join at rear. 2 black eyes* near front end. Mouth in center of underside, from which long, white feeding tube can be extended.

Habitat: Gills and legs of the Horseshoe Crab.

Range: Gulf of Maine to Florida and Texas.

Comments: This active worm can attach its sucker firmly to a Horseshoe Crab, where it feeds on food particles brought in by the host. When removed from a crab and kept in sea water, the worm will not feed and begins to shrink in size. The Limulus Leech lays eggs enclosed in a capsule fastened to the tissues of the Horseshoe Crab's gills by an anchorlike stalk.

217 **Oval Flatworm**
(*Allioplana californica*)
Order Polycladida

Description: 1½″ (38 mm) long, ¾″ (19 mm) wide. Oval, *thick, firm.* Bluish-green to pale olive; radiating brown branches of digestive tract may show through. *Nipplelike tentacles covered with eyespots;* 2 patches of eyespots over brain. Mouth in middle of underside.
Habitat: Under large rocks resting on gravel or coarse sand, and in crevices; near high-tide line.
Range: California and Baja California.
Comments: The Oval Flatworm feeds on tiny snails, whose tough, rasplike "tongues" can be found in its digestive tract.

218 **Oyster Leech**
(*Stylochus ellipticus*)
Order Polycladida

Description: 1″ (25 mm) long, ⅜″ (10 mm) wide. Oval, very thin and flat, with undulating margins. Pale yellowish, mottled with brown, gray, and dull green, frequently with a green-to-brown midline stripe toward the rear. *Row of tiny, black eyespots along front margin,* 2 clusters of eyespots over brain area, and on 2 tiny, retractable tentacles. Mouth in middle of underside.
Habitat: Among oysters and barnacles; on rocks.
Range: Bay of Fundy to Florida and Texas.
Comments: Oyster leeches insinuate themselves into open oysters in order to feed on their soft tissues. They also prey on barnacles. The Gray Oyster Leech (*S. frontalis*), which ranges from Florida to Texas, is 2″ (51 mm) long, 1″ (25 mm) wide, and gray, and has eyespots around the entire body margin. The slightly larger Red Oyster Leech (*S. oculiferus*) also ranges from Florida to Texas and is red with pink spots.

216 **Zebra Flatworm**
(*Stylochus zebra*)
Order Polycladida

Description: 1½″ (38 mm) long, ½″ (13 mm) wide. Oblong, with rounded head end and bluntly pointed rear end. *Yellowish to white, with many thin, dark brown crossbands,* some branching near the margin. Row of eyespots around margin, and short, stubby tentacles and brain area difficult to see because of worm's striking coloration. Mouth near middle of underside.

Habitat: In snail shells occupied by large hermit crabs.

Range: Cape Cod to Florida and Texas.

Comments: This worm is especially common in whelk shells occupied by the Flat-clawed Hermit Crab.

---

219 **Tapered Flatworm**
(*Notoplana acticola*)
Order Polycladida

Description: 2⅜″ (60 mm) long, ¾″ (19 mm) wide. Flat, tapered oval. Pale gray or tan, with darker spots along midlines; branches of digestive tract visible when full of food. *Widest near front,* tapering toward rear. No obvious tentacles, but position marked by *round clusters of eyespots; 25 eyespots in longitudinal bands* on each side of head over brain.

Habitat: Under rocks; between high- and low-tide lines.

Range: Entire coast of California.

Comments: This is one of the most common flatworms on rocky shores, and is an aggressive predator, eating animals half its size, including limpets and small barnacles. Most specimens are hermaphroditic. The Speckled Flatworm (*N. atomata*) is the most common flatworm on the rocky coast of New England, ranging from New Brunswick or farther north to Cape

Cod. It is 1½″ (38 mm) long, ¾″ (19 mm) wide, and elongate-oval. It is brownish-gray, mottled and streaky, and, like *N. acticola,* has 4 clusters of eyespots. It has no obvious tentacles.

---

214 **Horned Flatworm**
(*Eurylepta californica*)
Order Polycladida

Description: 1¼″ (32 mm) long, ½″ (13 mm) wide. Oval, with undulating margins. Pale gray, with white, ruffled margin; irregular white spots down middle; back has *sparse crisscross pattern of narrow, black lines ending in red* at margins. *2 long, flaring tentacles at front margin,* each with a cluster of eyespots and heavy black and red spot at base; eye clusters over brain form inverted V.
Habitat: Under rocks and among coralline algae; near low-tide line and below.
Range: C. California.
Comments: This flatworm's striking pattern makes it easy to identify. Its close relative, the Golden Horned Worm (*E. aurantiaca*), is yellowish-pink to salmon, with pinkish midstripe and many tiny, white dots on the back. It ranges from British Columbia to southern California.

---

213 **Monterey Flatworm**
(*Pseudoceros montereyensis*)
Order Polycladida

Description: 2″ (51 mm) long, 1″ (25 mm) wide. Elliptical, *margins deeply ruffled, marginal tentacles close together,* formed by folds of front margin. Pale grayish with black stripe down middle and near margins; scattered reddish and blackish spots.
Habitat: Under rocks in tidepools; near low-tide line.
Range: C. and s. California.

Comments: *P. montereyensis* swims by graceful undulations of its body margins, and can crawl upside-down on the surface film of quiet tidepools.

215 **Crozier's Flatworm**
(*Pseudoceros crozieri*)
Order Polycladida

Description: 2″ (51 mm) long, ¾″ (19 mm) wide. Elongate. Grayish-tan or green, with numerous fine lines of dark brown, thicker near margin, distributed irregularly across upper surface. *Pair of marginal tentacles, covered with eyespots,* set close together at head end, each formed by a fold of the body margin. Two clusters of eyespots over brain area behind tentacles.
Habitat: Among rocks and coral rubble; near low-tide line and in shallow water.
Range: Florida; Bahamas; West Indies.
Comments: This is the most common of the larger inshore flatworms in Florida.

232 **Leopard Flatworm**
(*Pseudoceros pardalis*)
Order Polycladida

Description: 1½″ (38 mm) long, ½″ (13 mm) wide. Elliptical. *Black, with rounded, yellow or tan spots on back.* Ruffled margin with *closely set white spots;* marginal tentacles close together, formed by fold of front margin.
Habitat: On coral reefs.
Range: Florida; Bahamas; West Indies.
Comments: Like many other members of the genus *Pseudoceros,* this worm is brilliantly colored, and especially beautiful when seen gliding over a coral head.

NEMERTEAN WORMS
(Phylum Rhynchocoela)

Most nemerteans are long, slender, and somewhat flattened. Many are highly colored—red, orange, yellow, brown, or green—some patterned above with stripes or spots, and paler underneath. Nemerteans range in size from less than an inch to several feet long. They are remarkably elastic and can stretch many times even their relaxed body length. They may be equipped with eyespots and sensory grooves, are covered with cilia, and consist internally of a solid mass of tissue, without a body cavity. Their nervous systems and excretory systems are like those of flatworms, and the digestive tract includes mouth and anus. The mouth is situated on the lower side near the nemertean's front end. The anus is located at the worm's rear tip. Soft-bodied and seemingly vulnerable, the nemerteans are predators. They have a long proboscis which can be thrust out to a distance greater than the animal's body length to entangle prey—usually small annelid worms and crustaceans—with abundant and sometimes paralyzing mucus. The proboscis is then retracted to deliver the prey to the mouth. The proboscis lies in a fluid-filled sac above the mouth. To extrude it, the worm contracts muscles in the sac wall, forcing the fluid to pop out the proboscis. A long muscle retracts the proboscis. In all species, the proboscis is coated with sticky mucus, which in some is toxic. Members of one class (Enopla) have one or more sharp stylets, like little harpoons, at the tip of the proboscis to penetrate the prey.

Sexes are separate in nearly all nemerteans. Some species shed their eggs directly into the water; others deposit them in masses of mucus. In most species, fertilization occurs externally. Eggs shed into the water develop into ciliated swimming larvae which eventually settle and assume adult body form. Those laid in mucus usually develop directly into minute replicas of the adult. Some nemerteans also reproduce asexually, regenerating whole individuals from fragments.

Classes of Rhynchocoeles: The Phylum Rhynchocoela is divided into two classes, the Anopla, or unarmed nemerteans (without stylets), and the Enopla, or armed nemerteans. The Anopla include many forms whose larvae are planktonic. Anoplans' sensory grooves, when present, consist of slits running lengthwise along each side of the head. Sensory grooves on enoplans curve from the side toward the midline above, demarking the head from the trunk. The young of all enoplans develop directly into adults. Field identification is based on body form, head shape, color, and presence and numbers of eyespots and sensory grooves. Color, however, is an undependable characteristic, varying greatly in many forms. Accurate identification may require tissue examination in a laboratory.

---

254 Six-lined Nemertean
(*Tubulanus sexlineatus*)
Class Anopla

Description: 8″ (20 cm) long, ⅛″ (3 mm) wide. In a thin, parchmentlike tube. Cylindrical, head round. *Upper surface brown or black. 6 white longitudinal stripes* and *150 white cross-stripes*. Lacks eyespots and sensory grooves.

Habitat: Among mussels, algae, and other growth on rocks and pilings; from low-tide line to a few feet below.

Range: Entire Pacific Coast.

Comments: This worm can stretch to lengths of more than 1 meter. The Tube

Nemertean (*T. pellucidus*) also forms a tube of parchment, but is much smaller, measuring only 1″ (25 mm) long and ⅛″ (3 mm) wide. It is slender, cylindrical, and white, sometimes with a pale orange stripe down its back, and ranges from Cape Cod to Florida and Texas, and from British Columbia to California.

259 **Red Lineus**
(*Lineus ruber*)
Class Anopla

Description: 8″ (20 cm) long, ⅛″ (3 mm) wide. Slender, slightly flattened, *head wider than adjacent part of body. Dark red, brownish, or greenish, pale at borders,* sometimes ringed with faint, white lines, with 4–8 black eyespots and longitudinal sensory groove on each side.

Habitat: Under rocks and shells, and among mussels and algal growth on both sand and mud bottoms; above low-tide line and below to shallow depths.

Range: Maine to Long Island Sound; Washington to c. California.

Comments: Some biologists regard the greenish form of this animal as a separate species. The Social Lineus (*L. socialis*), a gregarious worm frequently found in clumps, is about the same size, shape, and color as *L. ruber,* but is distinguished by its habit of coiling into a tight spiral when disturbed. It ranges from Maine to Florida and Texas. The Striped Lineus (*L. bicolor*), which ranges from Cape Cod to Cape Hatteras, is smaller, 2″ (51 mm) long and 1/16″ (2 mm) wide, and is greenish or brownish, with a white or pale yellow stripe down the back, and a pale undersurface.

Verrill's Nemertean
(*Micrura verrilli*)
Class Anopla

Description: 18″ (46 cm) long, ⅜″ (10 mm) wide. Slightly flattened toward the posterior end, with small tail; head rounded. *White, with rectangular blocks of reddish-purple down the back; upper surface of head orange.* Longitudinal sensory slits on each side, no eyespots.

Habitat: Under rocks in sandy mud, among roots of surfgrass and algal holdfasts; near low-tide line and below to depths of kelp holdfasts among which it occurs.

Range: Alaska to c. California.

Comments: This species is named for A. E. Verrill, a 19th century American naturalist. Leidy's Nemertean (*M. leidyi*) is similarly shaped, 12″ (30 cm) long and ¼″ (6 mm) wide. It is reddish-orange, rosy, or purplish-red and is paler underneath. It ranges from Maine to Florida and Texas, burrowing in sand and sandy mud near low-tide line and just below. The name honors Joseph Leidy, an early naturalist.

---

257 **Milky Nemertean**
(*Cerebratulus lacteus*)
Class Anopla

Description: 48″ (122 cm) long, ⅝″ (16 mm) wide. Cylindrical and firm at front end; flat, wide, and soft, *very thin at the edges over most of the length;* tail present, *head shaped like a spearhead,* wider than adjacent body. Milky white, yellowish, or pinkish in young forms, mature males red, females brownish. Deep longitudinal sensory grooves on head. No eyespots; *mouth an elongate slit.*

Habitat: Burrowing in sand or sandy mud, under rocks, in sheltered bays and estuaries; near low-tide line and below.

Range: Maine to Florida and Texas.

Comments: This nemertean is by far the largest ribbon worm on the Atlantic Coast of the United States. The California Swimming Nemertean (*C. californiensis*) is 36″ (91 cm) long and ½″ (13 mm) wide. It is similar in shape to *C. lacteus,* and its color varies from pale yellow, rosy cream, and buff to light brown and chocolate. It burrows in soft, sandy or muddy bottoms in protected bays from near low-tide line to water 150′ (46 m) deep, and ranges from Washington to California.

255, 256 **Chevron Amphiporus**
(*Amphiporus angulatus*)
Class Enopla

Description: 6″ (15 cm) long, ⅜″ (10 mm) wide. Thick, slimy. Reddish-brown to purplish above, whitish or pinkish below, rounded *head demarked by whitish sensory grooves, forming a rear-pointing chevron.* Pale area with 12 small eyespots on each side in front of chevron; 20 larger eyespots along each side of front margin, separated from other eyespots by a thin, pale line; proboscis thick, pinkish.

Habitat: Beneath rocks in sandy or gravelly places; from above low-tide line to water more than 450′ (137 m) deep.

Range: Maine to Cape Cod; Washington to s. California.

Comments: This species may be mistaken for a leech. There are at least 17 species of *Amphiporus* along the West Coast, making this the most common genus there. The Blood Nemertean (*A. cruentatus*), similar in shape to *A. angulatus,* is 1⅜″ (35 mm) long and ⅛″ (3 mm) wide, and is translucent yellow, pink, or orange, with 3 longitudinal vessels containing red blood. It ranges from Massachusetts to both coasts of Florida, and the entire Pacific Coast, and is found among hydroids, bryozoans, and algae, on rock and shell bottoms, from low-tide line to water 240′ (73 m) deep.

Wandering Nemertean
(*Paranemertes peregrina*)
Class Enopla

Description: 10″ (25 cm) long, ¼″ (6 mm) wide. Stout. Purplish, dark brown, or orange-brown above, yellow to white below. *Narrow, white, V-shaped line demarking head.*

Habitat: Among mussels, under rocks and coralline algae, and on protected mudflats; above low-tide line.

Range: Entire Pacific Coast.

Comments: An aggressive predator, the Wandering Nemertean can be found at low tide moving over wet rocks and mudflats, seeking annelid worms as prey. Anabaseine, the first nemertean toxin to be identified chemically, was extracted from this species.

Four-eyed Nemerteans
(*Tetrastemma* spp.)
Class Enopla

Description: 1¼″ (32 mm) long, ¼″ (6 mm) wide. Almost cylindrical. Pale green, yellow, pink, or white. Head demarked by shallow sensory grooves, with *4 eyespots arranged in a square.*

Habitat: On algae, eelgrass, hydroids, bryozoans, and other growth on rocks, pilings, and shelly bottoms; from low-tide line to water 45′ (14 m) deep.

Range: Maine to Florida and Louisiana; entire Pacific Coast.

Comments: Laboratory study is required to identify these tiny but abundant nemerteans as to species, since their color is so variable.

SEGMENTED WORMS
(Phylum Annelida)

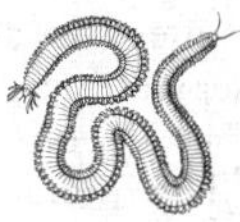

The Phylum Annelida includes about 9,000 known species belonging to three classes: the Hirudinea, Oligochaeta, and Polychaeta. The Hirudinea are the leeches, of which only a few species are parasitic on marine fishes. The Oligochaeta include the earthworms and most freshwater annelids, but very few marine species. The Polychaeta are nearly all marine, and include almost 6,000 species, two-thirds of all annelids. Our discussion will deal only with the polychaetes.

An annelid's body is usually elongate and more or less cylindrical, and consists of a series of *segments.* The body wall is covered with a thin, elastic *cuticle,* beneath which lie layers of circular and longitudinal muscles. These surround a fluid-filled body cavity (*coelom*) which is usually divided between the segments by cross-walls (*septa*). A series of contractions and elongations of the segments propels the worm forward. The more complex motions of wriggling and the undulatory swimming of certain marine worms depend on alternate contractions of left and right longitudinal bands, and on the movement of appendages present in many species.

The body plan is more advanced than that of phyla previously discussed. Annelids have a complete digestive system extending from the mouth on the first segment to the anus at the hind end of the body. Above the mouth is a lobe, the *prostomium:* a probing organ that often bears sensory structures and is useful in feeding and burrowing. Most annelids have well-developed circulatory, nervous, and excretory systems. Respiration is carried on through the cuticle, and some forms have specialized gills. All annelids are equipped with numerous glands under the cuticle that produce abundant mucus that helps keep the cuticle moist and is used in some species to catch food, build tubes, or form egg-masses. Most polychaetes have separate sexes. Mature eggs and sperm are liberated directly into the water. The rounded or top-shaped larva, or *trochophore*, has a mouth, digestive tract, and anus—and is thus able to feed—and a girdle of cilia for swimming. As it grows and elongates, new segments are added just ahead of the anus. When a certain number of segments and head features have formed, the little worm drops out of the plankton and assumes the adult life pattern of its species.

Subclasses of Polychaetes:

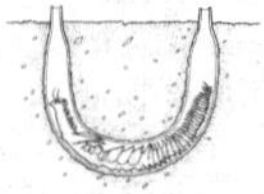

The Class Polychaeta is divided into two subclasses, the Errantia and the Sedentaria. The Errantia are worms that move about, while the Sedentaria remain in a tube or burrow. Errant polychaetes generally have well-developed, paired appendages (*parapodia*), trunk segments all similar to each other, good locomotory ability, and a head with eyespots and sensory appendages. They are predators, browsers, or bottom-dwelling deposit feeders. Sedentary polychaetes usually have a head without eyes or sensory appendages, but sometimes with abundant gills and feeding tentacles; a trunk divided into a thick *thorax* followed by a slender *abdomen*—the segments of each quite different; and appendages reduced and modified for adhering to the inside of the tube or burrow. They are generally bottom-dwelling deposit or filter feeders.

Milky Paddle Worm
(*Eteone lactea*)
Subclass Errantia

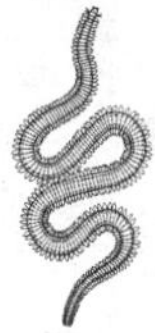

Description: 9″ (23 cm) long, ⅛″ (3 mm) wide. Elongate, slender, slightly flattened. Milky-white to pale yellow. 400 segments, *first segment with 2 pairs long tentacles;* lobe above mouth with *4 short antennae.* 2 small eyes. Appendages with *paddle-shaped or leaf-shaped lobes* on upper side.

Habitat: Burrowing in mud or sand, or among growth on rocks; from well above low-tide line to water 600′ (183 m) deep.

Range: Maine to Florida.

Comments: These worms are good swimmers, and come to the ocean surface at night in the summer months. The Varied-footed Paddle Worm (*E. heteropoda*), distinguished from *E. lactea* by differently shaped paddles, can be seen on mudflats at low tide.

247 **Leafy Paddle Worms**
(*Phyllodoce* spp.)
Subclass Errantia

Description: 18″ (46 cm) long, ⅜″ (10 mm) wide. Long, slender. Whitish, tan, brownish, greenish, or gray, some with dark band down middle of back, or with cross-stripes. Head with *4 pairs of long tentacles;* lobe above mouth heart-shaped, with *4 short antennae* and *2 prominent eyes.* Body segments with *large, leaflike, oval paddles* on upper side of appendages, smaller ones on lower side.

Habitat: Under rocks, among shells and gravel, and in algal holdfasts; from low-tide line to water 5000′ (1524 m) deep.

Range: Arctic to Florida and Texas; Alaska to Mexico.

Comments: These worms prey on other polychaetes, and on nemertean worms and other small creatures, and are themselves eaten by several species of fish, including cod, haddock, and plaice. Laboratory examination is usually necessary to determine the species of these worms.

248 **Green Paddle Worm**
(*Eulalia viridis*)
Subclass Errantia

Description: 6″ (15 cm) long, ⅛″ (3 mm) wide. Slender. *Pale to dark green.* 200 segments, first 3 with 4 pairs of tentacles equal in length; lobe above mouth oval, with 5 *antennae.* 1 pair large eyes. Appendages with *spearhead-shaped paddles* above.

Habitat: Under rocks in tidepools, in gravelly sand, on pilings and rocks among marine growth; from above low-tide line to water 500′ (152 m) deep.

Range: Arctic to New Jersey; California.

Comments: This worm was one of the most abundant animals found in an oceanographic study of organisms growing on ship hulls, floats, and harbor installations in New England. A closely related species, the Black-striped Paddle Worm (*E. aviculiseta*), comparable in size to *E. viridis* and common in California, is green with a black cross-stripe between adjacent segments.

Sea Mouse
(*Aphrodita hastata*)
Subclass Errantia

Description: 6″ (15 cm) long, 3″ (76 mm) wide. *Plump.* Yellow, bronze, blackish, occasionally iridescent. 40 segments. *Upper surface covered with numerous long bristles of various sizes and mucus and mud.*

Habitat: In soft mud bottoms; below low-tide line, from water 6′ (2 m) to more than 6000′ (1829 m) deep.

Range: Gulf of St. Lawrence to Chesapeake Bay.

Comments: The Sea Mouse, which is sometimes tossed ashore in large numbers during storms, has abundant bristles that give it a furry appearance, especially when the mud and mucus have been washed out.

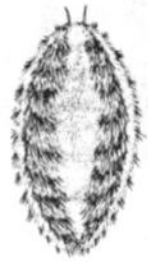

240 **Twelve-scaled Worm**
(*Lepidonotus squamatus*)
Subclass Errantia

Description: 2″ (51 mm) long, ⅝″ (16 mm) wide. Stout. Grayish, tan, or mottled brown. Covered above by *12 pairs of oval scales,* with tan, reddish, or greenish projections of several sizes; tentacles and antennae with dark bands, pointed tips.

Habitat: Under rocks, among marine growth, on pilings, and on gravel and shell bottoms; from above low-tide line to water more than 8000′ (2438 m) deep.

Range: Labrador to New Jersey; Alaska to California.

Comments: The name *Lepidonotus* means "scaly back," and when this worm is disturbed it rolls up like an armadillo into a scale-covered ball. It is tough and, unlike some of its relatives, does not easily lose its scales. The Commensal Twelve-scaled Worm (*L. sublevis*) is 1⅜″ (35 mm) long and ⅜″ (10 mm) wide. It is grayish, greenish, or reddish-brown, and has scales with low, conical projections of uniform size. It lives commensally in a shell with the Flat-clawed Hermit Crab, and among oysters and under rocks, from Cape Cod to Florida and Texas. The Variable Twelve-scaled Worm (*L. variabilis*) measures 1″ (25 mm) long and ¼″ (6 mm) wide, and has mottled gray or brown scales, each with its rear border edged in short hairs. It ranges from Florida to Texas and the West Indies, and is found among oyster shells, rocky rubble, and clumps of algae.

Pacific Scale Worm

(*Arctonoe vittata*)
Subclass Errantia

Description: 4″ (102 mm) long, ⅜″ (10 mm) wide. Back partially covered by more than *30 pairs of white, smooth, oval scales.* Exposed surface of back *tannish with dark cross-stripes,* heavier stripes on segments 7 and 9. 2 pairs of tiny eyes on frontal lobe; 1 pair of dark gray antennae.

Habitat: Free-living among rocks, or commensal on the Keyhole Limpet, the Gum Boot Chiton, and the Leather Star; from above low-tide line to water 800′ (244 m) deep.

Range: Entire Pacific Coast.

Comments: When living commensally in the mantle cavity of the Keyhole Limpet, this worm decidedly earns its keep. When the host limpet is attacked by the Ochre Sea Star, the worm bites the star's tube feet, causing it to retreat. If removed from the limpet, the worm can relocate its host and reestablish itself in the mantle cavity. Two other species of *Arctonoe* in Pacific waters are the Beautiful Scale Worm (*A. pulchra*), 2¾″ (70 mm) long and ¼″ (6 mm) high, found in the Giant Chiton, a giant keyhole limpet, and a sea cucumber; and the Fragile Scale Worm (*A. fragilis*), 3⅜″ (86 mm) long and ¼″ (6 mm) high, found on a sea star. Members of the genus tend to match the color of the host.

245 Eighteen-scaled Worm
(*Halosydna brevisetosa*)
Subclass Errantia

Description: 4⅜″ (111 mm) long, more than ⅜″ (10 mm) wide. Covered above with *18 pairs of grayish, brownish, or reddish-brown oval scales,* each scale with a pale dot. 1 pair of antennae, and 1 unpaired.

Habitat: Free-living among mussels, algal holdfasts, or growth on rocks and pilings; or commensal in the tubes of several other polychaetes and the snail shell occupied by Baker's Hermit Crab (*Paguristes bakeri*). From above low-tide line to water more than 1460′ (545 m) deep.

Range: Entire Pacific Coast.

Comments: The commensal forms of this scale worm are typically about twice as big as their free-living relatives. When several of these worms are put together in a container, however, they attack each other, biting off scales and bits of flesh.

239 Fifteen-scaled Worm
(*Harmothoe imbricata*)
Subclass Errantia

Description: 2½″ (64 mm) long, ¾″ (19 mm) wide. Thick, flattened. Covered above with *15 pairs of scales.* Reddish, orange, tan, brown, green, gray, black, speckled or mottled, or with black stripe down the back. 2 pairs of eyes on frontal lobe; front pair on lower side, rear pair on upper side.

Habitat: From open shores to very brackish estuaries; in rocky tidepools. Free-living under rocks, among marine growth; or commensal in tubes of other polychaetes, or in shells occupied by hermit crabs. From above low-tide line to water more than 11,000′ (3353 m) deep.

Range: Arctic to New Jersey; s. California.

Comments: This ubiquitous scale worm is tolerant

of great ranges of temperature, salinity, and depth. The Four-eyed Fifteen-scaled Worm (*H. extenuata*) is 3″ (76 mm) long, ¾″ (19 mm) wide, and likewise variable in color. It differs most obviously from *H. imbricata* in having both pairs of eyes on the upper surface of the frontal lobe. It has the same range as *H. imbricata.*

Burrowing Scale Worm
(*Sthenelais boa*)
Subclass Errantia

Description: 8″ (20 cm) long, ¼″ (6 mm) wide. Long, flattened. Gray, with mottled, brownish stripe down the back. *150 pairs of overlapping, kidney-shaped scales with fringed borders.*

Habitat: In mud, sandy mud, sandy gravel, and among eelgrass roots, and ranging into brackish water; from above low-tide line to water 480′ (146 m) deep.

Range: Cape Cod to Florida and Texas.

Comments: This worm is a rapid burrower, quickly making its way into mud or sand when disturbed. A related species, the Dusky Burrowing Scale Worm (*S. fusca*), which is greenish in color, is found from Washington to California, commonly among the roots of surfgrass. It measures 6″ (15 cm) long and ¼″ (6 mm) wide.

---

251 **Two-gilled Blood Worm**
(*Glycera dibranchiata*)
Subclass Errantia

Description: 15⅜″ (38 cm) long, ½″ (13 mm) wide. Long, round. Pink. *Lobe above mouth conical,* with 4 tiny antennae at tip. Pharynx everts as *long, bulbous proboscis with 4 black jaws at tip.* Each appendage with *red, fingerlike, non-retractable gill* on upper and lower side.

Habitat: In mud, sandy mud, and sandy gravel bottoms in bays and open waters; from near low-tide line to water 1322′ (403 m) deep.

Range: Gulf of St. Lawrence to Florida and Texas; c. California to Mexico.

Comments: This worm is sometimes called a Beak Thrower because it can suddenly and forcefully shoot out its proboscis. It uses this mechanism for burrowing, for ingesting prey, and for nipping the unwary person handling them. This species is used as fish bait, and is shipped all over the United States by bait diggers in Maine. The Tufted-gilled Blood Worm (*G. americana*) ranges from Cape Cod to Florida and Texas and along the entire Pacific Coast. It measures 14″ (36 cm) long and ½″ (13 mm) wide and differs from *G. dibranchiata* chiefly in having a retractable tuft of gills only on the upper side of each appendage.

Shimmy Worm
(*Nephtys bucera*)
Subclass Errantia

Description: 12″ (30 cm) long, ¾″ (19 mm) wide. Sturdy, somewhat flattened, especially at head end. Pale, with *brownish V-shaped bands* on upper side of each segment. *Lobe above mouth squarish, with 1 pair of short antennae at corners.* Appendages 2-lobed.

Habitat: In sand and sandy mud, ranging into brackish water; from above low-tide line to water 600′ (183 m) deep.

Range: Maine to Florida and Mississippi.

Comments: These worms swim by "shimmying" with a series of rapid undulations that progress from the tail end toward the head. The Leafy Shimmy Worm (*N. caeca*) ranges from the Arctic to Rhode Island, and from Alaska to northern California. It is whitish, greenish, or bronze, 8″ (20 cm) long, ⅝″ (16 mm) wide, and has leafy extensions behind each pair of appendages. The California Shimmy Worm (*N. californiensis*) is widespread along the Pacific Coast. It is 12″ (30 cm) long, ¾″ (19 mm) wide, and whitish or tawny, with a dark spread-eagle pattern on its frontal lobe. It inhabits open sandy beaches.

253 **Bat Star Worm**
(*Ophiodromus pugettensis*)
Subclass Errantia

Description: 1½″ (38 mm) long, ⅛″ (3 mm) wide. Robust. *Reddish-brown to purple or black.* Lobe above mouth with 3 antennae, 2 pairs of eyes; head with 6 *pairs of tentacles.* Appendages with long dorsal filaments.

Habitat: Free-living on mud bottoms or among marine growth; or commensal among the tube feet of sea stars, particularly the Bat Star; from above low-tide line out onto the continental shelf.

Range: Entire Pacific Coast.

Comments: These worms may live on sea stars or be free-living. Those removed from a sea star will readily return to it, and will also be attracted to a sample of water that the star once occupied, scenting its presence. The Bat Star host may have as many as 20 worms on it at once, and worms will leave one star for an approaching one. On the other hand, free-living worms are oblivious to sea stars and the water that holds them.

249 **Clam Worm**
(*Nereis virens*)
Subclass Errantia

Description: 36″ (91 cm) long, 1¾″ (44 mm) wide. Thicker in head region, tapered toward rear. 200 segments. Iridescent greenish, bluish, or greenish-brown above, usually with fine red, gold, or white spots; paler beneath; appendages red, showing blood vessels. Head with *4 pairs of tentacles of equal length;* a fleshy lip on each side of mouth, lobe above mouth broad, rectangular, with *pair of short tentacles;* proboscis with pair of strong, black jaws. 2 pairs of eyes; body appendages 2-lobed, *upper part of appendages broad and leaflike.*

Habitat: In sand, sandy mud, mud, clay, and various peat bottoms, among roots of eelgrass, in protected waters and in brackish estuaries; from near high-tide line to water more than 500′ (152 m) deep.

Range: Maine to Virginia; entire Pacific Coast.

Comments: The Clam Worm is a swift and voracious predator, feeding on other worms and invertebrates, carrion, and certain algae. It has a keen sense of smell and in captivity can readily locate bits of fresh clam meat. Another large nereid found in Pacific waters, *N. brandti,* is very difficult to distinguish from *N. virens* and indeed may intergrade with it.

250 Pelagic Clam Worm

(*Nereis pelagica*)
Subclass Errantia

Description: 6⅛″ (16 cm) long, ½″ (13 mm) wide. Long, thicker in head region, tapered toward rear. Iridescent greenish-brown, golden-brown, reddish-brown, or olive. Head with *4 pairs of tentacles of equal length;* each side of mouth with *fleshy lip longer than frontal lobe;* frontal lobe above mouth *tapered toward front end,* with pair of short antennae, 2 pairs of eyes; proboscis with pair of strong, black jaws. Body appendages with *3 whitish, bluntly conical knobs.*

Habitat: In sand or mud under rocks on open

shore; from low-tide line to water 3600′ (1100 m) deep.

Range: Arctic to Florida, south to Straits of Magellan; Alaska to Panama.

Comments: This worm does not tolerate reduced salinity, and hence is seldom found in bays and estuaries. During the breeding season it develops very large eyes, and appendages specialized for swimming, and leaves its burrow to mate in the open water.

---

270 **Plumed Worm**
(*Diopatra cuprea*)
Subclass Errantia

Description: 12″ (30 cm) long, ⅜″ (10 mm) wide. In a leathery tube. Front cylindrical, rear flattened and tapered. Reddish to brown, speckled with gray; appendages yellowish-brown; ripe males yellowish, females gray-green. Cuticle thick, wrinkled, iridescent. Lobe above mouth oval, short, with 1 pair of short, conical antennae in front, and *5 long antennae with ringed bases* on top. Proboscis with large jaws, segments 4 or 5 through 35 with *bushy gills on upper surfaces.*

Habitat: On protected mud- and sandflats, mixed with shell debris and gravel; from low-tide line to water 270′ (82 m) deep.

Range: Massachusetts to Florida and Louisiana.

Comments: Although the Plumed Worm dwells in a tube, it is an active predator, an unusual combination among polychaete worms. This worm can bite, so care in handling it is advised. The Ornate Plumed Worm (*D. ornata*), found below the low-tide line on rubbly bottoms in California, is so similar that only laboratory examination could reveal the difference between it and *D. cuprea.* Naturally, their ranges eliminate confusion in the field.

Fragile Worm
(*Lumbrineris fragilis*)
Subclass Errantia

Description: 15″ (38 cm) long, ½″ (13 mm) wide. Slender, cylindrical. Earthwormlike, with 340 segments. Iridescent; reddish-orange or brown, or yellowish with white bands. *Head without tentacles, antennae, or eyes; lobe above mouth short, conical;* proboscis with 3 pairs of black jaws. Paired appendages short, simple.

Habitat: Burrowing in mud, sand, or gravelly-mud bottoms; from low-tide line to water 11,000′ (3350 m) deep.

Range: Arctic to Virginia.

Comments: The Fragile Worm is an aggressive predator, preying on other bottom-dwelling invertebrates, and is itself eaten by cod, haddock, and other fishes. As its name suggests, it breaks readily, making it difficult to dislodge a specimen intact.

---

252 Opal Worm
(*Arabella iricolor*)
Subclass Errantia

Description: 24″ (61 cm) long, ¼″ (6 mm) wide. Long, slender. Reddish-brown, reddish-yellow, or greenish, with *brilliant metallic iridescence.* 500 segments; head without antennae or tentacles; lobe above mouth bluntly conical, with *4 black eyes in a row at rear margin.* Paired appendages simple, small.

Habitat: Burrowing in sand and sandy mud, in oyster and mussel beds, and among eelgrass roots, invading estuaries with low salinities; from low-tide line to water 275′ (84 m) deep.

Range: Massachusetts to Florida and Texas; entire Pacific Coast.

Comments: The Opal Worm, iridescent as its name implies, will first contract into a ball, but glistens brightly in the sunlight once cleaned of the sand that usually adheres to its abundantly secreted mucus.

242 **Orange Fire Worm**
(*Eurythoe complanata*)
Subclass Errantia

Description: 6″ (15 cm) long, 1/4″ (6 mm) wide. Moderately long, flattened. *Orange-yellow,* with tufts of *red gills* on upper surfaces of paired appendages, and abundant *white bristles* on sides. Head with 5 short tentacles; 2 pairs of eyes; *smooth, elongate, oval pad* from back of head to 4th segment.

Habitat: On the reef flat under rocks and old coral heads; from near low-tide line to water 50′ (15 m) deep.

Range: Florida, and throughout the West Indies and Gulf of Mexico.

Comments: **HIGHLY TOXIC.** This annelid merits the name "fire worm." If touched, it inflicts painful stings with bristles that pierce the skin, break off, and release a toxin. The pain and itching may last several days.

241 **Green Fire Worm**
(*Hermodice carunculata*)
Subclass Errantia

Description: 10″ (25 cm) long, 1/4″ (6 mm) wide. *Squarish in cross section. Greenish or reddish,* with tufts of *orange gills* on upper surface of each appendage, and abundant *white bristles* on the side. Head with 5 short tentacles; 2 pairs of eyes; large, *folded or wrinkled, lance-shaped pad* from back of head to 5th segment.

Habitat: Under rocks, in turtle grass beds, on coral reefs and flats; from low-tide line to water 50′ (15 m) deep.

Range: Florida; Bahamas; West Indies.

Comments: **HIGHLY TOXIC.** Like the Orange Fire Worm, this species inflicts a painful sting when handled.

243 Red-tipped Fire Worm
(*Chloeia viridis*)
Subclass Errantia

Description: 5″ (127 mm) long, 2″ (51 mm) wide. Elliptical and sturdy. Pale greenish or brownish, mottled, iridescent, with dark stripe down middle of back; side of each segment fringed with *large tuft of white bristles with red or orange tips.* Slender pad with pointed rear on back of head, reaching to segment 3.

Habitat: On sand or mud bottoms, among debris, on pilings and on floating objects; from below low-tide line to water more than 300′ (91 m) deep.

Range: Florida; Bahamas; West Indies.

Comments: **HIGHLY TOXIC.** This fire worm is an active predator that sometimes swallows a surprised fisherman's bait, hook and all. When handled, its abundant bristles sting severely.

246 Lug Worm
(*Arenicola cristata*)
Subclass Sedentaria

Description: 12″ (30 cm) long, 1″ (25 mm) wide. Firm and sturdy, thick in front, with tapering head and tail end; *skin coarse and checkered. Greenish-black.* Head without appendages or eyes; mouth with bulbous proboscis covered with short, fingerlike projections. Each segment with 5 rings, the thickest with tufts of long bristles above and ridged furrows with shorter hooks below.

Habitat: Burrowing in sandy mudflats in protected places; near low-tide line and just below.

Range: Cape Cod to Florida and Louisiana;

entire Pacific Coast.

Comments: To feed, the Lug Worm pumps water into its burrow, thus irrigating its gills and collapsing the muddy sand at the end of the burrow. It then eats that sand, from which it digests the organic matter. Periodically, the worm backs up to the surface to void the undigested sand and mud. Another species, the Northern Lug Worm (*A. marina*), 8″ (20 cm) long and ¾″ (19 mm) wide, found north of Cape Cod, extrudes its feces as a "casting," a rope of sandy mud lightly held together by mucus.

264, 271 Bamboo Worm
(*Clymenella torquata*)
Subclass Sedentaria

Description: 6″ (15 cm) long, ¼″ (6 mm) wide. In a sandy 10″ (25 mm) long tube. Cylindrical, slender. Brick-red with bright red joints, or green with green joints. 22 segments: head segment obliquely slanted, with bulbous, eversible proboscis; *flared collar at front end of 4th segment; 17 segments elongate and bearing short appendages* with bristles; the *last of 4 tail segments funnel-shaped, with fingerlike projections around the edge.*

Habitat: In vertical tube in sand or sandy-mud bottoms in protected places, ranging into brackish water; from near low-tide line to water more than 330′ (100 m) deep.

Range: Maine to North Carolina.

Comments: The long segments of these worms vividly recall sticks of bamboo. The green color of some populations of bamboo worms results from their eating 1-celled algae present in the mud. A green worm with several segments removed, placed in mud lacking those algae, will regenerate red segments. The most common closely related species on the West Coast is the Red-banded Bamboo Worm (*Axiothella rubrocincta*). It is the same size as *C. torquata,* is green with red bands on segments 4–8, and has a funnel-shaped terminal segment with unequal fingerlike projections around the edge. It frequently shares its vertical sandy tube with the commensal Bamboo Worm Pea Crab (*Pinnixa longipes*).

Sand Bar Worm
(*Ophelia denticulata*)
Subclass Sedentaria

Description: 3″ (76 mm) long, ¼″ (6 mm) wide. Spindle-shaped, with *deep groove on underside* from segment 10 to rear. Pink, reddish, or pale rosy-blue, iridescent. Head conical, without appendages; proboscis soft and button-shaped; tapering rear end terminates abruptly; anus surrounded by several short, fingerlike structures. *18 pairs of long, tapering, dark red gills* beginning with segment 10.

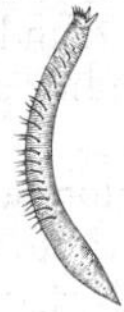

Habitat: Burrowing in clean sand bottoms; near low-tide line and below in shallow water.

Range: Maine to North Carolina.

Comments: This worm lives in the cleanest sandy bottoms, ingesting huge quantities of sand from which it nourishes itself by digesting the sparse film of microorganisms covering each grain, a very small percentage of the total swallowed. The Cosmopolitan Sand Bar Worm (*O. limacina*) is found in similar habitats on the West Coast of the United States. It is very similar to *O. denticulata* but has 23 pairs of gills.

269 **Polydora Mud Worm**
(*Polydora ligni*)
Subclass Sedentaria

Description: 1″ (25 mm) long, 1/16″ (2 mm) wide. In a mud-covered tube. Slender, cylindrical. Translucent, reddish. Head with *2 long antennae* (which are easily lost), lobe above mouth forked in front, with 4 eyes arranged in a rectangle; *5th bristle-bearing segment with large group of long bristles pointing upward;* tail somewhat flared. *14 pairs of gills,* usually beginning with segment 12.

Habitat: In soft, fragile tubes covered with mud and attached to hard objects in protected places on mud and clay bottoms; near low-tide line and in shallow water.

Range: Entire East Coast; California.

Comments: These worms are sometimes so abundant in oyster beds that they bury the oysters in several inches of mud tubes. There are many species in the genus *Polydora,* some of them boring into oyster shells or snail shells occupied by hermit crabs.

Sand Chimney Worm
(*Spio setosa*)
Subclass Sedentaria

Description: 3″ (76 mm) long, 1/8″ (3 mm) wide. Slender, cylindrical. In thick-walled but fragile tubes of sand that rise as chimneys about 1/4″ (6 mm) above the surface. Dull green or yellowish-green, with *red gills held erect over back on all trunk segments.* Head with *1 pair of long tentacles;* 4 eyes on top arranged in a square.

Habitat: In protected sandy places; near low-tide line and below in very shallow water.

Range: Maine to North Carolina.

Comments: These worms may be very abundant locally. The Long-horned Worm (*S. filicornis*) is found on both coasts of the

United States. It is smaller than *S. setosa*, 2″ (51 mm) long and ⅛″ (3 mm) wide, and greenish, with red gills on all segments, and can be distinguished from *S. setosa* with certainty only by microscopic examination of its bristles.

Parchment Worm

(*Chaetopterus variopedatus*)
Subclass Sedentaria

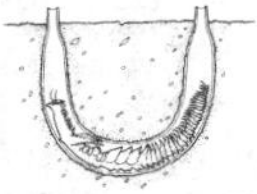

Description: 10″ (25 cm) long, 1″ (25 mm) wide. Soft. Divided into 3 distinct regions. In U-shaped, parchmentlike tubes more than 24″ (61 cm) long. Whitish, translucent. Front end has shovel-shaped head with pair of short tentacles, *pair of winglike appendages* on segment 12, *cup on segment 13* from which a food groove extends to the mouth; middle section with *3 prominent "fans"* on back of segments 14–16; rear section of 30 segments, each with pair of appendages loaded with sperm or eggs in the ripe adult.

Habitat: Buried in mud and sandy mudflats and eelgrass beds in protected bays and estuaries; from near low-tide line to water 25′ (8 m) deep.

Range: Cape Cod to Florida and Louisiana; c. and s. California.

Comments: The tube of this worm is distinctive enough to identify it. It is wide enough to accommodate the worm along most of its length, but each end tapers and rises about 1″ (25 mm) above the surface of the bottom. The worm cannot get out. Although the West Coast form seems to be the same species, its tube is not always U-shaped or buried, but is frequently irregular in shape and attached among growth of other organisms on wharf pilings and floats, or lying on the sandy bottom.

Florida Honeycomb Worm
(*Sabellaria floridensis*)
Subclass Sedentaria

Description: 3″ (76 mm) long, ¼″ (6 mm) wide. In sandy tube. Conical. Yellowish; 2 adjacent semicircular parts on top of head form *dark, disk-shaped lid edged by golden bristles* and surrounded by row of *reddish, retractable tentacles.* Tail slender, unsegmented, bent under body at sharp angle.

Habitat: Along open coast, attached to rocks, forming reefs of tubes of cemented sand grains; near low-tide line and below in shallow water.

Range: West coast of Florida.

Comments: Sand tubes of the Florida Honeycomb Worm form reefs more than 10′ (3 m) across. The worm extends its tentacles to trap fine plankton, organic particles, and sand. The food is moved to the mouth by cilia, and the sand is moved to the edge of the tube where it is cemented on by mucus. The California Honeycomb Worm (*S. cementarium*), which ranges from Alaska to southern California, is very similar, and forms large reefs along the open coast.

Pacific Black-bristled Honeycomb Worm
(*Phragmatopoma californica*)
Subclass Sedentaria

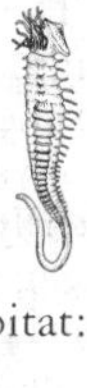

Description: 2″ (51 mm) long, ¼″ (6 mm) wide. In sandy tube. Conical. Cream-colored, with *cone-shaped lid of black bristles* on top of head, and semicircle of *lavender, retractable tentacles* in front of lid. Tail slender, not segmented, bent under body at sharp angle.

Habitat: On rocky shores, forming reefs of tubes of cemented sand grains; from above low-tide line to water 245′ (75 m) deep.

Range: C. California to Baja California.

Comments: The sandy reefs of this species are sometimes wider than 6′ (2 m) across. Reefs are attached to a large rock so situated that wave action swirls up sand which the worms catch and cement to the tube opening. The Atlantic Black-bristled Honeycomb Worm (*P. lapidosa*) is similar and builds large reefs on rocks and wharf pilings along the Atlantic coast of Florida.

161 **Eyed Fringed Worm**
(*Cirratulus cirratus*)
Subclass Sedentaria

Description: 4¾″ (121 mm) long, ⅛″ (3 mm) wide. In mud tubes. Cylindrical. Orange to yellowish. Head bluntly pointed, with *2–9 pairs of eyes* on top, arranged in an arc. *Cluster of long filaments on first bristle-bearing segment,* 1 or more pairs on most of the rest.

Habitat: Under rocks, mussel beds, and sponges; from near low-tide line to shallow depths.

Range: Maine to Cape Cod; entire Pacific Coast.

Comments: The Eyed Fringed Worm feeds by extending its filaments out of the tube and sweeping them over the bottom, picking up small organic particles for food. The filaments are fragile, and are easily broken off when the worm is handled.

158 **Large Fringed Worm**
(*Cirriformia grandis*)
Subclass Sedentaria

Description: 6″ (15 cm) long, ¼″ (6 mm) wide. Nearly uniform in diameter; head conical, no eyes; somewhat flattened toward rear. Yellowish, shading into red, green, orange, or brown; dark gut contents usually visible. Segments 1–7 without appendages; *segments 8 and 9 with large cluster of long, slender, reddish filaments; succeeding segments with 1 to several pairs of filaments.*

Habitat: Burrowing shallowly in soft mud, and under stones in muddy places; from near low-tide line to water 130′ (40 m) deep.

Range: Maine to North Carolina.

Comments: This worm apparently has a bad taste or contains toxic mucus, for if it is dropped into a tank of fish it is eagerly seized, then spit out. Captive fish quickly learn to avoid it while readily feeding on other worms.

---

160 **Luxurious Fringed Worm**
(*Cirriformia luxuriosa*)
Subclass Sedentaria

Description: 6″ (15 cm) long, ¼″ (6 mm) wide. Nearly uniform in diameter. Yellowish, shading to orange or red. Head conical, no eyes; segments in front ¼ each with *1 pair of red filamentous gills* without grooves; *segments 5–7 with 40 pairs of long, reddish-orange, grooved tentacles;* segments farther back with fewer tentacles.

Habitat: In crevices in tidepools, in mussel beds, and among roots of surfgrass; from above low-tide line to water 65′ (20 m) deep.

Range: C. California to Baja California.

Comments: This worm is named *luxuriosa* because of the abundance of its long tentacles. It burrows just below the surface of the sand or mud, and extends its gills and tentacles into the water for oxygen and small organic food particles.

Coralline Fringed Worm
(*Dodecaceria corallii*)
Subclass Sedentaria

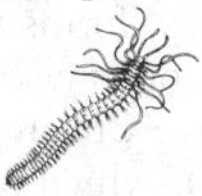

Description: ½″ (13 mm) long, 1/16″ (2 mm) wide. Cylindrical. Brown or greenish-brown; segmental bristles long, whitish. Head conical; *8 pairs of yellow or orange filaments* near head end.

Habitat: Burrowing into dead shells, dead corals, and encrusting coralline algae; from near low-tide line to water 165′ (50 m) deep.

Range: Maine to Florida and Gulf of Mexico.

Comments: These worms have the capacity to burrow galleries into limy substrates such as corals, shells, and coralline algae. Found along the Pacific Coast, Fewkes' Fringed Worm (*D. fewkesi*) is 1⅝″ (41 mm) long, ⅛″ (3 mm) wide, and dark brown or green to black, and has 11 pairs of dark filaments near the head end. Unlike *D. corallii,* it deposits limy tubes which may cluster in masses over 3′ (1 m) across.

268 **Ice Cream Cone Worm**
"Trumpet Worm"
(*Pectinaria gouldii*)
Subclass Sedentaria

Description: 1⅝″ (41 mm) long, ¼″ (6 mm) wide. *In cone-shaped, slightly curved tube of sand grains,* 2½″ (64 mm) long and ¾″ (19 mm) wide at front end. Conical; terminal segment flared. Creamy-pink mottled with red and blue. Head obliquely flattened, with 2 pairs of long, tapered antennae, *2 comblike sets of 15 large, golden bristles;* clusters of pale, flattened feeding tentacles extending forward from beneath mouth. 2 pairs of bright red gills on sides of head.

Habitat: In a vertical tube in sandy mud. Invading brackish estuaries with low salinities; from near low-tide line to water 90′ (27 m) deep.

Range: Maine to Florida.

Comments: Its tube proves this worm nature's stonemason *par excellence.* A wall made of a single layer of sand grains selected for size and fit, then firmly cemented together by dense mucus, forms the delicate, graceful tube, open at both ends. The Californian Ice Cream Cone Worm (*P. californiensis*), which ranges from Alaska to Baja California, is 2⅜" (60 mm) long and ⅜" (10 mm) wide. It has 2 comblike sets of 14 long, tapered bristles, but is otherwise similar to *P. gouldii.* The Coarse-grained Ice Cream Cone Worm (*Cistenides brevicoma*) occupies the same range and is of comparable size, but has short, blunt spines and uses coarse sand grains in its tube.

---

273 **Ornate Worm**
(*Amphitrite ornata*)
Subclass Sedentaria

Description: 15" (38 cm) long, ¾" (19 mm) wide. In a firm tube of sand, mud, or mucus. Thick thorax and more slender, tapered abdomen. Orange-pink to reddish or orange-brown. Head with *numerous long, yellowish-orange tentacles* and *3 pairs of branched, red gills;* appendages with bristles on *50 body segments.*

Habitat: In quiet, shallow bays, often brackish, in bottoms ranging from soft mud to firm muddy sand; near low-tide line and just below.

Range: Maine to North Carolina.

Comments: The tubes of this worm extend downward a foot or more. The opening is usually found in the center of a low hillock made of mud and sand excavated by the worm. South of Cape Cod, one can commonly find the Many-scaled Worm (*Lepidametria commensalis*) living commensally in the tube with the Ornate Worm.

159 **Johnston's Ornate Worm**
(*Amphitrite johnstoni*)
Subclass Sedentaria

Description: 10″ (25 cm) long, ½″ (13 mm) wide. In firm tube of sand or mud and mucus. Thick thorax and more slender, tapered abdomen. Orange-pink to red or orange-brown. Head with *numerous long, yellowish-orange tentacles,* and *3 pairs of branched, red gills;* appendages with bristles on *23–45 segments.*

Habitat: In tubes under rocks on open shores and bays; near low-tide line and below in shallow water.

Range: Arctic to New Jersey.

Comments: The tube of this worm is usually attached to the underside of a rock with the open end protruding, permitting the worm to extend its tentacles and gills. When covered with water, the tentacles sweep over the surface of the bottom, picking up fine organic particles for food and transporting them to the mouth in a ciliated groove on each tentacle.

Green Terebellid Worm
(*Eupolymnia crescentis*)
Subclass Sedentaria

Description: 6″ (15 cm) long, ½″ (13 mm) wide. Cylindrical, plump. *In parchmentlike tube.* Wider thorax and narrower abdomen. *Greenish,* with numerous tan tentacles and *3 pairs of reddish-brown gills* on head. Numerous eyespots behind tentacles. *17 segments with bristle-bearing appendages.*

Habitat: Attached to rocks in sandy-mud habitats; most common between high- and low-tide lines in protected bays; from near low-tide line to water 650′ (198 m) deep.

Range: Entire Pacific Coast.

Comments: A commensal *Pinnixa* crab is sometimes found living in this worm's rigid parchmentlike tube. Terebellids feed by extending long ciliated tentacles and picking up food particles from the bottom which are then moved to the mouth by ciliary action.

Crested Terebellid Worm

(*Pista cristata*)
Subclass Sedentaria

Description: 3½" (89 mm) long, ⅜" (10 mm) wide. Cylindrical, tapered toward rear. In tubes more than 6" (15 cm) long of coarse sand grains, pebbles, and debris. Wide thorax and narrower abdomen. Thorax red to orange-pink, abdomen pinkish-tan. Head with numerous long, white tentacles, *2 pairs of red gills, spirally branched around a long stem, 1 gill much larger,* 1 sometimes missing. *Lobes on sides of first 4 thoracic segments; 17 segments with bristle-bearing appendages.*

Habitat: In muddy sand and eelgrass beds, ranging into brackish water; near low-tide line and below in shallow water.

Range: Maine to Florida and Texas.

Comments: This species, in its sand-and-algae tube, can easily be identified by the single large gill which forms a dense pompom resembling a shaving brush. Two other species of *Pista* occur on the Pacific Coast: the Elongate Terebellid Worm (*P. elongata*) and the Pacific Terebellid Worm (*P. pacifica*). Both are much bigger worms. *P. elongata* is 8" (20 cm) long and ½" (13 mm) wide, is reddish or pinkish-tan, and has 3 pairs of branched gills. It builds a tube attached under rocks and crevices with a spongy network of fibers around the opening. It ranges from British Columbia to Panama. *P. pacifica* builds a vertical tube extending down into the protected sand or mud bottom, with an opening above ending in a triangular hood. The worm is 12¾" (32 cm) long, ⅝" (16 mm) wide, is reddish-brown, and has 3 pairs of branched gills. It ranges from British Columbia to southern California.

164 **Red Terebellid Worm**
(*Polycirrus eximius*)
Subclass Sedentaria

Description: 2¾″ (70 mm) long, ¼″ (6 mm) wide. Cylindrical, tapered toward rear. Wide thorax and narrower abdomen. *Blood-red.* Head with numerous tentacles, but *no gills* or eyes. *25 segments with appendages bearing bristles.*

Habitat: Without well-defined tubes; in burrows in soft mud bottoms, among eelgrass roots, or under rocks in muddy places, ranging into brackish water; from low-tide line to water 55′ (17 m) deep.

Range: Maine to North Carolina.

Comments: This worm has no circulatory system, but its body fluid is full of red blood cells. It extends its tentacles by forcing this fluid into them.

162, 165 **Curly Terebellid Worm**
(*Thelepus crispus*)
Subclass Sedentaria

Description: 11¼″ (28 cm) long, ¾″ (19 mm) wide. In a tough, membranous, sand-encrusted tube. Reddish-pink. Wide thorax, narrower abdomen. Head with numerous tentacles; *3 pairs of curly gills, each consisting of a cluster of red, unbranched filaments.*

Habitat: Under rocks on exposed, rocky shores; above low-tide line.

Range: Entire Pacific Coast.

Comments: This is the most common terebellid worm on the rocky shores of the West Coast, where its tough tube enables it to withstand the strong wave action.

163 **Pale Terebellid Worm**
(*Thelepus setosus*)
Subclass Sedentaria

Description: 8″ (20 cm) long, ½″ (13 mm) wide. In thin, debris-encrusted tube. Creamy-pink. Thorax wide, abdomen narrower. Head with numerous tentacles, *3 pairs of gills,* each consisting of a cluster of *red, unbranched filaments.*
Habitat: Under rocks on open shores; near low-tide line and below in shallow water.
Range: British Columbia to s. California.
Comments: This worm is widespread, also occurring in Japan and elsewhere in the Pacific, the Indian Ocean, the Red Sea, the Mediterranean Sea, and parts of the Atlantic Ocean.

143, 146, 147 **Giant Feather Duster**
(*Eudistylia polymorpha*)
Subclass Sedentaria

Description: 10″ (25 cm) long, ½″ (13 mm) wide. In sturdy, parchmentlike tubes. Body tannish, gills maroon, reddish, orange, or brown, usually with cross-bands of lighter and darker shades. Cylindrical, slightly flattened, tapered; collar at head end; *large plume of featherlike gills* 2½″ (64 mm) across when expanded.
Habitat: Attached to boulders or pilings, or wedged into crevices on open rocky shores; from near low-tide line to water 1400′ (427 m) deep.
Range: Entire Pacific Coast.
Comments: These worms frequently occur in large numbers in a tidepool, presenting a handsome sight that resembles a flower garden. They retract into their tubes with remarkable speed when touched, or even when a shadow passes over them. The gills have numerous eyespots which mediate this "shadow reflex."

140 **Slime Feather Duster**
(*Myxicola infundibulum*)
Subclass Sedentaria

Description: 8″ (20 cm) long, 1¼″ (32 mm) wide. *In a mass of slime.* Collar at head end, tapered toward rear. Cream-colored to tannish, with cream, yellow, greenish, bluish, or brownish *funnel-shaped whorl of gills interlaced almost to the tips* to form a membrane.

Habitat: In gravelly sand and among rocks; among mussels, tunicates, and other growth on rocks and pilings; from low-tide line to water 170′ (52 m) deep.

Range: Maine to New York; entire Pacific Coast.

Comments: Unlike other feather dusters, this one lives in a large, viscous mass of slime rather than a tube. *Myxicola* has in its longitudinal nerve cord a giant nerve fiber which reaches the length of the worm—one of the largest nerve fibers in the animal kingdom. The size of this fiber permits scientists to insert probes and make electronic measurements of the speed of nerve conduction.

138, 139 **Large-eyed Feather Duster**
(*Potamilla reniformis*)
Subclass Sedentaria

Description: 4″ (102 mm) long, ⅛″ (3 mm) wide. In thin, leathery tubes encrusted with sand, more than 6″ (15 cm) long. Slender, tapered toward rear. *60 segments.* Yellowish-green. Plume of about 20 gills, united at base, orange-red to reddish-brown, ringed with white, yellowish at tips, each with *1–8 large, red eyes, irregularly spaced.*

Habitat: On rocks and shells; from near low-tide line to water 340′ (104 m) deep.

Range: Maine to North Carolina.

Comments: This delicate and graceful worm is abundant on shelly bottoms on the south side of Cape Cod.

144 **Banded Feather Duster**
(*Sabella crassicornis*)
Subclass Sedentaria

Description: 2″ (51 mm) long, ⅛″ (3 mm) wide. In stiff, leathery tube 4″ (102 mm) long. Tapered. Creamy-pink to orange-tan. 4-lobed collar at head end, head with plume of *24 straight, feathery gills,* united at base, banded with various shades of red, each with *2–6 paired, dark red eyespots, evenly spaced* in rows across plume.

Habitat: Attached to rocks and shells; from near low-tide line to water 170′ (52 m) deep.

Range: Maine to Cape Cod; entire Pacific Coast.

Comments: Living in a dead-end tube presents a problem in voiding fecal pellets. Sabellids have a ciliated groove that runs the length of the body and carries the pellets to the top of the tube so the worm can dump them outside. The Black-eyed Feather Duster Worm (*S. melanostigma*), which also has evenly spaced rows of eyes on the plume and, in addition, a few wide bands of red, is about the same size as *S. crassicornis.* It is common in Florida and the Caribbean Sea.

---

141 **Magnificent Feather Duster**
(*Sabellastarte magnifica*)
Subclass Sedentaria

Description: 5″ (127 mm) long, ¾″ (19 mm) wide. In leathery tube. Sturdy, tapered. Tan to brown. Plume of 48 *featherlike gills,* each 4″ (102 mm) long, and with bands of brown and pale tan, chocolate and white, dark purple and brown, or mahogany-red and brown.

Habitat: In sand or gravel bottoms; on pilings, and among reef corals; below low-tide line in shallow tropical seas.

Range: Florida and Texas; Bahamas; West Indies.

Comments: This worm of exotic tropical beauty is the largest of the feather duster worms in American Atlantic waters. Its gills, like those of other sabellids, are used for filter feeding, and its feathery tentacles remove particles from the water.

83, 149 **Lacy Tube Worm**
(*Filograna implexa*)
Subclass Sedentaria

Description: ¼″ (6 mm) long, 1/25″ (1 mm) wide. Slender, threadlike. In slender, white, limy tube up to 2½″ (64 mm) long and ⅛″ (3 mm) wide. Pink to purple. *Head with plume of 8 gills and stalked, spoonlike lid* to close opening of tube.

Habitat: Numerous tubes aggregated into a lacy network 1′ (30 cm) or more across; attached to rocks; from below low-tide line to water 170′ (52 m) deep.

Range: Maine to Cape Cod.

Comments: These tiny worms will not be seen without a special effort, but their tubes can hardly be confused with those of any other species. Reproduction in *Filograna* is both sexual and asexual by binary fission, each worm dividing into 2 pieces and regenerating missing parts.

Atlantic Tube Worm
(*Hydroides uncinata*)
Subclass Sedentaria

Description: 3″ (76 mm) long, ⅛″ (3 mm) wide. In white, limy, twisted tube. Cylindrical, tapered toward rear. Translucent, yellowish or greenish, with green blood vessels visible. Head with solid or mottled purple, red, orange, yellow, tan, greenish, or white *plume of 18 pairs of gills, each pointed and without side branches at tip,* abundant branches near

base. *Lid stalked, funnel-shaped, edged with about 30 teeth.* Collar of 7 segments. Abdomen tapering.

Habitat: Attached to hard objects; from low-tide line to water 50′ (15 m) deep.

Range: Cape Cod to Florida and Texas.

Comments: These worms are common wherever there is an abundance of rocks or shells on which to settle. The tubes may be solitary or occur in tangled masses.

---

477 **Sinistral Spiral Tube Worm**
(*Spirorbis borealis*)
Subclass Sedentaria

Description: ⅛″ (3 mm) long, ⅛″ (3 mm) wide. In limy tube coiled like a snail shell, counterclockwise to left from opening toward center. Tiny worm inside. Whitish, translucent. *9 feathery gills;* stalked lid; collar of 3 segments.

Habitat: Attached to kelps, Irish moss, and other algae, and to rocks and shells; from above low-tide line to shallow depths.

Range: Maine to Cape Cod; entire Pacific Coast.

Comments: These worms are unusual among polychaetes in being hermaphroditic, the forward segments of the abdomen being female and the rear ones male. The tube of the Dextral Spiral Tube Worm (*S. spirillum*) coils to the right. The worm itself is similar to *S. borealis* in measurements, habitat, and distribution.

---

136 **Star Tube Worm**
(*Pomatostegus stellatus*)
Subclass Sedentaria

Description: 4″ (102 mm) long, ¼″ (6 mm) wide. In limy tube. Slender. Collar prominent, *2 sets of gills arranged in double-folded plume,* with alternating bands of yellow and orange, black and white, maroon and white. Rounded lid on thick stalk, with spiny border ending in a *star-shaped cluster of spines.* Thorax of 6 bristle-bearing segments.

Habitat: On reef tract, attached to coral rocks; below low-tide line in shallow seas.

Range: Florida, Caribbean Sea, and Gulf of Mexico wherever coral reefs occur.

Comments: These showy worms are common on the reef tract.

137, 142, 145 **Red Tube Worm**
(*Serpula vermicularis*)
Subclass Sedentaria

Description: 4″ (102 mm) long, ¼″ (6 mm) wide. In sinuous, limy tubes 4″ (102 mm) long. Collar prominent, abdomen tapered. Pinkish or red-orange; *plume of 40 pairs of gills* pink, orange, or red, with whitish bands. *Lid funnel-shaped, with 160 fine notches on the border.*

Habitat: Attached to rocks, pilings, floats, and shells in protected harbors and tidepools, and on open shores; from low-tide line to water more than 300′ (91 m) deep.

Range: Alaska to s. California.

Comments: This species was named in 1767 by the father of our modern system of classifying animals and plants, Carolus Linnaeus. Though he described it from specimens obtained in the North Atlantic, it does not occur on the eastern shores of the United States.

130, 131, 132, 133, 134, 135 **Spiral-gilled Tube Worm**
(*Spirobranchus giganteus*)
Subclass Sedentaria

Description: 4″ (102 mm) long, ⅜″ (10 mm) wide. In long, *limy tube* with a single spine on one side of opening. Bluish to tan. 200 segments. Prominent flared collar, *2 sets of yellow, orange, red, pink, blue, white, or tan spirally-wound gills,* each 1″ (25 mm) long, conical. *Rounded lid on a long stalk with a pair of antlerlike projections* with few short spikes at margins.

Habitat: On dead coral, or burrowed into living coral heads; below low-tide line in shallow seas.

Range: Florida, Bahamas, West Indies, and Gulf of Mexico wherever coral reefs occur.

Comments: These beautiful worms look like flowers on the surface of coral heads. A smaller species, the Spiny Spiral-gilled Tube Worm (*S. spinosus*), with tan, yellow, orange, reddish, brown, or blue gills, is 1″ (25 mm) long and ⅛″ (3 mm) wide, and is found burrowed into coralline algae in southern California.

ASCHELMINTHEAN WORMS
(Phylum Aschelminthes)

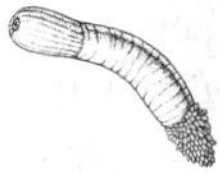

The Phylum Aschelminthes includes an assemblage of free-living and parasitic animals sharing a number of features of interest to the specialist, but not generally observable outside a laboratory. The worms' large, fluid-filled body cavity is enclosed by a body wall of epidermis covered by a tough cuticle and lined with body muscles. They have a tubular digestive system with a wall only 1 cell-layer thick, and devoid of muscles, which is comparable to the layer lining the gut of a higher animal. The worms moult a number of times until they reach the adult stage. At that time all cells are specialized and incapable of cell division.
Consequently, the adult is not only incapable of further enlargement, or regeneration of lost tissue, but even of repairing a wound.
The parasitic forms of these worms, all occurring inside the bodies of their plant or animal hosts, are not specifically marine, and hence are not considered in this book. Most of the free-living forms are microscopic, or so nearly so as to escape notice, and hence are also not included. The one exception is *Priapulus caudatus,* which is described below.

262 Tailed Priapulid Worm
(*Priapulus caudatus*)

Description: 3¼" (83 mm) long, ½" (13 mm) wide. Slender, cylindrical. Cream-colored, buff, yellowish, or tan. Trunk cylindrical, with numerous encircling, indented rings with tiny, wartlike nodules, especially on last few rings. Large, *completely retractable, club-shaped proboscis* ⅓ trunk length, with 25 *longitudinal rows of fine spines;* mouth at end, surrounded by 3 rings of *large, brown, inward-directed hooks.* Rear end with greatly stretchable tail appendage resembling bunch of grapes.

Habitat: Soft mud; from low-tide line to water 1650′ (500 m) deep.

Range: Arctic to Maine; Puget Sound to c. California.

Comments: While the rings on this worm's body look like segments, they are only superficial constrictions. *Priapulus* probes through soft mud by alternately extending and retracting its proboscis. When it encounters a prey organism, usually a polychaete worm, it grabs it with the hooks surrounding its mouth, retracts the proboscis, and swallows the prey whole.

PEANUT WORMS
(Phylum Sipuncula)

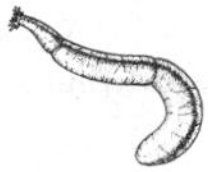

The small Phylum Sipuncula includes the peanut worms, a group with certain features in common with annelid worms, but several other characteristics that are unique. Both have similar developmental and reproductive patterns, a similar nervous system, and similar layers in the body and the walls of the digestive tract. In sipunculids, however, there is no sign of segmentation, and the digestive tract is U-shaped, doubling over and terminating in an anus in the upper midline, well toward the front end. These worms can extend and retract a large part of the front end into and out of the trunk and, when disturbed, contract into a plump, taut, elongate oval, in some species resembling a peanut kernel. The mouth and surrounding parts are the last to be seen when the worm extends fully, and the first to roll in and disappear as it retracts. The mouth is surrounded by tentacles. Animals with short tentacles feed on organic matter taken in with mud or sand; those with longer, branched ones filter organic particles from the water. A few of the more common sipunculids are described here.

---

263 Gould's Peanut Worm
(*Phascolopsis gouldii*)

Description: 12″ (30 cm) long, ½″ (13 mm) wide. Long and slender; no appendages. Whitish, creamy, pink, tan, or gray. Cylindrical, tapering to blunt point at rear, *front end more slender,* extending to a length equal to ½ contracted body length; mouth surrounded by ring of *short, unbranched tentacles.*

Habitat: Mud, and muddy or gravelly sand; near

low-tide line and below in shallow water.

Range: Bay of Fundy to Cape Hatteras.

Comments: This worm is one of the more common animals found on sandy mudflats on the south side of Cape Cod.

---

266 **Agassiz's Peanut Worm**
(*Phascolosoma agassizii*)

Description: 4¾″ (121 mm) long, ½″ (13 mm) wide. Long and slender; no appendages. Trunk light to dark brown, sometimes with purplish or brown spots; extensible front end paler, with *irregular dark rings.* Trunk rough, cylindrical, tapered to a blunt point at rear, covered with papillae, largest ones at rear end; extensible front end narrow, armed with *25 rings of small, toothlike spines,* mouth surrounded by ring of *small, slender tentacles.*

Habitat: In sand, under rocks, in roots of surfgrass, and in kelp holdfasts, mussel beds, and growth on pilings; above low-tide line.

Range: Alaska to Baja California.

Comments: This is the most common species of sipunculan on the West Coast. The Antillean Peanut Worm (*P. antillarum*) ranges from Florida through the West Indies. It is 2″ (51 mm) long and ¼″ (6 mm) wide, and light to chestnut-brown. It has trunk papillae, spines on the front end, and tiny tentacles like *P. agassizii.* It burrows into sand and soft rocks in shallow water.

Bushy-headed Peanut Worm
(*Themiste pyroides*)

Description: 8″ (20 cm) long, 2″ (51 mm) wide. Plump. Trunk rich brown with fine brown spots, extensible part paler, front tip reddish- or purplish-brown,

tentacles cream-colored or buff. Trunk thick, almost pear-shaped; middle region of extensible front end set with *many brown spines;* mouth surrounded by 4 highly-branched, *bushy tentacles.*

Habitat: Under rocks, in holes and cracks; near low-tide line and below in shallow water.

Range: British Columbia to Baja California.

Comments: From inside their hiding places these worms extend their bushy tentacles out to capture fine organic food particles on their mucus-covered branches. The Eelgrass Peanut Worm (*T. zostericola*) is similar, but more slender and without spines, and grows to 10″ (25 cm) long and 1¾″ (44 mm) wide. It is buff or grayish, with dark gray cross-lines around the front end.

ECHIURID WORMS
(Phylum Echiura)

The Phylum Echiura is a small group of worms, similar in many ways to the Annelida, but lacking any suggestion of segmentation. Their digestive and nervous systems, body wall structures, reproductive patterns, and even bristles (*setae*) are similar. Echiurids are plump-bodied, sausage-shaped, fluid-filled creatures with a non-retractable proboscis that has a deep groove or trough on the underside, leading into the mouth. Since the proboscis contains the brain and lies in front of the mouth, it is probably comparable to the lobe in front of the annelid mouth. The anus is at the rear tip of the trunk. One or more rings of bristles encircle the rear end of the worm, justifying the name Echiura, which means "spiny tail" in Greek.

261 Keyhole Urchin Spoon Worm
(*Thalassema mellita*)

Description: 1½" (38 mm) long, ¼" (6 mm) wide. Sausage-shaped, with proboscis. Brick-red; proboscis pale yellow. Trunk cylindrical, smooth, with *pair of large, hooklike bristles near head end* on underside; *no bristles around anus;* scooplike proboscis about ⅓ total length.

Habitat: In skeletons of dead keyhole urchins; below low-tide line in shallow water.

Range: Virginia to Florida.

Comments: These worms enter the skeletons of dead keyhole urchins when quite young, and become too large to escape. They feed by extending the scooplike proboscis, whose mucus-covered surface traps tiny organic particles.

265 Innkeeper Worm
(*Urechis caupo*)

Description: 7¼″ (18 cm) long, 1¾″ (44 mm) wide. Sausage-shaped. Yellowish-pink. Trunk smooth, cylindrical; *pair of hooklike bristles near head end,* circle of *11 strong bristles around anus. Short proboscis* at front end, shaped like spoon with sides turned up.

Habitat: Sandy mudflats.

Range: N. to s. California.

Comments: *Urechis* is called the Innkeeper Worm because its burrow is also inhabited by many commensals, including a goby fish, a scale worm, and 2 species of crabs. A tiny clam succeeds in living a foot or more down in the mud by the simple expedient of poking its siphons into the Innkeeper Worm's burrow, and drawing water from and expelling it into the burrow, rather than having to come to the surface.

MOLLUSKS
(Phylum Mollusca)

With the exception of insects, mollusks are probably more familiar to most people than any other group of invertebrates. They have been used as food since prehistoric times; clams, oysters, mussels, conchs, cockles, escargots, periwinkles, squids, and octopods have long been a significant part of the human diet.

In addition to the approximately 100,000 known living species, some 35,000 fossil mollusks have been described. The fossil record for this phylum extends over 600 million years and is extremely rich because the mineralized molluscan shell, or *valve*, fossilizes readily.

The body consists of three regions: a *head*, bearing the mouth and sense organs and containing the brain; a *visceral mass* surrounded by the body wall and containing most of the internal organs; and a *foot*, the muscular lower part of the body on which the animal creeps. A membranous extension of the body wall, the *mantle*, secretes the shell and encloses a *mantle cavity* containing the gills, anus, and excretory pores. The mouth in most groups is equipped with a long, tough, toothed, ribbonlike structure, the *radula*, by means of which the animal rasps food. Mollusks have well-developed organ systems—nervous, muscular, digestive, circulatory, respiratory, excretory, and reproductive—but, unlike annelids and arthropods, they lack body segmentation.

The patterns of sexual reproduction among mollusks vary greatly. Some are hermaphroditic, while others have separate sexes. Some lay large, yolky eggs enclosed in capsules; others spawn naked eggs which are fertilized and develop in the water. Some brood their young. Many marine species produce a characteristic swimming larva, the *veliger,* which has a ciliated, winglike structure on either side of the mouth and, seen head-on, is reminiscent of a microscopic butterfly.

Within the Phylum Mollusca there are 7 classes: Monoplacophora, Polyplacophora, Aplacophora, Gastropoda, Scaphopoda, Bivalvia or Pelecypoda, and Cephalopoda. Of these, the Monoplacophora, Aplacophora, and Scaphopoda are rare or deep-water animals and will not be mentioned further.

Class Polyplacophora:

Chitons are flattened, oval creatures with a row of 8 broad but short valves along the back. Surrounding the 8 valves, and in some cases covering them, is a margin of the mantle called the *girdle.* On the underside is a large foot with which the animal clings tenaciously to a solid surface; when pried loose, it usually curls into a ball like a pill bug or a hedgehog. On either side of the foot in the mantle cavity lie paired gills. At the front of the foot is the head, bearing the mouth with its rasp (*radula*), but without tentacles or eyes.

Most chitons feed by rasping fine algal growth from the surface of rocks and, in some cases, also prey on sedentary animals such as bryozoans, sponges, and protozoans.

Most species have separate sexes and spawn eggs or sperm. In a few species, the female broods eggs in the mantle cavity. Fertilized eggs of the spawning species develop into larvae similar to those of some polychaete annelids and, unlike the more abundant gastropods and bivalves, do not become veligers.

Classification of chitons is based on the structure of the 8 valves, the girdle, the gills, and the tooth pattern on the radula. Since the last element involves both dissection and microscopic examination, it will not be considered in this book. The Polyplacophora were formerly combined with the Aplacophora as the Class Amphineura.

Class Gastropoda: The Class Gastropoda contains about 80% of all living molluscan species, and includes the snails, limpets, abalones, garden slugs, sea slugs, and sea hares. Although most of these are found in the sea, many live in fresh water and some are land dwellers—the only mollusks to succeed in that environment.

The general body plan of a gastropod includes a spiral shell into which the animal can retract, a head equipped with tentacles, eyes, and mouth, and a large foot. Forms such as sea hares have only a vestigial, internal shell. Nudibranchs, or sea slugs, have shells in the larval stage, but lose them on maturing. Terrestrial slugs have no shell at any stage of the life cycle.

The snail shell is basically an elongate cone wound around an *axis.* Each turn around the axis is a *whorl.* The *body whorl* is the most recently formed and largest, and contains most of the snail's soft parts. The remaining whorls constitute the *spire,* which terminates in the *apex.* The apex is actually the original shell of the larval stage, and is called the *nuclear whorl* or *protoconch;* in most snails the nuclear whorl is worn off during growth. The inner side of the shell immediately adjacent to the axis about which it spirals is the *columella.* The columella may be hollow, opening in an *umbilicus* at the end opposite the apex, or may be closed over by a shell growth, the *umbilical callus.* The groove between adjacent whorls is called a *suture.* The outer surface of the whorls may be smooth or variously sculptured with small *beads,* larger *nodules,* sharp *spines, ribs* which run parallel to the axis, or heavy *cords,* finer *threads,* or very fine *grooves* or *lines,*

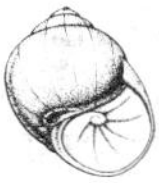

all running spirally. A prominent spiral ridge on a whorl is termed a *shoulder.* The surface may be coated with a tough, horny layer, the *periostracum.* The opening of the body whorl through which the head and foot are extended is the *aperture.* The aperture has an *outer lip* and an *inner lip.* In some snails the aperture may have a slender forward extension, the *siphonal canal,* in which lies a tubular fold of the edge of the mantle, the *siphon,* through which the animal takes in water. Many species have a horny, leathery, or limy lid, or *operculum,* attached to the side of the foot. When the head and foot are completely retracted, the operculum neatly seals off the aperture. By far the majority of snails are *dextral,* coiling to the right. A few species are *sinistral,* coiling to the left, and occasionally an individual of a normally dextral species will be sinistral. To make sure which "handedness" a snail exhibits, hold the shell with the apex up and aperture toward you. If the aperture is on your right, the snail is dextral; if it is on your left, the snail is sinistral. In the descriptions that follow, all species are considered dextral unless otherwise identified.

In cited measurements, the term "length" represents the length of the axis, even though anatomically it would be more correct to call this the height. In a few low, flattened forms, such as limpets, slipper snails, and abalones, the term "length" will designate true length, the distance from front to rear, and the term "height," the distance from top of shell to surface of substrate. The width is always the broadest dimension at right angles either to the axis or to the true length.

In an active marine snail the opening of the mantle cavity is above the head, not at the rear as the generalized molluscan body plan would indicate. This change in position is achieved by a process called *torsion,* a 180° counterclockwise rotation of the visceral mass during larval development. Having the mantle open forward improves water flow over the gills. Torsion also makes the snail asymmetrical internally. The paired nerve cords are twisted about each other, and organs from the original right side, such as the gill, auricle of the heart, and excretory opening, are diminished or eliminated. In some groups "detorsion" has occurred, and will be described later.

The Class Gastropoda contains 3 subclasses: the Prosobranchia (meaning "gill forward"), the Opisthobranchia ("gill behind"), and the Pulmonata ("with a lung"). The Prosobranchia make up the largest subclass and include a majority of the marine gastropods. These all possess a shell, have separate sexes, and demonstrate torsion.

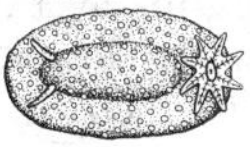

The Opisthobranchia include a variety of hermaphroditic gastropods with gills located toward the rear as a result of detorsion in development, following torsion. This subclass includes both shelled forms and shell-less animals such as sea hares and the colorful sea slugs, or nudibranchs.

In the Pulmonata, the mantle chamber is modified as a lung, enabling these mollusks to breathe air. Included here are all land snails and slugs, most of the freshwater snails, and a very few marine species.

Class Bivalvia: The bivalves known to most people are those commonly used as food: clams, oysters, mussels, cockles, and scallops.

The Class Bivalvia (also known as Pelecypoda) includes mollusks whose shells consist of two parts, or *valves,* hinged together along the upper midline. There are about 15,000 species of bivalves, more than 80% of them in the sea, and the others in fresh water.

The basic body plan of a bivalve is well adapted to a life of burrowing through the bottom. The two valves that enclose the rest of the body are convex, giving the animal a wedgelike or hatchetlike shape. (Pelecypoda means "hatchet foot" in Greek.) The foot is bladelike and can be protruded forward from between the valves. The mantle, which completely lines the shell, forms two openings at the rear end: a lower one, the *incurrent siphon,* to admit water into the mantle cavity, and an upper one, the *excurrent siphon,* for its escape. The siphons may be merely 2 slitlike openings between the two parts of the mantle, or 2 separate extensible tubes, or the 2 tubes may be united into a single *neck* which contains the 2 passages. The gills, meshworks of strands united into large sheets, are suspended on either side of the visceral mass within the mantle cavity. They are used both for respiration and for food collecting. The gills secrete mucus which traps food from the water—mostly 1-celled plants and fine particles of organic matter. The thin layer of food-laden mucus is moved down the gills, then forward along their edges, where long, flaplike *palps* transfer it to the mouth. The head of a bivalve is little more than a bump with a mouth and a pair of palps.

Locomotion in a bivalve is slow. In some bivalves the muscular foot is long and slender and protrudes forward between the valves, probing into the bottom and forming a knob at the tip for anchorage. As the foot muscles contract, the bivalve body is drawn forward a short distance. Other bivalves, however, such as scallops and file shells, have a quite different method of locomotion. They swim by a series of quick, clapping motions of the shells which jet water around the ends of the hinge line as the mollusk moves, open end forward. Still other bivalves lead completely sedentary lives attached to some solid or anchored in the bottom. Mussels have a gland on the foot which secretes a fluid that turns into a silken strand (*byssal thread*) upon contact with water. A mussel attaches firmly to a rock or piling by a cluster of byssal threads. Oyster larvae settle on some solid and attach the left valve by byssal threads. As the oyster grows, its mantle edge cements it to the solid with the same limy material that makes up the shell. Various boring bivalves may be found in clay, peat, sandstone, shale, limestone, shells, coral, or wood. Most marine bivalves have separate sexes and reproduce by spawning. Some are hermaphroditic, though these usually produce only eggs or sperm at any one time.

For purposes of identification, bivalve shells exhibit the most obvious and stable species characteristics. To determine the left valve from the right, hold the animal hinge-side upward and front end away from you. (The front end is the one from which the foot protrudes.) The right valve is then on your right and the left valve on your left. In some species you will find a gap between the valves for foot protrusion. Of course, if the live bivalve obligingly extends its foot, the question is settled. Alternatively, the protrusion of the siphons between the valves at the rear end lets you know indirectly which end is the front. For bivalves that typically lie with one valve down and one up, such as oysters, jewel boxes, and scallops, we refer to the upper and lower valves. Concentric growth lines can be seen on the shell surface in most bivalves. The center area surrounded by the lines is the oldest part of the shell, and is known as the *umbo* (plural *umbones*), or *beak*. It is usually located nearer the front end than the back, or bent forward. Many characteristics used by conchologists in identification of bivalves—such as *teeth,* by means of which the 2 valves interlock near the hinge line; *muscle scars,* where muscles that close the shells attach; and the *pallial line,* which marks the internal attachment of the mantle to the shell—are exhibited on the inner surface of the shell. Since these features cannot be observed on living animals, little reference will be made to them. In determining dimensions, the *length* of the shell is measured as a straight line between its front and rear edges. The *height* is the distance between uppermost (hinge line) and lowermost edges. However, in oysters the length is commonly considered as the distance from the umbo to the opposite end (the height by our previous definition).

Class Cephalopoda:

Members of the Class Cephalopoda, the squids, octopods, and nautiluses (the latter not included in this book), are specialized variations on the molluscan theme. The visceral mass is surrounded by a mantle enclosing gills and body openings. Cephalopods have a head—with a brain, sense organs, and a mouth with a radula—and a foot specialized into a number of arms surrounding the mouth and equipped with suckers. Some people refer to the arms as tentacles, but biologists reserve that term for a special pair of long appendages found in squids, and used for capturing prey. The mouth is equipped with a beak, shaped like that of a parrot, with which the animal kills and tears apart its prey. Some cephalopods also have salivary glands that secrete a paralyzing toxin. One small octopus of the South Pacific has a toxin so potent that its bite is lethal to a human.

Squids have thin internal shells, and octopods have none at all. Both use their *mantle cavities* and *siphons* for locomotion. Water is admitted into the mantle cavity through slits behind the head, and jetted with great force out the siphon beneath the neck. A squid can aim its siphon either forward or backward and jet itself in the opposite direction. Squids also have a pair of fins that serve both as stabilizers and as power sources for swimming by undulation. An octopus can only jet backwards with arms trailing, a mode of locomotion used mostly for swift escape. When on the bottom, it simply crawls along on its arms. An ink gland opening into the mantle cavity enables cephalopods, when threatened, to squirt out a cloud of ink and, in the confusion, make their escape. Squids living in ocean depths where the only light is "living light," or *bioluminescence,* confuse their enemies not with ink, but with a cloud of luminescent particles.

The octopus has the largest and most complex brain, as well as the best eyes, of any invertebrate. Experiments in animal behavior have shown these animals to have fine sensory discernment, both by touch and by vision, and a great capacity for learning.

Cephalopods have separate sexes and practice internal fertilization. Females deposit gelatinous masses containing large numbers of encapsulated, yolky eggs that hatch as recognizable miniatures of the parent. A female octopus guards her eggs until they hatch.

370, 374 **Mottled Red Chiton**
(*Tonicella marmorea*)
Class Polyplacophora

Description: 1½" (38 mm) long, ¾" (19 mm) wide. Oval to oblong. Mottled, usually with red and brown, occasionally with pink, orange, purple, blue, or green; inside surface tinted with rose. *Girdle leathery, shells dull, smooth,* slightly ridged down the middle; first valve flat or slightly convex. 25 gills on either side of broad foot.

Habitat: On rocks; from low-tide line to water 300′ (91 m) deep.

Range: Greenland to Massachusetts Bay.

Comments: The Mottled Red Chiton is an indiscriminate browser, feeding on small algae, protozoans, sponges, hydroids, bryozoans, and other sedentary growth. A similar species, ranging from the Arctic to Connecticut, is the Northern Red Chiton (*T. rubra,* formerly listed as *Ischnochiton ruber*), with which the Mottled Red Chiton can easily be confused. *T. rubra* is smaller, measuring ¾″ (19 mm) in length and ⅜″ (10 mm) in width, and is also mottled red, with valves ridged down the middle. Its girdle is covered with tiny scales.

371 **Lined Chiton**
(*Tonicella lineata*)
Class Polyplacophora

Description: 2″ (51 mm) long, 1″ (25 mm) wide. Oval to oblong. Mottled reddish-brown, with oblique zigzag lines of different color combinations, including light and dark red, light and dark blue, white, brown, or black; girdle greenish or yellowish, sometimes banded. *Valves low, smooth, rounded; girdle smooth,* leathery.

Habitat: On rocks covered with coralline algae; from low-tide line to water 180′ (55 m) deep.

Range: Alaska to s. California.

Comments: The color and disruptive line-pattern of this striking chiton make it hard to confuse with any other species and provide it with excellent camouflage in the dense growths of coralline algae on which it feeds.

375 **Rough-girdled Chiton**
(*Ceratozona squalida*)
Class Polyplacophora

Description: 2″ (51 mm) long, 1″ (25 mm) wide. Oval to oblong. Girdle leathery, brownish, densely *covered with brown to blackish hairs* ⅛″ (3 mm) long. Valves whitish with greenish mottling, surface rough; first valve with 10–11 radiating ribs, last one small, with 8–10 slits on rear edge. Gills as long as foot.

Habitat: On coral rocks; near low-tide line and below in shallow water.

Range: Florida; Bahamas; West Indies.

Comments: This rather dirty-looking chiton is abundant throughout its range. It withstands the battering surf to which it is exposed daily by settling tightly in a close-fitting depression which it has scraped in the soft coral rock.

380 **White Chiton**
(*Ischnochiton albus*)
Class Polyplacophora

Description: ½″ (13 mm) long, ¼″ (6 mm) wide. Oval to oblong. Girdle cream-colored to pale tan; *surface granular;* valves white, with faint tan or brownish growth lines, slight ridge down the middle, smooth; slightly concave, inside surface white; first valve large, almost semicircular. Gills beginning halfway back along the foot.

Habitat: On rocks; from low-tide line to water 25′ (8 m) deep.

Range: Arctic to Massachusetts Bay; Alaska.

Comments: This dainty little chiton is a cold-water form, very abundant in Passamaquoddy Bay and the Bay of Fundy. The Mesh-pitted Chiton (*I. papillosus*) ranges from Florida to Texas and to the Bahamas and West Indies. It, too, measures ½″ (13 mm) long and ¼″ (6 mm) wide, and has mottled olive-green shells and a narrow girdle with alternate bars of greenish-brown and white. Mertens' Chiton (*Lepidozona mertensii*) ranges from Alaska to Baja California. It is larger, measuring 1½″ (38 mm) long and ¾″ (19 mm) wide, and is usually yellowish with reddish-brown streaks and spots.

---

379 **Gum Boot Chiton**
(*Cryptochiton stelleri*)
Class Polyplacophora

Description: 13″ (33 cm) long, 5″ (127 mm) wide. Oblong, somewhat flattened. Brick-red to reddish-brown, thick, leathery, gritty; *girdle completely covering 8 white, butterfly-shaped valves;* 80 pairs of gills.

Habitat: On rocks in protected places near deep channels, and in kelp beds; from low-tide line to water 70′ (21 m) deep.

Range: Alaska to s. California.

Comments: This is the world's largest chiton, and the only one in American waters whose girdle completely covers its valves.

---

372 **California Nuttall's Chiton**
(*Nuttallina californica*)
Class Polyplacophora

Description: 2″ (51 mm) long, ⅝″ (16 mm) wide. Elongate, oval. Girdle mostly brown, with some white; wide, *mossy, with short, rigid spines.* Valves brown, olive, or dark gray, granulated on sides, with smooth central ridge. Inside surface bluish, sole of foot orange-yellow. 9 gills on either side of foot.

Habitat: On rocks in strong surf, in crevices, or among barnacles and mussels; from high-tide line to midtidal zone.

Range: Puget Sound to s. California.

Comments: This chiton feeds mostly on red and green algae, and its shells are commonly overgrown by these algae, which erode some of the dark surface.

377 **Hartweg's Chiton**
(*Cyanoplax hartwegii*)
Class Polyplacophora

Description: 2″ (51 mm) long, 1¼″ (32 mm) wide. Broadly oval. Dark brown, olive, or gray *girdle narrow, with fine granules;* valves paler than girdle, with scattered white spots near edges; low-arched, rounded, finely granulated, and interspersed with scattered larger, wartlike granules; without ridge down midline. Inside surface blue-green.

Habitat: On rocks protected from surf, under cover of brown algae, and in tidepools; between high- and low-tide lines.

Range: C. California to Baja California.

Comments: This chiton is nocturnal and nomadic, active mainly when the tide is out. It is less tenacious than other chitons and is found only in protected sites. Unlike some other chitons, it does not have an established territory, but occupies one site for several days and then moves on.

378 **Florida Slender Chiton**
(*Stenoplax floridana*)
Class Polyplacophora

Description: 1½″ (38 mm) long, ½″ (13 mm) wide. Long, oval, highly arched. Marbled, bluish-gray girdle *covered with dense, round scales;* wide. Valves whitish-green, with olive or gray markings and wavy, longitudinal, beaded ribs at the sides; end valves with concentric rows of beads; last valve with 9 slits at margin.

Habitat: On coral rocks; near low-tide line and below in shallow water.

Range: S.E. Florida; Bahamas; West Indies.

Comments: This chiton lays its eggs in coiled strings of jelly attached to rocks and shells. Heath's Chiton (*S. heathiana*), which ranges from central California to Baja California, is a larger species, reaching a length of 4⅜″ (111 mm) and a width of 2" (51 mm). Its girdle is buffy, and its valves cream-colored, mottled with gray or brown.

381 **Black Katy Chiton**
(*Katharina tunicata*)
Class Polyplacophora

Description: 4¾" (121 mm) long, 1¾" (44 mm) wide. Elongate, oval, thick. *Girdle black, thick, smooth, covering ⅔ of valve.* Valves gray, usually eroded; front valve with numerous tiny pits. Inner surface white, foot salmon to red.
Habitat: On rocks exposed to heavy wave action and full sun; between high- and low-tide lines and slightly below.
Range: Alaska to s. California.
Comments: The Black Katy is a tough chiton that can withstand both the rigors of crashing waves and the intense sun to which it is subjected at low tide.

376 **Mossy Chiton**
(*Mopalia muscosa*)
Class Polyplacophora

Description: 3⅝" (92 mm) long, 2¼" (57 mm) wide. Oval to oblong, flat. *Girdle covered by stiff mossy hairs,* dull brown, dark olive, or gray, wide, with slight notch at rear. Valves same color as girdle or paler, with flattened ridge down midline and fine, beaded longitudinal ridges in middle; crosswise-beaded ridges sometimes eroded on side. Valves often overgrown by algae or worm tubes. Inside surface bluish-green to lilac.
Habitat: On rocks protected from heavy wave action, and in tidepools; between high- and low-tide lines.
Range: Alaska to Baja California.
Comments: The Mossy Chiton tolerates a wide variety of environmental conditions,

including the reduced salinities of inland estuaries. The Hooked-bristled Chiton (*M. lignosa*) is somewhat smaller, 2¾″ (70 mm) long and 1⅞″ (48 mm) wide, variously combines blue, gray, green, brown, cream-color, and sometimes orange on its valves and girdle, and has a deep notch at the rear of the girdle.

373 **Veiled Chiton**
(*Placiphorella velata*)
Class Polyplacophora

Description: 2″ (51 mm) long, 1¾″ (44 mm) wide. Oval, flattened. Broad, beige to tan *girdle expanded in front as a head flap* with green and red undersurface, narrow at rear, with scattered hairlike scales. Veil of fleshy tentacles on head. Valves short, wide, reddish or brownish, with white, tan, or greenish spots and streaks.

Habitat: In crevices and on undersides of rocks covered with coralline algae; from low-tide line to water 50′ (15 m) deep.

Range: Alaska to Baja California.

Comments: The Veiled Chiton is the only chiton in American waters that captures active animal prey. Small crustaceans and worms are usually swallowed whole, larger ones shredded by the radula and then consumed.

Common Eastern Chiton
"Bee Chiton"
(*Chaetopleura apiculata*)
Class Polyplacophora

Description: ¾″ (19 mm) long, ½″ (13 mm) wide. Oval. Girdle mottled brown and cream-colored, narrow; valves whitish-cream, gray, or brownish, with slight ridge down middle; inner surface white or grayish. *Surface finely granulated with*

scattered short, transparent hairs; central region with *20 longitudinal rows of beaded ridges;* 3 sides sloping upward, meeting in a central apex, eroded in mature specimens. 24 gills.

Habitat: On rocks and shells; from low-tide line to water 90′ (27 m) deep.

Range: Cape Cod to west coast of Florida.

Comments: The species name, *apiculata,* is Latin for "ending in a point," or "with an apex." In calling this animal the Bee Chiton, the namer apparently confused the root-word *apex* with *apis,* Latin for "bee."

392, 393 Red Abalone
(*Haliotis rufescens*)
Class Gastropoda

Description: 12″ (30 cm) long, 9¼″ (24 cm) wide, 3″ (76 mm) high. Oval, somewhat flattened, *beret-shaped. Brick-red;* inner surface iridescent blue, green, and pinkish; head, tentacles, scalloped mantle, and mantle tentacles black. Prominent muscle scar. *4 holes in a row* on upper surface, occasionally fewer or more; older holes sealed off. Rough, with broad, wavelike ridges and fine spiral threads usually overgrown by algae and animal life, pitted by boring sponges and boring clams.

Habitat: On rocks and in crevices exposed to heavy surf; from low-tide line to water 540′ (165 m) deep.

Range: Oregon to Baja California.

Comments: The Red Abalone is highly prized as a delicacy and its meat commands premium prices in the fish market. Aside from humans, its chief predator is the sea otter, and in the otter's range the abalone is seldom found outside inaccessible crevices. The Japanese Abalone (*H. kamtschatkana*) is a smaller species, 6″ (15 cm) long and 4¼″ (108 mm) wide. It is highly arched, with 5 open holes and a corrugated surface. It ranges from Alaska to central

California. The Black Abalone (*H. cracherodii*) is also smaller, 6″ (15 cm) long and 4¼″ (108 mm) wide, with 8 open holes and a bluish or greenish-black shell. It is abundant and edible, but not fished commercially. It usually inhabits deep crevices in rocks, and occurs from between high- and low-tide lines to water 20′ (6 m) deep. It feeds mostly on large, brown algae.

384 Cayenne Keyhole Limpet
(*Diodora cayenensis*)
Class Gastropoda

Description: 2″ (51 mm) long, 1⅜″ (35 mm) wide. Oval, conical, widest at rear. Whitish, pinkish, buff, or gray; interior white or blue-gray, shiny; head and foot cream-colored. Keyhole opening just in front of apex; *deep pit behind apex*. Many ribs, *every 4th rib large,* crossed by concentric growth lines; margin scalloped. Head has pair of tentacles with eyes at base.
Habitat: On rocks and jetties; from well above low-tide line to moderately deep water.
Range: Virginia to Florida and Texas; Bahamas; West Indies.
Comments: This keyhole limpet moves about its territory feeding on algae, and always returns to a home base. Its shell is usually overgrown with algae, making it hard to see.

391 Rough Keyhole Limpet
(*Diodora aspera*)
Class Gastropoda

Description: 2¾″ (70 mm) long, 2″ (51 mm) wide. Oval, conical. Gray, with brownish radiating bands or with black and white radiating stripes; head and foot cream-colored. *Round opening in front of apex.* Coarse radial ribs, *every fourth rib large.*

Head has pair of tentacles at base. Shell surface rough.

Habitat: Under large rocks or growths of large algae; near low-tide line and below in shallow water.

Range: Alaska to Baja California.

Comments: The Rough Keyhole Limpet responds to an approaching predatory sea star by extending its mantle over its shell, making it difficult for the star's tube feet to attach. The Pacific Scale Worm, a commensal frequently to be found in the limpet's mantle cavity, earns its keep by biting the tube feet of an attacking sea star, making it retreat.

---

382 **Volcano Limpet**
(*Fissurella volcano*)
Class Gastropoda

Description: 1⅜″ (35 mm) long, ¾″ (19 mm) wide, ½″ (13 mm) high. Oval, conical, narrow at front end. Pink, with *reddish-brown rays;* interior pale green; mantle red-striped; foot yellow. Keyhole at apex, somewhat nearer front end, elongate, oval. Radiating ribs low, rounded, of various sizes. Margin slightly scalloped at larger ribs.

Habitat: On rocks; between high- and low-tide lines.

Range: N. California to Baja California.

Comments: Although this limpet has a symmetrical keyholed shell, its larva has a spiral shell with no keyhole. The tropical Atlantic Barbados Keyhole Limpet (*F. barbadensis*) has a similarly shaped shell, and is 1″ (25 mm) long and ¾″ (19 mm) wide, but its keyhole is almost round, with a red and green border. The shell is whitish to buffy-pink, with purplish lines between the ribs and occasional purple-brown blotches. It lives on wave-washed rocks.

388 **Giant Keyhole Limpet**
(*Megathura crenulata*)
Class Gastropoda

Description: 5⅛″ (130 mm) long, 3⅛″ (79 mm) wide, 1″ (25 mm) high. Body oval; longer and wider than shell. Buff to pink; interior white; sole of foot yellow to orange; *black to mottled gray mantle conceals most of shell in life.* Keyhole forward of center, oval, large.

Habitat: On rocks and breakwaters, in crevices and protected areas; near low-tide line and below.

Range: C. California to Baja California.

Comments: The Giant Keyhole Limpet feeds on algae and colonial tunicates. Native Americans used the conveniently perforated shell on wampum belts.

383 **Tortoise-shell Limpet**
(*Notoacmaea testudinalis*)
Class Gastropoda

Description: 1½″ (38 mm) long, 1⅛″ (28 mm) wide. Low, smooth, *cone-shaped, oval.* Banded irregularly with brownish or bluish; sometimes checkered; *inner surface dark brown,* shiny; mantle and head cream-colored to buff. 2 short tentacles; *gill visible* in mantle opening above head; foot with large sole.

Habitat: On rocks and kelps; above low-tide line and below in shallow water.

Range: Arctic to Long Island Sound.

Comments: This is the only common limpet in New England. When even slightly disturbed, it clamps down so tightly on a rock that the shell will break before it can be pried loose.

389 **Seaweed Limpet**
(*Notoacmaea incessa*)
Class Gastropoda

Description: ⅞″ (22 mm) long, ½″ (13 mm) wide, ⅝″ (16 mm) high. Narrowly oval, *higher than wide.* Deep brown, sometimes with white spot near apex; inner surface brown; soft parts brown. Apex slightly forward of center, *rear slope convex;* surface smooth, with fine radial lines.
Habitat: On the large, brown alga *Egregia;* near low-tide line and below.
Range: Alaska to Baja California.
Comments: This limpet feeds both on the small algae growing on the surface of *Egregia* and on that plant itself. The limpet cannot move from plant to plant, but spends its entire life on the alga where it settled as a larva, eating out a pit into which it fits perfectly.

385, 386 **Plate Limpet**
(*Notoacmaea scutum*)
Class Gastropoda

Description: 2½″ (64 mm) long, 2″ (51 mm) wide. Oval, cone-shaped. Brownish to greenish, with interrupted radial rows of whitish spots; inner surface whitish, with dark spot in center and dark brown rim. Apex slightly forward of center, *smooth, with about 40 low, flat-topped ridges; margin of shell notched* between ridges. Tentacles light brown.
Habitat: On rocks protected from strong surf action; between high- and low-tide lines.
Range: Alaska to s. California.
Comments: The Plate Limpet moves up and down on a rock as the tide rises and falls, apparently maintaining some ideal relationship with the water level.

387 **Shield Limpet**
(*Collisella pelta*)
Class Gastropoda

Description: 1⅝" (41 mm) long, 1" (25 mm) wide, ⅝" (16 mm) high. Oval, *apex slightly forward of center, slopes slightly convex.* Black, brown, or greenish irregular stripes on cream-colored background; sometimes checkered; inner surface bluish-white, with dark spot under apex, dark margin. Surface usually with 20–24 shallow, rounded ribs, but sometimes smooth.

Habitat: On rocks and holdfasts of brown algae, and in mussel beds; between high- and low-tide lines.

Range: Alaska to Baja California.

Comments: The Shield Limpet has a tidal activity rhythm and moves about only when wetted by waves or when under water. It feeds chiefly on several kinds of red and brown algae. The Rough Limpet (*C. scabra*) measures 1¼" (32 mm) long and ¾" (19 mm) wide. It has an apex which is lower than that of *C. pelta*, and forward of center, about 20 heavy ribs, and a scalloped margin. It is mottled gray to green, its inner surface white with brown radiating lines, and its soft parts cream-colored with small, black spots. It ranges from Oregon to Baja California. The Ribbed Limpet (*C. digitalis*), whose range is the same as that of *C. pelta*, is somewhat smaller, 1¼" (32 mm) long and ⅝" (16 mm) high, and its apex is almost at its front margin and overhanging it slightly.

---

390 **Owl Limpet**
(*Lottia gigantea*)
Class Gastropoda

Description: 3½" (89 mm) long, 2¾" (70 mm) wide, 1¼" (32 mm) high. Spotted brown and white; inner surface dark;

brown at margin; side of foot gray, sole of foot yellow to orange. *Owl-shaped muscle scar in middle of inner surface. Apex near front end.*

Habitat: On rocks exposed to heavy surf action; between high- and low-tide lines.

Range: Washington to Baja California.

Comments: The Owl Limpet is territorial, claiming an area of about 1 square foot (1000 square cm). It "bulldozes" other species of limpets out of its territory, and their absence permits the rapid growth of a film of algae from spores that settle in the cleared areas. The Owl Limpet feeds on the new algal growth in such areas, which can easily be spotted at some distance.

---

457 Purple-ringed Top Snail
(*Calliostoma annulatum*)
Class Gastropoda

Description: 1½" (38 mm) long, 1¼" (32 mm) wide. Nearly conical. *Golden-yellow, with purplish stripe at suture;* apex pink, head and foot yellow-orange with brown spots. *5–9 beaded spiral threads,* spire of 5 whorls, no umbilicus.

Habitat: Offshore in forests of giant kelp; inshore on rocks; near low-tide line and below.

Range: Alaska to Baja California.

Comments: This strikingly-colored mollusk is swift for a snail. In an experiment where marked animals were released at the base of kelps, they ascended as far as 30′ (9 m) in 24 hours. The Blue Top Snail (*C. ligatum*) is a slightly smaller, chocolate-brown species, 1⅛" (28 mm) long and 1" (25 mm) wide, with tan spiral ridges, a blue shell lining, and a brown foot with an orange sole. It ranges from Alaska to southern California and lives in the kelp forest closer to the ocean bottom. The Channeled Top Snail (*C. canaliculatum*), which is a little larger, 1⅝" (41 mm) long and 1⅜" (35 mm) wide, is white to gray, with brown spiral stripes and a tan, brown-spotted foot, and lives near the top of the kelp forest.

458 Pearly Top Snail
(*Calliostoma occidentale*)
Class Gastropoda

Description: ½" (13 mm) long, ½" (13 mm) wide. Nearly conical. *Pearly white,* aperture iridescent. *Spire of 5–6 whorls,* 3–4 spiral cords, *no umbilicus.* Lip fragile.
Habitat: Below low-tide line to water 5525′ (1684 m) deep.
Range: Nova Scotia to Massachusetts.
Comments: Its iridescent, pearly appearance makes this beautiful snail easy to recognize.

468 Greenland Top Snail
(*Margarites groenlandicus*)
Class Gastropoda

Description: ⅜" (10 mm) long, ½" (13 mm) wide. Thin, spire somewhat flattened. *Rosy red or creamy-tan,* head and foot cream-colored. Aperture round; operculum thin and horny. 2 long, white tentacles with eye at base. 5 whorls. *Numerous fine spiral lines,* funnel-shaped umbilicus, 12 mantle tentacles.
Habitat: On kelps and other algae, and on rock bottoms; from low-tide line to water 800′ (244 m) deep.
Range: Greenland to Massachusetts Bay.
Comments: This delicate and handsome little snail has two color phases: rosy red and creamy-tan. It feeds on fine algae growing on kelps and other substrates. The Smooth Top Snail (*M. helicinus*) shares its habitat, range, and size, but differs in having a smooth, shiny, brownish and iridescent shell, and a dark gray head, foot, and tentacles.

459 **Red Top Snail**
(*Astraea gibberosa*)
Class Gastropoda

Description: 1¾″ (44 mm) long, 3″ (76 mm) wide. Spire conical. Brown periostracum covering red shell; reddish-brown mouth; light brown head and foot, with meshwork of dark brown lines. *5–6 strong spiral cords of nodules;* base flat, aperture oval, operculum oval, smooth, limy.

Habitat: On rock bottoms; from low-tide line to water 240′ (73 m) deep.

Range: British Columbia to Baja California.

Comments: The Red Top Snail feeds by scraping small algae and other growth from the surface of rocks. It can keep aquarium walls free of such growth. In older literature it is listed as *A. inaequalis.*

---

460 **Black Turban Snail**
(*Tegula funebralis*)
Class Gastropoda

Description: 1″ (25 mm) long, 1¼″ (32 mm) wide. Heavy. *Dark purple or black; pearly white below at columella;* head and foot black. Umbilicus closed, or merely a dimple. 2 nodules at columella base. 4 slightly convex whorls, body whorl with weak spiral lines.

Habitat: On rocky shores; between high- and low-tide lines.

Range: British Columbia to Baja California.

Comments: This species is one of the most abundant snails on the Pacific Coast. At low tide it may be found in crevices and protected places. Large individuals may live to be 20 to 30 years old. The Brown Turban Snail (*T. brunnea*) is similar, but is light chestnut-brown, has 1 nodule at the base of the columella, and a dark brown foot. It ranges from Oregon to Santa Barbara Island, and is more commonly found below the low-tide line and on kelps.

Bleeding Tooth
(*Nerita peloronta*)
Class Gastropoda

Description: 1½″ (38 mm) long, 1¾″ (44 mm) wide. Thick, globular. *Whitish, with black and red zigzag markings, 1 or 2 white teeth on blood-red inner lip.* 4–5 whorls; flattened spire; umbilicus closed; aperture semicircular. Operculum horny, brownish-green.

Habitat: On open rocky shores; near low-tide line and below in shallow water.

Range: S. Florida; West Indies.

Comments: The teeth on the red inner lip of this species remind one of the milk teeth of children as they are being lost. The Checkered Nerita (*N. tessellata*) is a smaller species, ¾″ (19 mm) long and ⅞″ (22 mm) wide, with 12 spiral cords and a black-and-white spotted or checkered pattern. It ranges from Florida to Texas and the West Indies.

474 Chink Snail
(*Lacuna vincta*)
Class Gastropoda

Description: ⅜″ (10 mm) long, ¼″ (6 mm) wide. Thin, spire cone-shaped. Brownish, often banded; operculum yellow, horny; head and foot dark gray; tentacles white. 4–5 convex whorls, sutures deep, apex pointed; *umbilicus a small, narrow chink* beside columella.

Habitat: On eelgrass, kelps, and other seaweeds; from low-tide line to water 72′ (22 m) deep.

Range: Arctic to New Jersey.

Comments: This little snail is abundant on seaweeds in northern New England. In summer it lays a gelatinous mass of eggs in the form of a diminutive white doughnut attached to the algae.

472 **Common Periwinkle**
(*Littorina littorea*)
Class Gastropoda

Description: 1" (25 mm) long, ¾" (19 mm) wide. Bluntly conical. Grayish; *outer lip thick, black; inner lip white;* head and foot dark gray; sole cream-colored. 6–7 whorls, spire sharp, squat. Aperture oval, no umbilicus. Adults smooth, immatures have fine spiral threads.

Habitat: On rocky shores, mussels, gravel, and coarse sand; between high- and low-tide lines.

Range: Labrador to Maryland; c. California.

Comments: The Common Periwinkle is by far the most abundant snail in New England. It was introduced from Europe some time before the middle of the 19th century. This snail is eaten in Europe (it is delicious when steamed and served with drawn butter) and can be found for sale in Italian fish markets in our Eastern cities. A Pacific species, the Eroded Periwinkle (*L. planaxis*), occupies the same kind of habitat, higher on the rocks than any other West Coast mollusk. It grows to be ¾" (19 mm) long and ½" (13 mm) wide, is gray-brown, sometimes with pale spots, and possesses a smooth, white columella and frequently an eroded shell. It ranges from Oregon to Baja California. The Sitka Periwinkle (*L. sitkana*) ranges from the Bering Sea to Oregon on protected rocky shores. It is ¾" (19 mm) long and ½" (13 mm) wide, has 10–12 strong spiral threads on the body whorl, and is dark gray to rust-brown, sometimes with 2–3 white spiral bands. In Florida, the Gulf of Mexico, and the West Indies, the Zebra Periwinkle (*L. ziczac*) occupies the same general habitat. It grows to be 1" (25 mm) long and ⅜" (10 mm) wide, and is distinguished by its many narrow zigzag lines of chestnut to purplish-brown against a white or bluish-white background.

473 **Marsh Periwinkle**
(*Littorina irrorata*)
Class Gastropoda

Description: 1″ (25 mm) long, ⅝″ (16 mm) wide. Conical; thick. *Whitish, usually with spiral rows of reddish-brown streaks.* 4–5 whorls; many fine spiral grooves; apex sharp. Aperture yellowish, white inside. Outer lip stout, sharp. Operculum brown, horny.

Habitat: Among grasses in tidal marshes.

Range: New Jersey to Florida and Texas.

Comments: The Marsh Periwinkle can be found at low tide on the stems of marsh grass, with its operculum sealed tightly shut. It remains that way until liberated by the rising tide. The Angulate Periwinkle (*L. angulifera*), which ranges to Bermuda, southern Florida, the Bahamas, and the West Indies, inhabits mangrove swamps, where it is frequently seen well above the high-tide line. It reaches a length of 1¼″ (32 mm) and a width of ⅝″ (16 mm), has 6 convex whorls, a sharp apex, a body whorl with 30–40 spiral lines, and a pattern of elongate dark spots which form oblique stripes on a background varying from bluish-white through dull yellow to grayish-brown.

469 **Northern Yellow Periwinkle**
(*Littorina obtusata*)
Class Gastropoda

Description: ½″ (13 mm) long, ½″ (13 mm) wide. Globular, smooth. Whitish, yellow, or orange; solid or banded brown. *Spire rounded, flattened;* no umbilicus. Columella white, head and foot yellowish, sole white. Operculum yellow to brownish, horny.

Habitat: On rockweed and other large algae; near low-tide line and below.

Range: Labrador to New Jersey.

Comments: This periwinkle is less able than most

to withstand drying, and it survives low tides by hiding in piles of the seaweeds on which it lives.

470 **Rough Periwinkle**
(*Littorina saxatilis*)
Class Gastropoda

Description: ⅝″ (16 mm) long, ½″ (13 mm) wide. Bluntly conical. Whitish, yellowish, gray, or brown, sometimes with spiral stripes. Valve oval, *4–5 rounded whorls, sutures deep,* apex sharp; surface with smooth, fine spiral cords. Operculum horny, dark brown.
Habitat: On rocks; near and above high-tide line.
Range: Arctic to New Jersey.
Comments: This species is found higher on the New England shorelines than any other periwinkle. It is even found above high-tide line in the splash zone, where it can remain wet and feed only for a very short time each tidal cycle. When the tide ebbs, it seeks out a crevice or the shady side of a rock, closes its operculum (which is then sealed by mucus as it dries), and awaits the next wetting.

471 **Checkered Periwinkle**
(*Littorina scutulata*)
Class Gastropoda

Description: ½″ (13 mm) long, ⅜″ (10 mm) wide. Moderately conical. Brownish to nearly black, with lighter spots in *irregularly checkered pattern;* edge of lip with whitish spots against brown interior. 5 *whorls,* surface smooth, shiny. *Columella narrow.* Operculum horny, brownish.
Habitat: On rocky shores; between high- and low-tide lines.
Range: Alaska to Baja California.
Comments: This periwinkle tolerates exposure less

well than the Eroded Periwinkle, which shares the same habitat. At low tide it hides in crevices and among algal holdfasts.

399 Boring Turret Snail
(*Turritella acropora*)
Class Gastropoda

Description: 1″ (25 mm) long, ¼″ (6 mm) wide. Long, conical. Yellowish to brownish-orange. *10–12 flat-sided whorls,* sutures indistinct; many fine spiral threads; *aperture squarish.* Operculum brown, horny.

Habitat: On sand or mud bottoms; below low-tide line in shallow water.

Range: North Carolina to Florida and Texas; West Indies.

Comments: This long, slender snail feeds on organic matter that it separates from the mud and sand. Its relative of south Florida, the Bahamas, and the West Indies, the Variegated Turret Snail (*T. variegata*), is much larger—4″ (102 mm) long and 1″(25 mm) wide—and is white to purplish with reddish vertical streaks.

476 Scaled Worm Snail
(*Serpulorbis squamigerus*)
Class Gastropoda

Description: 5″ (127 mm) long, ½″ (13 mm) wide. *Attached to solid substrate.* Irregularly shaped, open end upturned. Gray or pinkish, *scaly longitudinal cords,* no operculum.

Habitat: Attached to rocks and pilings in protected places; from between high- and low-tide lines to water 65′ (20 m) deep.

Range: C. California to Baja California.

Comments: Like other worm snails, this animal feeds by spinning strands of mucus into the water to trap microscopic organisms and organic particles. It then extends its radula to pull in and eat the mucus.

---

475 Common Worm Snail
(*Vermicularia spirata*)
Class Gastropoda

Description: 6″ (15 cm) long, ¼″ (6 mm) wide. Tubular, loosely spiraled. Yellowish to amber. First few whorls from apex tightly spiraled, smooth; *all whorls but first few loose and irregularly winding.* Tubular, with spiral lines. Head cream-colored, with tentacles and eyes. Operculum horny.

Habitat: On sand or mud bottoms; below low-tide line in shallow water.

Range: S. Massachusetts to Florida; West Indies.

Comments: Though commonly called worm snails, these animals are more closely related to the turret snails. The pattern of shell growth begins like that of a turret, but then becomes irregular and wormlike.

---

433 Common Sundial
(*Architectonica nobilis*)
Class Gastropoda

Description: ¾″ (19 mm) long, 2″ (51 mm) wide. Circular, flattened. Cream-colored, with brownish spots adjacent to sutures. 6–7 whorls, with *4–5 beaded spiral cords; umbilicus round and deep,* bordered by heavily beaded cords. Operculum brown, horny.

Habitat: On sand bottoms; below low-tide line in shallow water.

Range: North Carolina to Florida and Texas; West Indies.

Comments: The scientific name of this mollusk means "noble master builder" in Latin —a tribute to the elegant construction of the animal's shell.

396 **Costate Horn Snail**
(*Cerithidea costata*)
Class Gastropoda

Description: ½" (13 mm) long, ⅛" (3 mm) wide. Sharply conical. Translucent, yellowish-brown; *9–12 convex whorls, curved ribs* on each, fading on last 2 or 3. Sutures deep, aperture round. Operculum round, brownish, horny.

Habitat: On mud bottoms; near low-tide line and below in shallow water.

Range: Florida; West Indies.

Comments: These animals can be seen by the thousands on mudflats, where they are eaten by birds. The Ladder Horn Snail (*C. scalariformis*) is larger, 1¼" (32 mm) long and ⅜" (10 mm) wide, and is red-brown to violet, usually with several whitish spiral bands, 10–13 whorls with closely set ribs, and distinct sutures. It ranges from South Carolina to Florida and the West Indies and lives in the same habitat as the Costate Horn Snail. The California Horn Snail (*C. californica*) is a still larger species, 1¾" (44 mm) long, and ⅝" (16 mm) wide, and is dark brown, with 10–12 ribbed whorls. It is very common in bays and estuaries from central California to Baja California. All these snails feed on the organic matter of the mud surface.

---

401 **Black Horn Snail**
(*Batillaria minima*)
Class Gastropoda

Description: ¾" (19 mm) long, ¼" (6 mm) wide. Sharply conical. Black, brown, or gray, sometimes banded. 6–8 whorls, apex sharp, sutures distinct, *ribs low, knobby.* Aperture oval, *siphonal canal short, turned to left;* operculum round, horny, with many spiralled lines.

Habitat: On mud bottoms; near low-tide line and below in shallow water.

Range: S. Florida; West Indies.

Comments: This snail is abundant on mudflats and is highly variable in its markings, some animals solidly colored, others strikingly banded. The California False Cerith (*B. attramentaria*), a species introduced from Japan and found from Washington to central California, is larger, 1⅞″ (48 mm) long and ⅝″ (16 mm) wide, and varies from black to tan, with white spots or bands. It is frequently seen crawling upside down on the surface film of shallow pools on a mudflat.

---

402 **Florida Cerith**
(*Cerithium floridanum*)
Class Gastropoda

Description: 1½″ (38 mm) long, ⅜″ (10 mm) wide. Sharply conical. Sturdy. *Whitish or gray, with spiral pattern of brown spots.* Apex sharp, 10–13 whorls with *several spiral rows of 18–20 regularly spaced beads,* with fine, granular spiral threads between. Sutures distinct. Aperture oval, oblique; outer lip scalloped; siphonal canal well developed. Operculum brown, horny, with few spirals.

Habitat: On algae, sea grasses, and rocks; near low-tide line and below in shallow water.

Range: North Carolina to Florida and Texas.

Comments: These snails are usually abundant wherever they occur. The Lettered Cerith (*C. literatum*) is shorter and stockier, 1″ (25 mm) long and ½″ (13 mm) wide, and is whitish, with spiral rows of reddish to black squares. It ranges from Florida to the Bahamas and the West Indies. The Ivory Cerith (*C. eburneum*) has about the same range as *C. literatum* and is comparable in size and shape to *C. floridanum,* but has 4–6 spiral rows of beads per whorl, with the beads of the middle row slightly larger.

It may be all white, or cream-colored, or have irregular reddish-brown spots. The Dwarf Cerith (*C. variabile*) also ranges from Florida through the West Indies, but is not more than ½″ (13 mm) long and ⅛″ (3 mm) wide, and is dark brown, usually speckled with white.

405 **Alternate Bittium**
(*Bittium alternatum*)
Class Gastropoda

Description: ¼″ (6 mm) long, 1/16″ (2 mm) wide. Sharply conical. 6–8 rounded whorls *crosshatched by spiral and vertical lines.* Slate-gray to brown-black, sometimes speckled. Sutures distinct; aperture rounded, oblique. Outer lip sharp, flared; operculum horny.

Habitat: On eelgrass, algae, or sand bottoms in quiet water; from low-tide line to water 180′ (55 m) deep.

Range: Gulf of St. Lawrence to Virginia.

Comments: Young snails of this species, which are reddish-brown, sometimes occur in such numbers that they make the sand seem alive. The Variable Bittium (*B. varium*) ranges from Maryland to Florida and Texas. It is the same size and lives in comparable habitats, but is grayish and has a thickened rib on the body whorl. The Threaded Bittium (*B. eschrichtii*) ranges from Alaska to Baja California, is larger, ¾″ (19 mm) long and ¼″ (6 mm) wide, whitish to gray, sometimes with brown spiral bands, and has 8–9 whorls with 5–6 spiral threads. It lives in coarse sand or gravel, or under rocks.

466 Common Purple Sea Snail
(*Janthina janthina*)
Class Gastropoda

Description: ¾″ (19 mm) long, 1½″ (38 mm) wide. Globular, thin, fragile. *Pale violet above, deep violet-purple below.* 3–4 sloping whorls, aperture large, no umbilicus or operculum.
Habitat: Floating on ocean surface.
Range: Cape Cod to Florida and Texas; California.
Comments: This snail is remarkable in its adaptation to life as a drifter. It manufactures a raft of mucus bubbles to which it clings upside down. If detached from its float, it sinks to the bottom and dies.

395 Angulate Wentletrap
(*Epitonium angulatum*)
Class Gastropoda

Description: 1″ (25 mm) long, ⅜″ (10 mm) wide. Tall, conical. Glossy white. 6–8 convex whorls separated by deep sutures, each whorl with 9–10 *vertical, bladelike ribs, each with a blunt angle above;* ribs all vertically aligned. Aperture almost circular; lips thick; no umbilicus. Operculum brown, horny.
Habitat: Around sea anemones; from low-tide line to water 150′ (46 m) deep.
Range: Long Island to Florida and Texas.
Comments: This species is one of the more common and widely distributed representatives of a very large group of Atlantic wentletraps. The Tinted Wentletrap (*E. tinctum*) grows to 1¼″ (32 mm) long and ⅜″ (10 mm) wide, and ranges from British Columbia to Baja California. It has 9–13 ribs per whorl, and frequently has a brown or purple spiral stripe. The name "wentletrap" is derived from Dutch and means "spiral staircase."

394 **Greenland Wentletrap**
(*Epitonium greenlandicum*)
Class Gastropoda

Description: 2″ (51 mm) long, ⅝″ (16 mm) wide. Tall, conical. Dull yellowish-white. *11–12 convex whorls* separated by deep sutures; whorl with *12 stout, flat, vertical ribs,* all vertically aligned, and *8–9 rounded spiral cords.* Aperture circular; lips thick; no umbilicus. Operculum brown, horny.

Habitat: On rock, sand, or mud bottoms; in water 60–567′ (18–173 m) deep.

Range: Arctic to Long Island; Alaska to British Columbia.

Comments: This showy wentletrap is one of the largest in the North Atlantic, and is covered with prominent spiral sculpture.

---

462 **Common Slipper Snail**
"Boat Snail"
(*Crepidula fornicata*)
Class Gastropoda

Description: 2″ (51 mm) long, 1″ (25 mm) wide. Oval; somewhat convex; curved to fit site of attachment. Apex turned to one side. Whitish to tan, mottled with brown; interior white; small head and tentacles pale buff. White, *shelflike deck on underside covers ½ of aperture;* large foot.

Habitat: Attached to almost any hard object; near low-tide line and below in shallow water.

Range: Gulf of St. Lawrence to Florida and Texas; introduced into c. California.

Comments: This snail remains attached by its foot to one place for much of its life; its shape conforms to the surface of its substrate. The Eastern White Slipper Snail (*C. plana*) reaches a length of 1½″ (38 mm) and a width of ¾″ (19 mm), is white, and usually flat or slightly concave or convex, depending on the shape of the substrate. It ranges from Nova Scotia to Florida and Texas. The Hooked Slipper Snail (*C. adunca*) ranges from British Columbia to Baja California. Its shell is 1″ (25 mm) long and ¾″ (19 mm) wide, and its apex is high and curved in a hook to the rear.

435 **Queen Conch**
(*Strombus gigas*)
Class Gastropoda

Description: 12″ (30 cm) long, 8″ (20 cm) wide. Spindle-shaped. Thick, heavy. Yellowish, covered with thick, horny periostracum, *lining rosy pink.* Short, conical spire, 8–10 whorls with *blunt spines on shoulder;* large body whorl. Aperture long, narrow; outer lip thick, widely flared in adult; operculum clawlike. Head with large proboscis and tentacles, whitish mottled with gray, large eye at base of each tentacle.

Habitat: In shallow water; below low-tide line.

Range: S. Florida; West Indies.

Comments: The Queen Conch is heavily fished and has become uncommon in Florida. Its meat is the main ingredient of conch chowder, and its big, showy shell a much-sought souvenir. The Fighting Conch (*S. alatus*), which ranges from South Carolina to Florida and Texas, is smaller, 4″ (102 mm) long and 2½″ (64 mm) wide, and is reddish-brown, sometimes mottled, with short spines on the shoulders.

450 **Four-spotted Trivia**
(*Trivia quadripunctata*)
Class Gastropoda

Description: ¼″ (6 mm) long, ¼″ (6 mm) wide. Coffee-bean-shaped, plump, oval. Bright pink, with 4 dark reddish-brown spots on the back; mantle and foot scarlet, mantle covering shell entirely when expanded. Long, *slitlike aperture crossed by 24 fine ribs.*

Habitat: Near low-tide line and below in shallow water.

Range: S. Florida; West Indies.

Comments: This little snail is abundant, and its faded shell is often found on beaches, the dots worn away. Spire, sutures, and columella are all overgrown by the body whorl and cannot be seen from the outside.

444 **Chestnut Cowrie**
(*Cypraea spadicea*)
Class Gastropoda

Description: 2⅝″ (67 mm) long, 1⅜″ (35 mm) wide. *Oval, polished, spire covered by body whorl.* Chestnut-brown above, edged by gray or white; white underneath; mantle rosy brown with dark spots; edge fringed; 2 long, reddish tentacles, with black eye at base. Long, slitlike aperture edged by 20–23 teeth; no operculum.

Habitat: Among seaweeds and surfgrass; from low-tide line to water 160′ (49 m) deep.

Range: S. California to Baja California.

Comments: Cowries are primarily tropical snails, prized by collectors for their beauty. The Chestnut Cowrie is the only one common in California waters. The Atlantic Gray Cowrie (*C. cinerea*) is found in southeast Florida and the West Indies. It is plump, 1½″ (38 mm) long and ¾″ (19 mm) wide, and grayish to orange-brown with dark specks above, and cream-colored to lilac below. It is common under rocks on reefs. The Atlantic Deer Cowrie (*C. cervus*) grows to be 5″ (127 mm) long and 3″ (76 mm) wide, is light brown, with white spots and a white line down the middle of the back, and is lighter below. Its aperture is edged by 35 brown-edged teeth. It ranges from southern Florida through the West Indies and to Texas.

449 **Flamingo Tongue**
(*Cyphoma gibbosum*)
Class Gastropoda

Description: 1″ (25 mm) long, ½″ (13 mm) wide. Elongate, oval. Solid, polished; creamy-orange to buff. *White to pinkish mantle, with black-edged, squarish, yellow spots,* covers shell completely when extended; white siphon with black rim, yellow tentacles with black stripe and white tip. *Hump or ridge across middle.* Aperture narrow, as long as shell; foot long.

Habitat: On sea fans and other soft corals; in shallow water.

Range: North Carolina to Florida; West Indies.

Comments: The Flamingo Tongue feeds on the soft tissues of sea fans and occasionally on other soft corals. During the day it can be found clinging to the base of these colonies.

467 **Shark Eye**
(*Polinices duplicatus*)
Class Gastropoda

Description: 2½″ (64 mm) long, 3¾″ (95 mm) wide. Smooth. Slate-gray to tan; dull gray mantle covers shell when extended. Periostracum thin; shell globe-shaped; 4–5 whorls; *spire low and rounded.* 2 tentacles yellow with black stripe. *Umbilicus almost covered by brown callus; deep.* Columella white. Aperture oval, flattened at inner lip; foot large, with plowlike front end; dull gray. Operculum horny, thin, brown.

Habitat: On sand and sandy-mud bottoms; near low-tide line and below in shallow water.

Range: Cape Cod to Florida and Texas.

Comments: The Shark Eye plows through the sand in search of clams or other mollusks, surrounds them with its large foot, and drills a neat, round hole through the shell, scraping out the soft parts with its radula. Lewis' Moon Snail (*P. lewisii*) ranges from British Columbia to Baja California. It grows to be 3¼″ (83 mm) long and 5″ (127 mm) wide, and is similar in color to the Shark Eye, but has a shallow groove between the shoulder and adjacent suture.

461 Northern Moon Snail
(*Lunatia heros*)
Class Gastropoda

Description: 4½″ (114 mm) long, 3½″ (89 mm) wide. Globe-shaped. Whitish to brownish, with thin, yellow periostracum. Apex somewhat flattened; 5 convex whorls; *umbilicus large, round, deep.* Aperture large, oval, whitish inside, sometimes with tan or purplish spots. Foot large, gray, with plowlike front end; mantle gray, tentacles yellow. Operculum thin, horny, light brown.

Habitat: On sand or sandy-mud bottoms; from low-tide line to water more than 1200′ (366 m) deep.

Range: Labrador to North Carolina.

Comments: Named for its round, whitish shell, this moon snail lays its eggs in a curved ribbon, or egg collar, made of sand and mucus. The Spotted Moon Snail (*L. triseriata*) has a comparable range and is much smaller, measuring 1″ (25 mm) long and ¾″ (19 mm) wide, and its shell usually has 3 brownish broken spiral bands on a buffy background. The foot and mantle are white and the tentacles black.

465 **Common Baby's Ear**
(*Sinum perspectivum*)
Class Gastropoda

Description: 2″ (51 mm) long, 1¼″ (32 mm) wide. Ear-shaped, wide, flattened. Mantle covers entire shell when extended. White, with thin, yellowish periostracum. Apex on same plane as body whorl; 3 whorls; *many fine spiral lines crossed by growth lines.* Aperture large; columella curved, white.
Habitat: On sand bottoms; below low-tide line in shallow water.
Range: Virginia to Florida and Texas; West Indies.
Comments: This moon snail secretes large amounts of mucus when handled.

434 **Emperor Helmet**
(*Cassis madagascariensis*)
Class Gastropoda

Description: 9″ (23 cm) long, 8″ (20 cm) wide. Massive, heavy, triangular. Cream-colored. Spire almost flat; body whorl with *3 rows of large, blunt knobs,* 1 or more much bigger than others. Aperture long, narrow; *inner and outer lips flattened in same plane,* thick, deep salmon with white, riblike teeth, brown bands between them. Upturned siphonal canal at front end. Operculum semicircular, horny.
Habitat: In shallow water; below low-tide line.
Range: S.E. Florida; West Indies.
Comments: Despite its species name, the Emperor Helmet is a tropical Atlantic animal and does not occur around Madagascar. A subspecies, *C. madagascariensis spinella,* is the most common helmet in the Florida Keys. It reaches 14″ (36 cm) in length and 12½″ (32 cm) in width, and its knobs are all about the same size.

454 **Scotch Bonnet**
(*Phalium granulatum*)
Class Gastropoda

Description: 4″ (102 mm) long, 3″ (76 mm) wide. Moderately heavy. Cream-colored, with spiral bands of regularly spaced, yellow to brown square spots. Periostracum thin. 5 whorls; short, cone-shaped spire; body whorl with *20 spiral grooves crossed by deep growth lines,* giving it a beaded appearance. Aperture large; outer lip heavy, toothed inside; inner lip broad, bumpy on lower part. Siphonal canal upturned; operculum horny, crescent-shaped, light brown.

Habitat: In shallow, warm water; below low-tide line.

Range: North Carolina to Florida, Texas, and Brazil; West Indies.

Comments: The handsome Scotch Bonnet is a popular item in shell shops. Empty, in the wild, it is frequently occupied by the Striped Hermit Crab; alive, it preys on sand dollars and sea urchins.

419 **Atlantic Hairy Triton**
(*Cymatium pileare*)
Class Gastropoda

Description: 5½″ (140 mm) long, 2¾″ (70 mm) wide. Rugged; covered with hairy, matted, golden-brown periostracum. Pale brown banded with gray and white. Spire pointed; 5–6 whorls with spiral cords crosshatched by lines. *Shoulder bumpy; heavy longitudinal ridge on either side* of large, brown aperture. Wrinkled, white teeth on both lips; siphonal canal upturned; operculum oval, horny, brown.

Habitat: In shallow water around reefs; below low-tide line.

Range: S.E. Florida and Texas; West Indies.

Comments: The Atlantic Hairy Triton and its close relatives are predatory on other mollusks and on sea stars. Its empty shells are sometimes occupied by hermit crabs. Older literature lists this snail as *C. martinianum.*

418 **Oregon Hairy Triton**
(*Fusitriton oregonensis*)
Class Gastropoda

Description: 5″ (127 mm) long, 2½″ (64 mm) wide. Evenly tapered from apex. Light brown; covered with *gray-brown, heavy, bristly periostracum;* operculum brown, thick, horny. 6 whorls, each with *16–18 longitudinal ribs crossed by many paired spiral lines.* Aperture large, oval, white inside; prominent siphonal canal.
Habitat: On rock bottoms in shallow water; near low-tide line and below.
Range: Alaska to s. California.
Comments: The Oregon Hairy Triton is an aggressive predator on mollusks and echinoderms, especially sea urchins. It rasps off the outer tissue and bores holes through the shell plates to get at the internal organs. Sometimes urchins are found with blackish scars resulting from encounters with this snail.

451 **Unicorns**
(*Acanthina* spp.)
Class Gastropoda

Description: 1–1⅝″ (25–41 mm) long, ¾–1″ (19–25 mm) wide. Spindle-shaped. Blue-gray, with rows of small, dark dots, or checkered black and white. *5–6 whorls* with slight shoulder, and spiral cords or threads, *shallow suture. Outer lip with sharp spine,* siphonal canal short. Operculum oval, horny, brownish.
Habitat: On rocks and pilings, among mussels and barnacles, in protected places; between high- and low-tide lines.
Range: Puget Sound to Baja California.
Comments: Among the various species of unicorns

are the Angular Unicorn (*A. spirata*) which measures 1⅝″ (41 mm) long and 1″ (25 mm) wide, has a ridged shoulder, and is pale blue-gray with rows of dots, and the Checkered Unicorn (*A. paucilirata*), a southern species ranging from southern California to Baja California, which is smaller than A. *spirata,* but is distinguished by its pattern of large, black and white checks.

---

409 **Spotted Unicorn**
(*A. punctulata*)
Class Gastropoda

Description: 1″ (25 mm) long, ¾″ (19 mm) wide. Spindle-shaped. Solid, rather smooth. Light tan, with spiral rows of dark spots. *Whorls with rounded shoulders,* and fine spiral lines. Aperture oval, brownish-white inside; *lower part of outer lip with sharp spine.* Siphonal canal short. Operculum oval, horny, brownish.

Habitat: On rocks, among barnacles and mussels; between high- and low-tide lines.

Range: C. California to Baja California.

Comments: At low tide these snails may be found on exposed rocks, but many collect in tidepools, or keep wet by nestling among contracted anemones. They feed on barnacles and periwinkles, and occasionally on other snails encountered as they move up and down with the tides.

---

436 **Leafy Hornmouth**
(*Ceratostoma foliatum*)
Class Gastropoda

Description: 4″ (102 mm) long, 2¼″ (57 mm) wide. Spindle-shaped, irregular. White, gray, or yellow-brown. Body whorl with *3 large, longitudinal flanges,* one on either side of aperture, one on upper side.

Aperture oval, white; outer lip with strong tooth near front end. Siphonal canal sealed over into a tube, turned up and to the right; operculum brown, horny.

Habitat: On rocks; from low-tide line to water 200′ (61 m) deep.

Range: Alaska to s. California.

Comments: The unusual shape of the shell, with its 3 winglike flanges, enables this snail to land foot-downward when it is dislodged from a rock and tumbles through the water.

437 **Apple Murex**
(*Phyllonotus pomum*)
Class Gastropoda

Description: 4½″ (114 mm) long, 2¼″ (57 mm) wide. Spindle-shaped. Sturdy, rough. Yellowish-tan with brown mottlings. 5–6 whorls with *revolving ridges and sharp ribs. 3 prominent longitudinal ridges on each turn.* Sutures indistinct; aperture large, rounded, ivory to orange, with brown spot at upper end; outer lip thick, rough. Siphonal canal about as long as width of aperture, curving upward. Operculum horny, yellowish.

Habitat: On gravel bottoms, in turtle grass beds, and on reef flats; below low-tide line in shallow water.

Range: North Carolina to Florida and Texas; West Indies.

Comments: The Apple Murex is a very common shallow-water species. Its eggs are laid in tough, leathery capsules attached to rocks.

438 **Lace Murex**
(*Chicoreus florifer*)
Class Gastropoda

Description: 3″ (76 mm) long, 1½″ (38 mm) wide. Irregular, spiny. Brownish-black to light brown. Sturdy, rough, *7 whorls, 3 prominent spiny ridges;* aperture small, round, with *8–10 large, leaflike spines* along outer lip, long siphonal canal. Operculum horny, blackish.

Habitat: Among mangroves, on rocks, mud, or sand bottoms; below low-tide line in shallow water.

Range: Florida; Bahamas; West Indies.

Comments: This is the most common murex in southern Florida. Young specimens are almost entirely pink, the darker pigments developing as the snail grows.

414 Emarginate Dogwinkle
(*Nucella emarginata*)
Class Gastropoda

Description: 1⅝″ (41 mm) long, 1″ (25 mm) wide. Plump, stout. White, gray, yellow, orange, brown, or black, with or without spiral stripes. Spire conical, *4–5 convex whorls with 5–6 beaded spiral cords; sutures shallow.* Aperture oval; outer lip thick, flared, toothed; inner lip smooth, siphonal canal short. Operculum horny, brownish.

Habitat: On rocks, among mussels and barnacles; between high- and low-tide lines.

Range: Alaska to Baja California.

Comments: The female of this species lays clusters of horny, flask-shaped capsules, each containing 500–600 eggs. Most of the eggs are sterile and are eaten by larvae developing from the fertile eggs. The Channeled Dogwinkle (*N. canaliculata*), which ranges from Alaska to central California, is about the same size, and mottled white to dark orange. It lacks beaded cords, but has 14–16 small, smooth spiral cords on the whorl, and has deeper sutures.

410 **Poulson's Rock Snail**
(*Roperia poulsoni*)
Class Gastropoda

Description: 2⅜" (60 mm) long, 1⅛" (28 mm) wide. Spindle-shaped. Whitish, with many narrow, brown spiral stripes. Periostracum thin, brownish. Spire tall, apex sharp; 5–6 whorls, each with 8–9 *bumpy ribs* crossed by *fine spiral lines and 4–5 large cords,* the latter making the bumps on the ribs. Aperture oval, long, white inside; outer lip thick, with 7–8 teeth; siphonal canal moderately long, nearly closed.

Habitat: On rocks and pilings; in tidepools; near low-tide line and below in shallow water.

Range: S. California to Baja California.

Comments: This snail, which feeds on barnacles, mussels and other bivalves and snails by rasping out the soft parts, is, in turn, eaten by the Pacific Rock Crab.

415 **Rock Snail**
(*Thais haemastoma*)
Class Gastropoda

Description: 5" (127 mm) long, 2½" (64 mm) wide. Shell solid, spindle-shaped. Light gray to yellowish, with irregular brown spots, inside salmon-pink. Apex pointed; 5–6 convex whorls, sutures deep, *fine spiral lines with 2 rows of bumps,* sometimes large; aperture large, oval, *outer lip thick, with small teeth,* inner lip smooth over columella, siphonal canal short, oblique; operculum brown, horny.

Habitat: On rocks and other solid objects, between high- and low-tide lines.

Range: North Carolina to Florida and Texas; West Indies; Mexico.

Comments: These snails are predators on barnacles and on oysters and other mollusks. Mollusk specialists divide this species into two subspecies, the Florida Rock

Snail, which is smaller, somewhat more bumpy, but less deeply sutured, and ranges from North Carolina through the West Indies; and the larger Hays' Rock Snail, which is smoother, but more deeply sutured, and ranges from western Florida to Texas and Mexico.

456 **Atlantic Dogwinkle**
(*Thais lapillus*)
Class Gastropoda

Description: 2″ (51 mm) long, 1⅛″ (28 mm) wide. Spindle-shaped. White, pink, lavender, yellow, orange, brown, or black, sometimes with spiral stripes. 5–6 *whorls, spire short, apex bluntly pointed;* surface with rounded spiral ridges, sometimes scaly; older specimens smoother. *Aperture oval, outer lip thick,* siphonal canal short, open. Operculum brown, horny.

Habitat: On rocks, among mussels and barnacles; between high- and low-tide line.

Range: Labrador to Long Island Sound.

Comments: The color and surface sculpture of the shell in this species are so variable that one might readily assume extreme specimens to be different species. The snail feeds either on barnacles or on small mussels, and if the young begin by feeding on one food species, they tend to prefer it over the other throughout life.

407 **Atlantic Oyster Drill**
(*Urosalpinx cinerea*)
Class Gastropoda

Description: 1¼″ (32 mm) long, ⅝″ (16 mm) wide. Plump, spindle-shaped. Gray. Spire moderately tall, sutures deep. 5–6 whorls, *9–12 low, rounded vertical folds per whorl.* Many strong spiral cords;

aperture oval, *purple inside;* outer lip sharp. Siphonal canal short, flaring; foot and siphon cream-colored; operculum dark, horny.

Habitat: On rocks, oyster beds, and wharf pilings; from above low-tide line to water 50′ (15 m) deep.

Range: Gulf of St. Lawrence to Florida; introduced from British Columbia into s. California.

Comments: This snail is the enemy of oystermen because it feeds primarily on young oysters, drilling holes in their shells and sucking out the meat. People who culture oysters on strings of empty shells suspended from rafts avoid the snail by towing the rafts up river mouths to areas of low salinity tolerated by the oysters but not the snail.

---

455 **Mottled Dove Snail**
(*Columbella mercatoria*)
Class Gastropoda

Description: ¾″ (19 mm) long, ½″ (13 mm) wide. Solid. Mottled or encircling bars of white and brown on pink, orange, or yellow background. Spire blunt, sutures moderately deep. 5–6 whorls with many spiral lines. *Aperture long and narrow; both lips toothed.* Operculum horny, small.

Habitat: On sand, mud, and rock bottoms; in shallow water.

Range: S.E. Florida, West Indies.

Comments: These carnivorous snails are most often found under rocks, where they prey upon small bivalves and crustaceans. Their shells are commonly found on Florida beaches.

406 **Greedy Dove Snail**
(*Anachis avara*)
Class Gastropoda

Description: ½″ (13 mm) long, ¼″ (6 mm) wide. *Spindle-shaped.* Mottled brownish-yellow. Spire tall, sutures shallow. 6–7 *whorls bearing spiral lines and vertical folds.* Aperture narrow; outer lip thick, sharp-edged, with tiny teeth; inner lip smooth, white. Siphonal canal short; operculum oval, horny, brown.
Habitat: On seaweeds, grassflats, rocks, and mud; near low-tide line and below.
Range: Cape Cod to Florida and Texas.
Comments: This snail is called greedy because of its voracious habits—it feeds both in the daytime and at night. It is especially abundant in the southern part of its range.

---

453 **Lunar Dove Snail**
(*Mitrella lunata*)
Class Gastropoda

Description: ¼″ (6 mm) long, ⅛″ (3 mm) wide. Spindle-shaped. Glossy, yellowish, or tan, with *brown, crescent-shaped or zigzag markings.* Spire tall, 6 whorls with fine spiral threads; sutures shallow. Aperture narrow, oval; 4 teeth on inner margin of outer lip. Foot, siphon, and tentacles white with few black specks; operculum horny, brown.
Habitat: On seaweeds and other growth, and on mud bottoms; near low-tide line and below in shallow water.
Range: Maine to Florida and Texas; West Indies.
Comments: The name *Mitrella* means "little miter" in Latin, and was probably suggested by the similarity between the shell's shape and that of a bishop's hat.

408 **Waved Whelk**
(*Buccinum undatum*)
Class Gastropoda

Description: 4″ (102 mm) long, 2″ (51 mm) wide. Conical, sturdy. Grayish or tannish, with tones of pink or yellow. Periostracum thick, gray; spire tall. Apex sharp. 5–6 convex whorls with *9–18 wavy longitudinal folds on upper half;* 5–8 spiral lines. Sutures deep; aperture large, flared, whitish inside. Lips thick. Siphonal canal short; head and foot white blotched with black; operculum horny, pale brown.
Habitat: On rock bottoms; from low-tide line to water 600′ (183 m) deep.
Range: Arctic to New Jersey.
Comments: The Waved Whelk is chiefly a scavenger of animal matter and is quickly attracted to any dead or damaged animal. It invades lobster pots to feed on the bait, probing into the fish carcasses with a long proboscis.

424 **Stimpson's Whelk**
(*Colus stimpsoni*)
Class Gastropoda

Description: 4″ (102 mm) long, 1½″ (38 mm) wide. Stout, spindle-shaped. Chalky-white, covered with thin, brownish periostracum. Spire tall. *7–8 flattish whorls separated by shallow sutures;* many spiral lines. Aperture ½ length of shell, oval, white inside. Siphonal canal large, moderately long; head and foot whitish with sparse black speckles; operculum teardrop-shaped, horny, thick, brown.
Habitat: On rock bottoms; from below low-tide line to water 2800′ (853 m) deep.
Range: Labrador to Cape Hatteras.
Comments: Stimpson's Whelk is found in tidepools from Maine north, wherever tidal fluctuations are great. Farther south its shells are found by diving and

dredging, washed ashore by storms, and brought into shallow water by large hermit crabs.

420, 425 **Corded Neptune**
(*Neptunea lyrata*)
Class Gastropoda

Description: 5″ (127 mm) long, 2¼″ (57 mm) wide. Spindle-shaped, stout. Dull white, with reddish-brown or brown cords, aperture white or pale tan. Shell thick, with 6–7 whorls, *7–10 thick, raised spiral cords* on body whorl, faint spiral threads between; aperture large, ½ body length, siphonal canal short, open; head and foot pale gray, speckled with black; operculum thick, horny, brown.

Habitat: On rocky shores; from low-tide line to water 300′ (91 m) deep.

Range: Alaska to Puget Sound; Nova Scotia to Cape Cod.

Comments: The shells of these whelks are frequently brought into shallow water by large hermit crabs. The Atlantic form of the Corded Neptune was formerly listed as *N. decemcostata,* but is now considered a subspecies of *N. lyrata.*

417 **Channeled Whelk**
(*Busycon canaliculatum*)
Class Gastropoda

Description: 7″ (18 cm) long, 3½″ (89 mm) wide. Pear-shaped. Pale buffy-gray. Periostracum yellowish-brown, intact in young specimens, mostly eroded away in older ones. Spire short. *5–6 whorls with square shoulder bearing small, low knobs;* fine spiral lines, *deeply channeled suture.* Aperture and siphonal canal ⅔ body length, yellowish inside.

Outer lip flared. Head and foot whitish. Operculum horny, brown.

Habitat: On sand and sandy-mud bottoms; from low-tide line to water 60′ (18 m) deep.

Range: Cape Cod to n. Florida; introduced into San Francisco Bay.

Comments: The Channeled Whelk feeds mostly on bivalves, which it digs out of the bottom. The Knobbed Whelk (*B. carica*) shares the Channeled Whelk's range and habitat. It grows to 9″ (23 cm) in length and 4½″ (114 mm) in width, has a series of large knobs on the shoulder, and a black head and foot. Its shell is yellowish-gray with a reddish-orange interior.

427, 428 **Lightning Whelk**
(*Busycon contrarium*)
Class Gastropoda

Description: 16″ (41 cm) long, 7″ (18 cm) wide. Long, heavy, spindle-shaped. White or cream-colored, with longitudinal zigzag streaks of brown. *Spirals to left. Spire low,* shoulder with *rows of triangular knobs.* Aperture long and wide, merging gradually into siphonal canal. Foot black.

Habitat: On sand, mud, and shell bottoms; below low-tide line in shallow water.

Range: North Carolina to Florida and Texas; Yucatán.

Comments: This large whelk is unusual in spiralling to the left. Its size and beauty make it an attractive souvenir.

---

429 **Crown Conch**
(*Melongena corona*)
Class Gastropoda

Description: 5″ (127 mm) long, 3″ (76 mm) wide. Sturdy; extremely variable in size, shape, and spine pattern. Cream-colored, with spiral bands of black, brown, yellow, or bluish. *Shoulder and side of whorl with 1–4 sharp semi-tubular spines* pointing upward or outward; many spiral lines. Suture deep; aperture wide, oval. Siphonal canal wide, short; operculum horny, clawlike.

Habitat: In shallow water; below low-tide line.

Range: Florida to Mexico.

Comments: The variability of this conch's shell has led to the naming of a number of subspecies, but since a large collection of these shells shows a complete gradation from one type to the other, the subspecies are probably not valid.

452 **Mud Dog Whelk**
(*Nassarius obsoletus*)
Class Gastropoda

Description: 1″ (25 mm) long, ⅜″ (10 mm) wide. Bluntly spindle-shaped. Moderately tall spire usually *eroded at apex. Purplish-black.* 6 whorls with many diverse spiral lines crossed by weak folds and growth lines. Sutures shallow. Aperture oval; outer lip thin, inner lip brown and gray, shiny. Siphonal canal no more than a notch; columella with oblique fold; inner lip glossy. Head and foot pale gray, heavily mottled with black. Operculum horny, blackish.

Habitat: On mudflats; above low-tide line and just below.

Range: Gulf of St. Lawrence to Florida; introduced from British Columbia into c. California.

Comments: The shell of this snail is usually coated with mud, since it buries itself as the tide goes out. It feeds both on carrion and on organic matter in the mud, and uses its keen sense of smell to locate animal carcasses.

411 **Giant Western Nassa**
(*Nassarius fossatus*)
Class Gastropoda

Description: 2″ (51 mm) long, 1″ (25 mm) wide. Spindle-shaped. Gray, or orange-brown to brown, with orange inner lip. *Spire tall, apex sharp;* 6–7 whorls with rounded ribs crossed by *15 heavy spiral lines;* sutures well defined. Aperture oval, *outer lip toothed,* siphonal canal short; inner lip glossy, *columella with oblique fold.* Operculum horny, blackish.

Habitat: On mudflats; above low-tide line and just below.

Range: British Columbia to Baja California.

Comments: This is the largest and one of the most common mud snails on the Pacific Coast.

412 **New England Dog Whelk**
(*Nassarius trivittatus*)
Class Gastropoda

Description: ¾″ (19 mm) long, ⅜″ (10 mm) wide. Spindle-shaped. Whitish to yellowish-gray, sometimes with brownish spiral stripes. Spire tall, with 6–7 whorls, *deep sutures;* 10–12 strong spiral lines crossed by ribs of equal size, creating a *checkered pattern.* Aperture oval, outer lip sharp, thin, scalloped; siphonal canal short. Operculum horny, brown.

Habitat: On sand or sandy-mud bottoms; from low-tide line to water 270′ (82 m) deep.

Range: Nova Scotia to South Carolina.

Comments: Unlike most members of this genus, the New England Dog Whelk prefers a sandy to a muddy habitat.

413 **Mottled Dog Whelk**
(*Nassarius vibex*)
Class Gastropoda

Description: ½″ (13 mm) long, ¼″ (6 mm) wide. Spindle-shaped. Whitish, mottled with gray or brown. Shell thick and heavy. Spire moderately tall, *apex bluntly pointed,* sutures well defined; whorls with *strong vertical folds crossed by indistinct spiral lines,* making surface corrugated. Aperture oval, *outer and inner lips thick.* Operculum horny and brown.

Habitat: On sand and mudflats; near low-tide line and below in shallow water.

Range: Cape Cod to Florida and Texas; Bahamas; West Indies.

Comments: This is the most widely distributed of the Atlantic dog whelks. It feeds by sorting out organic particles from the mud or sand, and by scavenging larger dead matter.

426 **Florida Horse Conch**
(*Pleuroploca gigantea*)
Class Gastropoda

Description: 24″ (61 cm) long, 10″ (25 cm) wide. Heavy, spindle-shaped. Chalky-white to salmon, with heavy, brown periostracum; salmon to tan inside. Columella has 3 oblique folds. Apex usually somewhat eroded; 8–10 convex whorls with *strong spiral cords irregularly spaced, fine lines between;* obvious longitudinal growth lines. Shoulder has large, low bumps; aperture oval; inner lip has 3 wrinkled folds. Siphonal canal long, turned up; head and foot scarlet. Operculum thick, horny, brown.

Habitat: Offshore and in inlets on sand bottoms; below low-tide line in shallow water.

Range: North Carolina to Florida and Texas; Mexico.

Comments: This is the largest snail on the Atlantic Coast of the United States, and one of the biggest in the world. The shells of older specimens are usually well populated with bryozoans, barnacles, tube worms, and other sedentary invertebrates.

421 **True Tulip Snail**
(*Fasciolaria tulipa*)
Class Gastropoda

Description: 10″ (25 cm) long, 4″ (102 mm) wide. Spindle-shaped. Grayish, pinkish, or orange-red, with *brown spots and many broken spiral stripes.* Apex sharp. Spire moderately high; *8–9 smooth convex whorls,* with fine spiral threads and growth lines. Few wrinkles below sutures; aperture oval, long; outer lip with fine teeth; thin inner lip with 2 oblique folds. Columella has 2 spiral ridges. Head and foot black with white speckles; sole of foot reddish-brown. Operculum horny, brown.

Habitat: On sand and grass bottoms; from low-tide line to water 30′ (9 m) deep.

Range: North Carolina to Florida and Texas; West Indies.

Comments: The orange-red color phase of this tulip snail is most common in the Florida Keys and the Yucatán peninsula.

422 **Banded Tulip Snail**
(*Fasciolaria hunteria*)
Class Gastropoda

Description: 3″ (76 mm) long, 1⅜″ (35 mm) wide. Smooth, spindle-shaped. Mottled bluish, greenish-gray, or orange-brown, with vertical whitish patches and *widely spaced, unbroken, brown spiral lines.* Shell smooth, 7–8 whorls, apex sharp, *sutures indistinct.* Aperture long, oval; *columella with 2 spiral ridges;* siphonal canal short, open. Foot black with white speckles. Operculum horny, brown.

Habitat: On sand and grass bottoms; near low-tide line and below in shallow water.
Range: North Carolina to Florida and Texas; Bahamas; West Indies, Yucatán.
Comments: This handsome snail is an aggressive predator that feeds on many other species of snails. It, in turn, is eaten by the Florida Horse Conch.

441, 442 **Lettered Olive**
(*Oliva sayana*)
Class Gastropoda

Description: 2½″ (64 mm) long, ⅞″ (22 mm) wide. *Long, oval, almost cylindrical. Glossy,* cream-colored or tan, with many brown or purplish-brown zigzag markings. 5–6 whorls with deep sutures; large body whorl, low spire. Aperture narrow, as long as body whorl; outer lip thin, purplish inside, inner lip white, with wrinkled folds. Siphonal canal only a notch. No operculum.
Habitat: On sand bottoms; near low-tide line and below in shallow water.
Range: North Carolina to Florida and Texas.
Comments: These snails are called olives because their shape resembles that of an olive pit. They are commonly used in shellcraft.

440 **Netted Olive**
(*Oliva reticularis*)
Class Gastropoda

Description: 1½″ (38 mm) long, ⅝″ (16 mm) wide. Oval, almost cylindrical, glossy. Pale gray, with netlike pattern of purple-brown lines. *Body whorl large; low, conical spire* of 5–6 whorls. Aperture narrow, as long as body whorl; outer lip thin; *inner lip with wrinkled spiral folds;* siphonal canal only a notch. No operculum.
Habitat: On sand bottoms; below low-tide line

to water more than 300′ (91 m) deep.

Range: S. Florida; West Indies.

Comments: The handsome, shiny shell of this species makes it an attractive collector's item.

447 **Purple Dwarf Olive**
(*Olivella biplicata*)
Class Gastropoda

Description: 1¼″ (32 mm) long, ⅝″ (16 mm) wide. *Oval, sturdy. Glossy,* white, lavender, grayish-black, rarely yellow or orange. Spire low, conical. 4–5 smooth whorls; sutures not indented. Aperture about ⅔ the length of shell; inner lip with 2 raised spiral folds; siphonal canal only a notch; foot and mantle white to cream-colored.

Habitat: On sand bottoms; from low-tide line to water 150′ (46 m) deep.

Range: British Columbia to Baja California.

Comments: Dwarf Olives, like their larger relatives, have a plowlike front end on the foot with which, during the day, they generally burrow under the sand, leaving a track behind. The Variable Dwarf Olive (*O. mutica*) is smaller, ½″ (13 mm) long and ¼″ (6 mm) wide. Its colors are also variable: gray, brown, yellow, whitish, sometimes banded with brown and white. It ranges from North Carolina to Florida and Texas, and to the West Indies.

403 **Beaded Miter**
(*Mitra nodulosa*)
Class Gastropoda

Description: 1″ (25 mm) long, ⅜″ (10 mm) wide. *Long, glossy,* brownish-orange. Apex sharp, spire tall, sutures distinct. 9–10 whorls with slight shoulders; *surface checkered* with longitudinal ribs crossed by deep spiral lines. Aperture short;

inner lip with 4 folds, siphonal canal only a notch.

Habitat: On rock bottoms; near low-tide line and below in shallow water.

Range: North Carolina to Florida; West Indies.

Comments: Miters have a specialized radula with long barbs which is useful for spearing the worms and other soft-bodied invertebrates that they eat. They are commonly found in protected crevices under rocks at low tide.

400 **Ida's Miter**
(*Mitra idae*)
Class Gastropoda

Description: 3¼″ (83 mm) long, 1⅛″ (28 mm) wide. Long, spindle-shaped. Shell dark brown under thick, black periostracum. Apex sharp, spire tall, 7–8 whorls; *body whorl ½ total length;* surface with fine lines crossed by fine longitudinal ribs. *Aperture long, narrow,* inner lip with *3 spiral ridges.* Head and foot white.

Habitat: Among rocks, in tidepools, and in kelp beds; from low-tide line to moderately deep water.

Range: N. California to Baja California.

Comments: Ida's Miter is uncommon in tidepools, but fairly common in kelp beds. The striking contrast of its black shell and white body makes it easy to identify.

423 **Junonia**
(*Scaphella junonia*)
Class Gastropoda

Description: 6″ (15 cm) long, 2½″ (64 mm) wide. Spindle-shaped. *Cream-colored, with spiral rows of squarish red-brown spots.* 5–6 smooth, convex whorls; sutures distinct, aperture ⅔ length of shell. Narrow outer lip thin, inner lip with 4 oblique folds near short siphonal canal. No operculum.

Habitat: Below low-tide line in water 6–180′ (2–55 m) deep.
Range: North Carolina to Florida and Texas.
Comments: The beautiful shell of this animal is much desired by collectors and commands high prices. It is dredged in large numbers by shrimp trawlers in the Gulf of Mexico.

---

416 **Common Nutmeg**
(*Cancellaria reticulata*)
Class Gastropoda

Description: 1¾″ (44 mm) long, 1″ (25 mm) wide. Spindle-shaped. Sturdy. Cream-colored, with reddish-brown bands and mottlings. *Surface meshlike with longitudinal ribs crossed by spiral lines of same size; whorls convex;* spire moderately tall, sharp; sutures moderately deep. Aperture long, oval, white inside. Outer lip thin, finely scalloped; inner lip with 2 twisted folds, upper one much larger; siphonal canal short, turned up. No operculum.
Habitat: In shallow water; below low-tide line.
Range: North Carolina to Florida and Texas; West Indies.
Comments: Although Common Nutmegs do not normally occur close to the beach, they may be washed up in large numbers during a storm.

---

439 **Common Marginella**
(*Prunum apicinum*)
Class Gastropoda

Description: ½″ (13 mm) long, ⅜″ (10 mm) wide. Plump. *Glossy;* cream-colored, yellowish, golden, or orange-brown, with 4 reddish spots on outer lip. 3–4 whorls; spire short; *body whorl large, convex.* Aperture long, narrow; outer lip smooth; *inner lip with 4 strong folds.* Siphonal canal only a notch. No operculum.

Habitat: In grass beds; at low-tide line and below in shallow water.

Range: North Carolina to Florida and Texas; West Indies.

Comments: These little snails are extremely abundant in turtle grass beds. They are carnivorous scavengers, readily attracted to carcasses.

---

445 California Cone
(*Conus californicus*)
Class Gastropoda

Description: 1⅝″ (41 mm) long, ¾″ (19 mm) wide. Cone-shaped. Tannish-gray, with dark brown periostracum. *Spire conical and slightly concave; body whorl long, conical, with fine spiral lines. Aperture long and slitlike,* outer lip thin, siphonal canal only a notch. Operculum small, horny, brown.

Habitat: On rocks and sand bottoms; from low-tide line to water 100′ (30 m) deep.

Range: C. California to Baja California.

Comments: This little cone is the only member of its genus on the Pacific Coast. It feeds on snails, bivalves, polychaetes, and even small fish, killing them with its harpoon and injected toxins, then swallowing the prey whole.

---

430 Alphabet Cone
(*Conus spurius*)
Class Gastropoda

Description: 3″ (76 mm) long, 1½″ (38 mm) wide. Cone-shaped, smooth. *Creamy-white, with spiral rows of squarish orange-brown spots* or with spots merging into mottlings. 9–10 whorls, first 4–5 forming a low spire, others flat, with prominent shoulder. Aperture long, narrow. Siphonal canal only a notch; outer lip thin. Operculum horny, small.

Habitat: Frequently near shore; in shallow water around reefs and reef flats.
Range: Florida; Gulf of Mexico; West Indies.
Comments: **Mildly toxic.** Cones are all predatory and are equipped both with poison glands and with a radula that has detachable, dartlike teeth. A live cone held in the hand might harpoon the holder—a danger to be considered in collecting them!

431 **Mouse Cone**
(*Conus mus*)
Class Gastropoda

Description: 1½" (38 mm) long, ¾" (19 mm) wide. Cone-shaped. Mottled yellowish-gray to reddish-brown, usually with pale band in middle of body whorl, and white spots on shoulder. Periostracum thick, velvety, yellowish, or greenish-brown. *Spire elevated, conical; shoulder with low knobs;* 6–7 whorls; large, conical body whorl. *Aperture long, slitlike;* outer lip thin; siphonal canal only a notch. Operculum small, horny, brown.
Habitat: Around coral reefs and on reef flats; near low-tide line and below in shallow water.
Range: S. Florida; Bahamas; West Indies.
Comments: The Mouse Cone is one of the most common shallow-water cones in Florida. Its name probably refers both to its size and its color.

432 **Stearns' Cone**
(*Conus stearnsi*)
Class Gastropoda

Description: 1" (25 mm) long, ⅜" (10 mm) wide. Cone-shaped. Gray, with spiral rows of brown and white dots, sometimes mottled reddish-brown. *Spire high, apex sharp, conical,* 7–8 whorls with concave tops; *body whorl long, with spiral lines at*

	base. Aperture long, slitlike; outer lip thin; siphonal canal only a notch. Operculum small, horny, brown.
Habitat:	On sand bottoms; from low-tide line to water 30′ (9 m) deep.
Range:	North Carolina to Florida and Texas; Bahamas; West Indies; Yucatán.
Comments:	This little shell is sometimes collected by sifting sand through a ¼″ mesh screen sieve. It is one of the smallest American cones.

397 Concave Auger
(*Terebra concava*)
Class Gastropoda

Description:	1″ (25 mm) long, ¼″ (6 mm) wide. Long, tapering to sharp point. Yellowish-gray. *Spire long, ⅘ total length, 12 slightly concave whorls,* sutures shallow; *beaded cord below sutures* with 20 beads per whorl, 4–5 fine lines on surface of whorl. Aperture oval, small; outer lip thin; siphonal canal short.
Habitat:	On sand bottoms; near low-tide line and below in shallow water.
Range:	North Carolina to both coasts of Florida.
Comments:	These slender little snails burrow in the sand where they prey on polychaetes, and perhaps other invertebrates, which they immobilize with a mild toxin on the radula.

398 Common Atlantic Auger
(*Terebra dislocata*)
Class Gastropoda

Description:	2″ (51 mm) long, ⅜″ (10 mm) wide. Long, tapering to a sharp point. Pinkish-gray to pale orange. 15 whorls, with fine spiral lines crossing *25 longitudinal ribs per whorl; beaded spiral band below suture.* Aperture small, square to oval. Outer lip thin; inner lip

glossy. Columella has 2 fused spiral folds near base. Short siphonal canal, bent upward. Operculum thin, horny, yellow-brown.

Habitat: On sand bottoms; near low-tide line and below in shallow water.

Range: Virginia to Florida and Texas; West Indies.

Comments: Augers are carnivorous, but this species has neither poison glands nor radula, and we do not know precisely what it eats. Because it is found among burrows of acorn worms, and because a related Pacific species is known to feed on acorn worms, these are probably its prey.

---

404 **Oyster Turret**
(*Crassispira ostrearum*)
Class Gastropoda

Description: 1″ (25 mm) long, ¼″ (6 mm) wide. Spindle-shaped, solid. Blackish-, reddish-, or yellow-brown. Spire tall, apex sharp. 7–8 whorls; *a smooth cord below suture line;* about 20 low, beaded longitudinal ribs per whorl crossed by 16–20 spiral threads. Aperture long, oval, with U-shaped notch at suture line; outer lip thin, wavy; siphonal canal short. Operculum horny, dark.

Habitat: From low-tide line to water 500′ (152 m) deep.

Range: North Carolina to Florida; Cuba.

Comments: The Oyster Turret is one of the most common Atlantic turrets in the United States, and is a predator on polychaete worms and small mollusks.

---

443 **Common West Indian Bubble**
(*Bulla occidentalis*)
Class Gastropoda

Description: 1″ (25 mm) long, ⅝″ (16 mm) wide. Oval, rotund, thin-shelled. Pale reddish-gray, with purplish-brown spots or vague bands. *Apex sunken, body whorl large, bulgy;* surface smooth, shiny. *Aperture longer than shell axis,* wide at front end, rounded at both ends; columella white.

Habitat: Grass or sand mudflats; near low-tide line and below in shallow water.

Range: North Carolina to Florida; Bahamas; West Indies.

Comments: This snail is reported to eat small mollusks whole, crushing them with the hard plates in its stomach lining.

---

464 California Bubble
(*Bulla gouldiana*)
Class Gastropoda

Description: 2¼″ (57 mm) long, 1½″ (38 mm) wide. Oval, rotund, *thin, smooth. Apex sunken,* reddish- or grayish-brown, mottled with dark and light V-shaped markings. Aperture wide at front end, narrow at rear. Head, foot, and mantle orange to yellowish-brown, with pale speckles.

Habitat: In mud and sandy-mud bottoms; just below low-tide line.

Range: S. California to Baja California.

Comments: This opisthobranch snail is the largest of the California bubble snails. It lives in quiet estuaries, buries itself in mud when the tide is out, and is most active at night.

---

463 White Paper Bubble
(*Haminoea vesicula*)
Class Gastropoda

Description: ⅝″ (16 mm) long, ⅜″ (10 mm) wide. Oval, fragile. Whitish or greenish-yellow. Thin, yellow-orange, or rust periostracum. Large, *barrel-shaped body whorl, spire sunken.* Aperture long, wide at front end; *outer lip flared.*

Habitat: On mudflats, sandy-mud bottoms, and marina floats; in bays; near low-tide line and below in shallow water.

Range: Alaska to Baja California.

Comments: This snail, whose body is too large to fit inside its shell, burrows in sand or mud just below the surface. It is eaten by the predatory Navanax, a large gastropod. The Solitary Paper Bubble (*H. solitaria*), which ranges from Cape Cod to North Carolina, grows to be ½" (13 mm) long and ¼" (6 mm) wide. It is smooth and lustrous, and ranges in color from bluish-white to yellowish-brown. It lives in a similar habitat.

---

211 **Warty Sea Cat**
(*Dolabrifera dolabrifera*)
Class Gastropoda

Description: 6" (15 cm) long, 2" (51 mm) wide. Oval, soft, rough. Gray to gray-green; sole of foot light green with pale dots. Head with *2 pairs of tentacles,* pair of black eyes between tentacle pairs. *Rear half of body wide,* rounded, plump; *surface rough, warty.*

Habitat: On algal growth; near low-tide line in shallow water.

Range: Florida; Bahamas; West Indies to Brazil.

Comments: Also called the Green Sea Hare, this species is well-camouflaged among the algae on which it feeds.

---

210 **Spotted Sea Hare**
(*Aplysia dactylomela*)
Class Gastropoda

Description: 5" (127 mm) long, 2½" (64 mm) wide. Plump, with long foot. Yellow or yellowish-green, with irregular violet-black circles. Head with *1 pair of antennae below, near mouth, 1 larger pair above, farther back,* with eyes in front of base. *Pair of long, winglike flaps* along upper side of body. Shell small, internal. Foot extends from head to beyond body mass.

Habitat: In turtle grass beds and protected sand-flats, and on reef flats; below low-tide line in shallow water.

Range: Florida and Texas; West Indies.

Comments: Like other sea hares this species is hermaphroditic and lays long, sticky strings of a million or more eggs, which are entangled in seaweeds.

---

209 California Sea Hare
(*Aplysia californica*)
Class Gastropoda

Description: 16″ (41 cm) long, 8″ (20 cm) wide, 8″ (20 cm) high. Plump, soft, with *long, winglike flaps near the top on either side; shell small, internal.* Reddish, brownish, or greenish, with mottled white and dark spots and lines; young usually reddish. Head with 1 pair of antennae low, near mouth, 1 larger pair above, farther back, eyes in front of them. Foot extends length of animal from head to beyond body mass.

Habitat: In sheltered places with few waves; from low-tide line to water 60′ (18 m) deep.

Range: N. California to Baja California.

Comments: Sea hares are so called because of the fancied resemblance of the second pair of antennae to a hare's long ears, and the similarity of the animal's general shape to that of the crouched hare. Like their namesakes, sea hares are herbivorous, eating a variety of red, green, and brown algae—from whose pigment they derive their color—and eelgrass. The California Black Sea Hare (*A. vaccaria*) is perhaps the world's largest gastropod. It attains a length of 30″ (76 cm), a width of 15″ (38 cm), and weight of 35 lbs (16 kg). This animal is reddish-brown to black, with white speckles, and ranges from southern California to Baja California.

148 **Ragged Sea Hare**
(*Bursatella leachi*)
Class Gastropoda

Description: 4″ (102 mm) long, 1½″ (38 mm) wide, 1½″ (38 mm) high. Thickest in rear ⅓, soft. Gray-green to olive, with white speckles. *Many ragged, gray to tan filaments on surface;* 2 pairs of ragged antennae difficult to distinguish from filaments. No shell in adult.

Habitat: In shallow water where algae are abundant.

Range: S. and w. Florida and Texas; West Indies.

Comments: These creatures are not easy to see because their color and ragged appearance camouflage them against the background of plant growth on which they feed.

234 **Navanax**
(*Chelidonura inermis*)
Class Gastropoda

Description: 8″ (20 cm) long, 2″ (51 mm) wide, 2″ (51 mm) high. Torpedo- or cigar-shaped. Velvety-brown or black, with violet sheen and numerous yellowish or white speckles sometimes arranged in streaks; margin edged in bright blue. *Head rounded at sides,* with *1 pair of stumpy antennae;* long flaps fold up over the back; *rear end of foot with deep notch.*

Habitat: In protected bays and on mudflats; from low-tide line to water 25′ (8 m) deep.

Range: C. California to Baja California.

Comments: An active predator that follows the mucus trails of its prey, this animal is carnivorous, feeding mostly on bubble snails, but also on other opisthobranchs such as nudibranchs and sea hares.

212 **Common Lettuce Slug**
(*Tridachia crispata*)
Class Gastropoda

Description: 1½" (38 mm) long, ¼" (6 mm) wide. Long. Greenish or bluish-green, with white speckles. Head wide, rounded at sides, with 1 pair of knobby antennae. *2 curly folds of mantle run length of back.*
Habitat: In quiet places such as grass beds; near low-tide line and below in shallow water.
Range: S. Florida and the West Indies.
Comments: The Common Lettuce Slug is herbivorous, and is equipped with a specialized radula that permits it to puncture cells of algae and suck out the juice.

222 **Hairy Doris**
(*Acanthodoris pilosa*)
Class Gastropoda

Description: 1¼" (32 mm) long, ⅝" (16 mm) wide. Oval, convex. Pale lemon-yellow or purplish-brown. *Covered with soft, slender, conical projections.* Antennae comblike, nearly equal in size, bent backward. *Ring of 7–9 plumelike gills* on back at rear end.
Habitat: On seaweeds, with a heavy growth of bryozoans; in shallow water, near low-tide line and below.
Range: Arctic to Connecticut; Alaska to Puget Sound.
Comments: The Hairy Doris feeds on the rubbery Porcupine Bryozoan, which is usually found growing on the base of rockweed and knotted wrack. It lays eggs in the form of white ribbons which attach onto the seaweed or the rock to which it is fastened.

228 Sea Lemon
(*Anisodoris nobilis*)
Class Gastropoda

Description: 10″ (25 cm) long, 3″ (76 mm) wide. Elongately oval, *covered on top with short, rounded projections.* Orange to pale yellow, whitish at tips of projections, *black spots between but not on projections.* Antennae comblike, short. Ring of 6 *white-edged, frilly gills* on back near rear end.

Habitat: From low-tide line to water 750′ (229 m) deep.

Range: British Columbia to Baja California.

Comments: This nudibranch is one of the largest on the Pacific Coast. It is known to feed on several different kinds of sponges and on dead organic matter. The Sea Lemon emits a strong, fruity odor which seems to discourage predators.

---

205 Atlantic Ancula
(*Ancula gibbosa*)
Class Gastropoda

Description: ½″ (13 mm) long, ⅛″ (3 mm) wide. Plump, elongated and tapered toward rear. Transparent, whitish. *Tentacles with 2 extra projections;* ring of gills on middle of back surrounded by ring of clublike projections; antennae and all projections lemon-yellow at tips.

Habitat: On various large seaweeds; near low-tide line and below.

Range: Arctic to Massachusetts.

Comments: The Atlantic Ancula's yellowish internal organs can be seen through the transparent whitish body wall. When out of the water, the soft body collapses, and some say it resembles a lightly fried egg. The Pacific Ancula (*A. pacifica*), which ranges from British Columbia to southern California, is about the same in size and appearance, but has 3 yellow stripes running the length of the body.

224 **White Knight Doris**
(*Archidoris odhneri*)
Class Gastropoda

Description: 8″ (20 cm) long, 3″ (76 mm) wide. Oval, convex. Pure white, rarely pale yellow. Covered with *low, rounded or conical projections.* Comblike antennae large. *Ring of 7 white, feathery gills* on back near rear end.

Habitat: On rocks; from low-tide line to water 75′ (23 m) deep.

Range: Alaska to s. California.

Comments: This is perhaps the largest nudibranch in American waters. Although it is commonly 2–3″ (51–76 mm) long, scuba divers unofficially report specimens 12″ (30 cm) or more in length. This mollusk feeds on several species of sponges.

231 **Monterey Doris**
(*Archidoris montereyensis*)
Class Gastropoda

Description: 2″ (51 mm) long, ¾″ (19 mm) wide. Oval. *Covered with cone-shaped projections.* Bright yellow to yellow-orange, with *dark spots both on and between projections.* Comblike antennae short and thick; ring of 7 yellowish, feathery gills on back near rear end.

Habitat: On rocky shores; from low-tide line to water 160′ (49 m) deep.

Range: Alaska to s. California.

Comments: The Monterey Doris feeds chiefly on the Crumb of Bread Sponge. It lays its eggs in a long, yellow ribbon which may contain 2 million eggs, and which is attached in a spiral to a rock.

225 **Yellow-edged Cadlina**
(*Cadlina luteomarginata*)
Class Gastropoda

Description: 3¼″ (83 mm) long, 1⅜″ (35 mm) wide. Oval, convex. White or pale yellow, bright yellow band around margin. *Comblike antennae; ring of* 6 *white, feathery gills* near rear end; body covered with low, rounded, *yellow-tipped projections.*

Habitat: Under rocks, in tidepools; from low-tide line to water 65′ (20 m) deep.

Range: British Columbia to Baja California.

Comments: This nudibranch's back feels gritty to the touch because supporting sharp spicules project through its surface. It feeds on several species of sponges.

223 **White Atlantic Cadlina**
(*Cadlina laevis*)
Class Gastropoda

Description: 1″ (25 mm) long, ⅜″ (10 mm) wide. Oval. Slightly convex. *White, semitransparent, with opaque white specks and larger, lemon-yellow spots* near the margin of the back. Comblike antennae and ring of feathery gills, usually yellow at tips.

Habitat: On rock bottoms; near low-tide line and below.

Range: Arctic to Massachusetts.

Comments: Although the populations of slime sponges on which it feeds are abundant on rocky New England shores, the White Atlantic Cadlina is found only locally, in widely separated populations.

221 **Ringed Doris**
(*Diaulula sandiegensis*)
Class Gastropoda

Description: 3⅝″ (92 mm) long, 1½″ (38 mm) wide. Oval. *Creamy-white to gray or light brown, with a few black circles or spots* of varying size on the back. Comblike antennae, ring of 6–7 gills on back near rear end. Velvety in appearance, gritty to the touch.

Habitat: On rocky shores; from low-tide line to water 110′ (34 m) deep.

Range: Alaska to Baja California.

Comments: The Ringed Doris lays a white spiral ribbon of eggs, attaching it to an overhanging rock ledge. Such a ribbon may contain 16 million eggs. Specimens of this animal in the northern part of its range generally have more rings and spots than those farther south.

---

227 **Salted Doris**
(*Doriopsilla albopunctata*)
Class Gastropoda

Description: 2¾″ (70 mm) long, 1″ (25 mm) wide. Oval, somewhat flattened; back covered with low projections. Bright yellow to reddish-brown, *small white dot on each projection.* Comblike antennae usually reddish; ring of 5 whitish, feathery gills on back near rear end.

Habitat: On rocky shores; from low-tide line to water 150′ (46 m) deep.

Range: N. California to Baja California.

Comments: This nudibranch has no radula, but is able to feed on sponges by secreting a fluid from its mouth to soften the surface of the sponge, and then sucking up the soft matter. The white dots on the animal's back look like a sprinkling of salt, hence the common name.

199 **Hopkins' Rose**
(*Hopkinsia rosacea*)
Class Gastropoda

Description: 1¼″ (32 mm) long, ½″ (13 mm) wide. Oval. Deep rosy-pink. *Covered by long, fingerlike projections,* with comblike antennae and ring of feathery gills on back near rear end.
Habitat: On rocky shores; near low-tide line and just below.
Range: Oregon to Baja California.
Comments: The long, fingerlike projections on the back of this beautifully colored nudibranch almost conceal the antennae and gills. Hopkins' Rose feeds on a pink encrusting bryozoan, and lays a pink spiral ribbon of eggs.

---

233 **Blue-and-gold Nudibranch**
(*Hypselodoris californiensis*)
Class Gastropoda

Description: 2⅝″ (67 mm) long, ⅝″ (16 mm) wide. Long, oval; *royal blue,* with paler blue edge of mantle. *4 longitudinal rows of bright yellow spots. Rear end of foot not covered by mantle.* Comblike antennae darker blue, as is ring of gills on back near rear end.
Habitat: On rocky shores; from low-tide line to water 95′ (29 m) deep.
Range: C. California to Baja California.
Comments: Like many nudibranchs, this handsome species secretes a substance which protects it from predators. It feeds on several kinds of sponges.

229, 230 **Rough-mantled Doris**
(*Onchidoris bilamellata*)
Class Gastropoda

Description: 1″ (25 mm) long, ¾″ (19 mm) wide. Mixed pattern of chocolate- to rusty-brown and cream-color. Broadly oval.

Back covered with many *short, thick, knobby projections.* Comblike antennae; *16–32 simple featherlike gills arranged in 2 half-rings* on back near rear end.

Habitat: On rocks and pilings near mud bottoms; from well above low-tide line to water 25′ (8 m) deep.

Range: Bay of Fundy to Rhode Island; Alaska to Baja California.

Comments: This nudibranch feeds on acorn barnacles. In New England it can frequently be found in large numbers, 20 or more per rock, under barnacle-covered boulders in quiet estuaries. Formerly known as *O. fusca.*

---

226 Crimson Doris
(*Rostanga pulchra*)
Class Gastropoda

Description: 1¼″ (32 mm) long, ¾″ (19 mm) wide. Broadly oval. *Yellow-red to scarlet,* with tiny dark spots sprinkled over the surface. Comblike antennae short, stout; ring of *6–9 short, feathery gills* on back near rear end. Velvety in appearance.

Habitat: On rocky shores with overhanging ledges or large boulders; from low-tide line to water 30′ (9 m) deep.

Range: British Columbia to Baja California.

Comments: This nudibranch feeds on several kinds of red sponges, whose pigments it incorporates into its own body. *Rostanga* matches the sponge in color, as does the spiral ribbon of eggs it lays on the sponge surface.

---

206 Sea Clown Nudibranch
(*Triopha catalinae*)
Class Gastropoda

Description: 6″ (15 cm) long, ¾″ (19 mm) wide. Almost cylindrical. White or grayish-tan, with *orange-red tips to all projections, antennae, and gills.* Head broad, with *8–12 short, branching projections directed forward; stumpy, branched projections scattered over back.* Comblike antennae and ring of 5 feathery gills on rear half of back.

Habitat: In tidepools and kelp beds; from above low-tide line to water 110′ (34 m) deep.

Range: Alaska to Baja California.

Comments: This white Sea Clown with orange spots is a spectacular sight. When stranded in a tidepool with small fish it is not attacked, perhaps because of some substance it secretes. It feeds on bryozoans. The species *catalinae* was formerly known as *T. carpenteri.* The Spotted Triopha (*T. maculata*), which ranges from northern California to Baja California, is similar in size and structure, but is red to reddish-brown or black, with numerous white to bluish spots and all appendages tipped bright red. The larger forms of this species were formerly known as *T. grandis.*

---

208 Bushy-backed Sea Slug
(*Dendronotus frondosus*)
Class Gastropoda

Description: 4⅝″ (117 mm) long, 1″ (25 mm) wide. Widest in middle, tapered to point at rear end. Grayish-brown to rusty-red mottled with white spots, or pure white. Head blunt, with 6 branched projections extending forward. Comblike antennae set in sheaths with whorl of branched projections; *2 rows of 5–8 bushy projections* along back.

Habitat: On rocks and among seaweeds; from low-tide line to water 360′ (110 m) deep.

Range: Arctic to New Jersey; Alaska to California.

Comments: The Bushy-backed Sea Slug is commonly found wherever there is an abundance of the hydroids on which it feeds. It is also known to browse on bryozoans and colonial tunicates.

203, 204 **Salmon-gilled Nudibranch**
(*Coryphella salmonacea*)
Class Gastropoda

Description: 1¾" (44 mm) long, ⅜" (10 mm) wide. Long, narrow; rear end tapered to sharp point. Translucent white, with opaque white stripe down middle of back. *100 long, fingerlike projections with salmon-colored cores rising in clumps* from each side of back. Head with 2 pairs of long antennae. Front end of foot with *sharp, hooklike projection* bent backward on each side.
Habitat: On rocks, among algal growth; near low-tide line and below to deep water.
Range: Greenland to Massachusetts Bay.
Comments: This cold-water species is found in late winter and spring in the southern part of its range, and into late summer in the northern part. It feeds on several kinds of hydroids.

202 **Red-gilled Nudibranch**
(*Coryphella rufibranchialis*)
Class Gastropoda

Description: 1¼" (32 mm) long, ⅜" (10 mm) wide. Long, narrow; rear end tapered to sharp point. Translucent white, with opaque white line down middle of back. *100 long, fingerlike projections* with bright red, occasionally brown, core; *projections rise in clumps* from each side of back, *opaque white ring near tip.* Head with 2 pairs of long antennae. Front end of foot with *sharp, hooklike extension* on each side.
Habitat: Among seaweeds and hydroids; near low-tide line and below to deep water.
Range: Arctic to New York; British Columbia.

Comments: Although this animal is called the Red-gilled nudibranch, the projections on its back are not gills. Each contains a slender extension of the gut. When the nudibranch feeds on certain kinds of hydroids, it can incorporate the stinging cells from the digested prey into the tissues of these projections so that they can explode on contact just as they would in the hydroid's tentacles.

Maned Nudibranch
(*Aeolidia papillosa*)
Class Gastropoda

Description: 4″ (102 mm) long, 1½″ (38 mm) wide. Thick and stubby. Whitish, gray, or tawny-brown, with pale speckles. Back covered with *hundreds of slender, fingerlike projections; bare area down midline.* 2 pairs of antennae on squarish head; rear end tapered to blunt point.

Habitat: On rocks, pilings, and mudflats; from low-tide line to water 2200′ (671 m) deep.

Range: Arctic to Maryland; Alaska to s. California.

Comments: This sturdy nudibranch feeds chiefly on sea anemones, with a preference for the Frilled Anemone. The animal tends to take on the color of the anemones on which it preys.

---

200 **Elegant Eolid**
(*Flabellinopsis iodinea*)
Class Gastropoda

Description: 3⅝″ (92 mm) long, ½″ (13 mm) wide. Higher than wide. *Vivid purple. 2 pairs of antennae, the second pair bright red;* back covered with *numerous orange, fingerlike projections.*

Habitat: On pilings and rocks, and in kelp beds; from low-tide line to water 110′ (34 m) deep.

Range: British Columbia to Baja California.

Comments: This gaudily colored nudibranch has flattened sides, and can drop from a surface and swim with a series of quick, alternate, U-shaped bends of its body.

201 **Hermissenda Nudibranch**
(*Hermissenda crassicornis*)
Class Gastropoda

Description: 3¼″ (83 mm) long, ⅜″ (10 mm) wide. Broadest just behind head, tapering to a fine point at the rear. Bluish-white, with orange line down middle of back; *margins with pale electric-blue lines;* first pair of tentacles with blue lines, second pair bluish with raised rings. Numerous fingerlike projections, swollen in middle, with *orange stripe just below white tip of each projection,* in 2 clusters on each side of back.

Habitat: In tidepools, and on rocks, pilings, and mudflats; from low-tide line to water 110′ (34 m) deep.

Range: Alaska to Baja California.

Comments: This is one of the most abundant nudibranchs on the Pacific Coast. It feeds on hydroids and other invertebrates, and is even cannibalistic; when two individuals meet they frequently fight, biting chunks of tissue from each other.

220 **Lion Nudibranch**
(*Melibe leonina*)
Class Gastropoda

Description: 4″ (102 mm) long, 1″ (25 mm) wide. Long, thick, big-headed. *Large, rounded hood more than 2″ (51 mm) wide around the mouth, edged with 2 rows of long, slender tentacles.* Pale yellowish or greenish-brown. Antennae on hood rounded, leaflike; 5–6 pairs of rounded, leaflike projections on back.

Habitat: In eelgrass beds; near low-tide line and below, and in kelp beds in deeper water.
Range: Alaska to Baja California.
Comments: This remarkable nudibranch uses its huge hood to trap small crustaceans that either swim or creep into its reach. It will also eat small mollusks and larvae of other invertebrates by raising and expanding its hood, and then bringing it down abruptly to scoop up any hapless creature that has come too close.

446, 448 **Salt-marsh Snail**
(*Melampus bidentatus*)
Class Gastropoda

Description: ½″ (13 mm) long, ⅜″ (10 mm) wide. Shaped like a top: oval, thin, smooth. Greenish-olive to brown, sometimes banded. *5 whorls; low, cone-shaped spire;* sutures shallow. Aperture long, narrow, widest at head end. Inner lip white, with 2 folds across lower part; head and foot dark gray.
Habitat: In salt marshes and backwaters; well above low-tide line in the debris among marsh grasses.
Range: Gulf of St. Lawrence to Florida and Texas.
Comments: The Salt-marsh Snail is one of the few air-breathing snails in the marine habitat. The Olive Ear Snail (*M. olivaceus*) ranges from southern California to Baja California. It is similar in color but a little larger, ¾″ (19 mm) in length and ½″ (13 mm) in width. It lives near the high-tide line among the pickleweed at the edge of mudflats.

329 **Atlantic Nut Clam**
(*Nucula proxima*)
Class Bivalvia

Description: ⅜″ (10 mm) long, ¼″ (6 mm) high. Obliquely oval, plump. Whitish, with thin, olive-green periostracum. Surface smooth, with faint radiating lines; *umbo prominent, bent forward; lower margin with fine teeth;* hinge line bent, foot square at end.

Habitat: In mud and sandy-mud bottoms; below low-tide line in shallow water.

Range: Gulf of St. Lawrence to Florida and Texas.

Comments: Unlike most bivalves, this clam feeds by probing into the surrounding mud with the tentacles which branch from its palps. It, in turn, is eaten by many bottom-feeding fish and by diving ducks.

301 **Thin Nut Clam**
(*Nuculana tenuisulcata*)
Class Bivalvia

Description: ¾″ (19 mm) long, ⅜″ (10 mm) high. Thin, flattened; *rear end long, tapered, curved slightly upward.* Pale yellowish-brown. Surface with *closely spaced concentric grooves;* hinge line bent; siphons small, short, close together; foot square at end.

Habitat: In mud bottoms; from just below low-tide line to water more than 800′ (244 m) deep.

Range: Arctic to Rhode Island.

Comments: Like other nut clams, this species feeds by transferring organic particles adhering to its mucus-coated tentacles into its mouth.

300 **File Yoldia**
(*Yoldia limatula*)
Class Bivalvia

Description: 2½″ (64 mm) long, 1¼″ (32 mm) high. Thin, rounded at front end, *sloping and narrowing toward rear;* smooth, flattened. *Greenish-tan to chestnut-brown.* Umbones small, halfway between ends; siphons small, close together; foot square at end.
Habitat: In mud bottoms; from below low-tide line to water 70′ (21 m) deep.
Range: Gulf of St. Lawrence to New Jersey; Alaska to s. California.
Comments: The File Yoldia is an active clam, capable of leaping by sudden thrusts of its foot when disturbed.

Veiled Clam
(*Solemya velum*)
Class Bivalvia

Description: 1½″ (38 mm) long, ¾″ (19 mm) high. Oblong, with rounded ends; fragile, smooth. *Brownish periostracum with scalloped edges overhanging shell.* 15 yellowish lines radiating from umbo; foot whitish, sole flattened with saw-toothed edges.
Habitat: In sandy-mud bottoms; above low-tide line and below in shallow water.
Range: Nova Scotia to Florida.
Comments: This active little clam can swim by squirting sudden jets of water out of its siphons.

Turkey Wing
(*Arca zebra*)
Class Bivalvia

Description: 4″ (102 mm) long, 2″ (51 mm) high. Oblong, stout, with *gape at both ends.* Yellowish or tannish, with *reddish-brown zebra stripes across 35 strong*

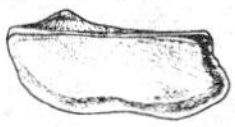

radiating ribs. Periostracum brown, shaggy; mantle with numerous eyes. Umbones well toward front end, far apart; hinge line straight.

Habitat: Attached to rocks and shells; near low-tide line and below in shallow water.

Range: North Carolina to Florida and Texas; Bahamas; West Indies to Brazil.

Comments: The Turkey Wing is one of those mollusks with gapes between the valves for the siphons and for the byssal threads that moor it. Its colorful shells are abundantly used in the Florida shell industry.

---

323 White-bearded Ark
(*Barbatia candida*)
Class Bivalvia

Description: 3″ (76 mm) long, 1¾″ (44 mm) high. Obliquely rectangular, rounded at front end; slanting to blunt rear end. Lower margin shallowly indented, white, with *soft, beardlike, yellow-brown periostracum.* Fine ribs crossed by growth lines, producing *beaded surface.* Gape for foot.

Habitat: Attached to rocks; below low-tide line in shallow water.

Range: North Carolina to Florida and Texas; Bahamas; West Indies to Brazil.

Comments: This ark attaches itself to rocks with its tuft of byssal threads.

---

369 Blood Ark
(*Anadara ovalis*)
Class Bivalvia

Description: 2¼″ (57 mm) long, 2″ (51 mm) high. Thick, solid, *roundish oval.* White, with hairy, brownish periostracum covering lower half; *soft parts blood-red. 26–35 strong ribs.* Umbones prominent, close together. Lower edge fluted by ribs; no gape.

Habitat: In sand bottoms; below low-tide line

in shallow water.

Range: Cape Cod to Florida and Texas; Bahamas; West Indies.

Comments: The red, oxygen-transporting pigment hemoglobin, found in relatively few mollusks, occurs in this ark's blood and tissues, hence their color.

---

367 **Ponderous Ark**
(*Noetia ponderosa*)
Class Bivalvia

Description: 2½″ (64 mm) long, 2¼″ (57 mm) high. Heavy, convex, almost triangular, with oblique slope toward rear end; *heart-shaped from end view.* Creamy-white, covered by dark brown, furry periostracum. *Prominent umbones curved downward toward hinge line. 30 flattened, squarish ribs* crossed by fine concentric lines on lower part.

Habitat: In sand bottoms; at low-tide line and below in shallow water.

Range: Virginia to Florida and Texas.

Comments: Shells of this species are sometimes washed up on beaches as far north as Cape Cod, Nantucket, and Long Island, but these shells are thought to be fossils; *N. ponderosa* is common in Miocene and more recent fossil beds far north of its present range.

---

365 **Comb Bittersweet**
(*Glycymeris pectinata*)
Class Bivalvia

Description: 1¼″ (32 mm) long, 1¼″ (32 mm) high. *Almost circular, shallowly convex.* Whitish or pale yellow, with yellowish-brown spots or bands. Mantle edged with small eyes. *20 or more raised, rounded, radiating ribs* crossed by fine growth lines. Umbones curved toward hinge line. Foot crescent-shaped.

Habitat: In sand and gravel bottoms; below low-tide line in shallow water.

Range: North Carolina to Florida; Bahamas; West Indies.

Comments: The pretty shells of this species lose their color readily if pounded by surf and tossed up on the beach.

294 **Horse Mussel**
(*Modiolus modiolus*)
Class Bivalvia

Description: 6″ (15 cm) long, 3½″ (89 mm) high. Heavy, oblong-oval; front end short and narrow, rear end wider, rounded. Pinkish-white, covered by *dark brown periostracum, hairy at lower margin.* Soft parts orange. *Umbones near but not at front end.* Slitlike siphons dark, fringed.

Habitat: In crevices among rocks; from low-tide line to water 500′ (152 m) deep.

Range: Arctic to n. Florida; Alaska to California.

Comments: The shells of older horse mussels are usually encrusted with coralline algae, and sometimes with holdfasts of kelp. After storms in New England, it is common to find large kelps washed ashore, their holdfasts still gripping horse mussels pulled free from their moorings. The Fat Horse Mussel (*M. capax*) ranges from central California to Peru. It is 4″ (102 mm) long and 1¾″ (44 mm) high. Its shell is broad and orange-brown under a coarse, hairy, brown periostracum, and its soft tissues are orange like those of *M. modiolus.* It is found attached to rocks and pilings in protected bays, from above low-tide line to water 160′ (49 m) deep.
The Straight Horse Mussel (*M. rectus*), a more slender form, reaches a length of 9⅜″ (24 cm) and a height of 3¼″ (83 mm). Its shell is bluish, covered with a glossy brown periostracum, and beaded at the rear end. The beak is set well back from the front end. Its soft tissues are yellowish-orange. It lies in a sand,

gravel, or mud bottom with only the rear edge of its shell projecting above the surface, its byssal threads attached to a buried rock. It ranges from British Columbia to Baja California.

295 **Ribbed Mussel**
(*Ischadium demissum*)
Class Bivalvia

Description: 4″ (102 mm) long, 1¾″ (44 mm) high. Oblong-oval, narrow and round at front end, long and wide toward rear; lower margin indented. Yellowish-brown to brownish-black, glossy. Soft tissues lemon-yellow. *Many strong radiating ribs. Umbones near but not at front end;* gape for byssal tuft, 2 slitlike siphons with frilled edges.

Habitat: In salt marshes and brackish, muddy estuaries; between high- and low-tide line.

Range: Gulf of St. Lawrence to Florida and Texas; introduced into San Francisco Bay, Newport Bay, and Los Angeles Harbor.

Comments: The Ribbed Mussel is a particularly rugged species, able to tolerate water temperatures up to 133°F (56.1°C) and salinities twice that of normal sea water.

293 **Blue Mussel**
(*Mytilus edulis*)
Class Bivalvia

Description: 4″ (102 mm) long, 2″ (51 mm) high. *Long, rounded triangle. Blue-black to black,* with shiny periostracum. Edge of black, fringed mantle forming 2 short siphons. *Umbones form apex at front end;* numerous growth lines; gape for byssal tuft. Foot slender, brownish.

Habitat: Attached to rocks, pilings, and almost any solid object; between high- and low-tide lines.

Range: Arctic to South Carolina; Alaska to Baja California.

Comments: The Blue Mussel usually occurs in dense masses. This is an edible mussel, cultured and used more often as food by Europeans than by Americans, but becoming common in American fish markets and restaurants in recent years.

---

292 California Mussel
(*Mytilus californianus*)
Class Bivalvia

Description: 10″ (25 cm) long, 4″ (102 mm) high. Long, rounded triangle. Bluish-black. Umbones form apex at front end, rear end broad and rounded. Surface with *strong radial ribs and irregular growth lines;* frequently eroded. Gape for byssal tuft. Foot slender, orange.

Habitat: On rocks, wharf piles, and unprotected shores; from well above low-tide line to water 80′ (24 m) deep.

Range: Alaska to Baja California.

Comments: These mussels occur in massive growths, frequently intermixed with the Leaf Barnacle. This species, one of the most abundant animals on the West Coast, is eaten by sea stars, predatory snails, crabs, shorebirds, and sea otters, and is collected as food by humans.

---

357 Flat Tree Oyster
(*Isognomon alatus*)
Class Bivalvia

Description: 3″ (76 mm) long, 3″ (76 mm) wide. *Upper shell flat, fanlike; lower shell deeper.* Grayish outside; pearly inside. Outside *surface flaky,* with prominent growth lines. Hinge line with *8–12 prominent grooves;* gape for byssal tuft near hinge line at front margin.

Habitat: Attached to rocks, reefs, pilings, and

mangrove roots; between high- and low-tide lines and below to water 5′ (15 m) deep.

Range: S. Florida and Texas; Bahamas; West Indies to Central America.

Comments: This oyster is commonly found in large, compact clumps, especially on mangrove roots.

---

346 **Atlantic Pearl Oyster**
(*Pinctada radiata*)
Class Bivalvia

Description: 3″ (76 mm) long, 3″ (76 mm) wide. *Roundish,* with *2 short wings* at ends of *straight hinge line;* valves flattish, equal in size. Outside tan or greenish-brown, with mottlings or rays of purplish-brown or black; inside pearly. Periostracum brown. Surface with irregularly concentric rows of scaly spines. Gape for byssal tuft at front end.

Habitat: Attached to rocks and stems of sea fans; below low-tide line in shallow water, and on reefs in water to 65′ (20 m) deep.

Range: S. Fiorida to Texas; Bahamas; West Indies.

Comments: The shell of this species has a beautiful mother-of-pearl lining and the oyster itself is a source of pearls.

298, 354 **Stiff Pen Shell**
(*Atrina rigida*)
Class Bivalvia

Description: 12″ (30 cm) long, 6½″ (17 cm) wide. Thick, *triangular, or wedge-shaped;* upper margin straight, lower margin rounded, gape at rear end. Light to dark brown, or purplish-brown. Ridge in middle of valves; *15 or more radiating ribs,* fading toward umbones, decorated with *high tubular spines.* Gape for byssal tuft on lower margin near umbo.

Habitat: In sand bottoms; at low-tide line and below in shallow water.
Range: North Carolina to Florida; Bahamas; West Indies.
Comments: The Stiff Pen Shell produces handsome black pearls. Pen shells lie partially buried in the bottom, with only the rear edge of the shell above the surface, posing a danger to the barefoot wader.

---

299 **Saw-toothed Pen Shell**
(*Atrina serrata*)
Class Bivalvia

Description: 10″ (25 cm) long, 4″ (102 mm) high. Fragile, *triangular, or wedge-shaped.* Translucent, grayish-tan, or greenish-brown, sometimes mottled. *30 radiating ribs set with sharp spines,* making *rear margin saw-toothed;* lower front surface smooth. Tuft of byssal threads at front end.
Habitat: In mud; near low-tide line and below in shallow water.
Range: North Carolina to Florida and Texas; Mexico; West Indies.
Comments: This species is harvested in Mexico, where its adductor muscles are canned and sold as scallops. It is very abundant in some regions, and is washed ashore in large numbers by storms.

---

361 **Kitten's Paw**
(*Plicatula gibbosa*)
Class Bivalvia

Description: 1″ (25 mm) long, 1″ (25 mm) wide. Solid, *fan-shaped.* White to grayish, with red-brown radiating lines. Valves with interlocking wavy lower margins, lower (right) valve cemented to rock. *5–7 high, wavy ribs.*
Habitat: Cemented to rocks; from above low-tide line to water 300′ (91 m) deep.
Range: North Carolina to Florida and Texas;

Bahamas; West Indies.

Comments: The shape of this little shell indeed resembles the paw of a kitten. When it is washed up on the beach, the colors quickly fade in the sunlight.

355 **Iceland Scallop**
(*Chlamys islandicus*)
Class Bivalvia

Description: 4″ (102 mm) long, 3⅞″ (98 mm) wide. Oval, upper valve slightly more convex than lower. Gray, cream-colored, yellow, orange, peach, red, pink, or purple, sometimes with concentric light and dark bands. Mantle cream-colored, with many tentacles and black eyes. Upper valve with about *50 radiating ribs,* sometimes in irregular groups, with small, erect scales. Wings at hinge line unequal, rear (left) one only half as long as front (right) one.

Habitat: On sand and gravel bottoms; in water from 6–1000′ (2–305 m) deep.

Range: Greenland to Massachusetts; Alaska to Puget Sound.

Comments: The upper valves of most scallops, including this one, are commonly encrusted with tube worms, bryozoans, coralline algae, and other growth, hiding the shell color.

353 **Atlantic Bay Scallop**
(*Argopecten irradians*)
Class Bivalvia

Description: 2⅞″ (73 mm) long, 3″ (76 mm) wide. Almost circular. Drab gray to yellowish-brown or reddish, often purple near the hinge. *30–40 bright blue eyes* on mantle margin. *17–18 low, rounded ribs; wings large, of equal size.*

Habitat: In eelgrass beds and sandy-mud bottoms; near low-tide line and below in shallow water.

Range: Cape Cod to Florida and Texas.

Comments: Unlike most bivalves, a scallop has well-developed eyes set in rows along the edge of its mantle. Each eye has a lens, retina, and optic nerve, and is quick to see even slight movement nearby. Some species of scallops are actually capable of short migrations, swimming by clapping their valves as though taking a bite out of the water, and propelled forward 3 feet at a time by the strong jets of water forced out as the valves snap shut. The scallops that range from Florida to Louisiana are considered a separate subspecies, as are those from Louisiana to Texas. The Calico Scallop (*A. gibbus*) ranges from North Carolina to Texas and the north coasts of Puerto Rico and Cuba. It grows to be 2″ (51 mm) wide, has about 20 rounded radiating ribs, and is quite plump. Its color is highly variable and mottled, combining white, red, purple, brown, orange, and yellow.

351 Giant Rock Scallop
(*Hinnites giganteus*)
Class Bivalvia

Description: Free-swimming juvenile: 1″ (25 mm) long, 1″ (25 mm) wide. Almost circular. Yellow to orange. Mantle with many blue eyes. *16–18 strong ribs which extend as spines* beyond shell margin; about *6 fine lines between ribs. Wings unequal,* rear (left) one larger. Attached adults: 10″ (25 cm) long, 10″ (25 cm) wide. Thick, irregular, taking shape of rock or shell to which it attaches. *Juvenile pattern still visible* near hinge line. Brownish; *mantle orange;* eyes blue.

Habitat: Among rocks, on pilings, and on floats; from low-tide line to water 150′ (46 m) deep.

Range: British Columbia to Baja California.

Comments: This species develops and behaves like other scallops and is free-swimming until its valves grow to about 1″ (25 mm). At that time it settles on a solid object and its mantle secretes a limy material, cementing down the lower (right) shell and making the animal sedentary for the rest of its life.

352 **Lion's Paw**
(*Nodipecten nodosus*)
Class Bivalvia

Description: 6″ (15 cm) long, 6″ (15 cm) wide. Strong and heavy, somewhat convex. Orange, bright red, or maroon. 7–9 *large, coarse ribs* with several concentric rows of *large, hollow projections;* many fine cords between ribs. Wings unequal, rear (left) one larger.
Habitat: On sand and rubble bottoms; below low-tide line in shallow water.
Range: Cape Hatteras to Florida and Texas; Bahamas; West Indies.
Comments: These large, handsome scallops are prized by shell collectors. The hollow bumps along the ribs are reminiscent of the knuckles on the toes of a lion.

356 **Atlantic Deep-sea Scallop**
(*Placopecten magellanicus*)
Class Bivalvia

Description: 8″ (20 cm) long, 8″ (20 cm) wide. Almost circular, slightly convex. Upper valve yellowish, purplish-gray, or whitish; lower valve white. Mantle white splotched with tan; *eyes steely gray.* Surface with *fine radiating lines and grooves. Wings about equal in size.*
Habitat: On sand and gravel bottoms; in water 12–400′ (4–122 m) deep.
Range: Labrador to Cape Hatteras.
Comments: This large scallop is the one commonly sold in fish markets. Only the large central muscle that closes the valves is used as food.

349 **Atlantic Thorny Oyster**
(*Spondylus americanus*)
Class Bivalvia

Description: 5½″ (140 mm) long, 5½″ (140 mm) wide. Almost circular, stout. Whitish, yellow, orange, or reddish-brown, sometimes mottled. Right (lower) valve larger, attached to a solid object; left (upper) valve smaller, with *many radiating ribs with scattered spines 2″ (51 mm) long.*

Habitat: On rocks, coral reefs, and wrecks; from low-tide line to water 150′ (46 m) deep.

Range: North Carolina to Florida and Texas; Bahamas; West Indies.

Comments: When specimens of this handsome species are found washed up on the beach, their spines have frequently been broken off by surf action.

350 **Rough File Shell**
(*Lima scabra*)
Class Bivalvia

Description: 1⅞″ (48 mm) long, 3¾″ (95 mm) high. Oval, moderately thick. Whitish, with thin, brown periostracum; mantle orange-red with *long tentacles the same color as mantle.* Upper end narrower, with small umbones projecting beyond hinge line; lower margin rounded. *Many radiating ribs with short, pointed, overlapping scales.*

Habitat: In crevices attached to rocks; from low-tide line to water 425′ (130 m) deep.

Range: South Carolina to Florida; Bahamas; West Indies.

Comments: File shells nestle in rocky crevices and attach themselves with byssal threads to the substrate. When disturbed, however, they can quickly detach from the byssus and swim erratically away by clapping movements of the valves. Hemphill's File Shell (*L. hemphilli*) ranges from central California to

Mexico. It grows to 1⅛″ (28 mm) high and 2″ (51 mm) wide, and has many fine, threadlike ribs crossed by concentric threads, but has a creamy-white mantle and tentacles. The marginal tentacles are sticky, and may be shed if the individual is attacked.

325 **Antillean File Shell**
(*Lima pellucida*)
Class Bivalvia

Description: 1½″ (38 mm) long, 2¼″ (57 mm) high. Oval, thin-shelled. *Translucent white.* Periostracum thin, tan, usually lost. Upper end narrow, with small umbones projecting beyond hinge line; valves somewhat oblique, lower margins rounded. *Many fine radiating ribs, unequal in size, making lower margin saw-toothed.*

Habitat: In crevices among rocks; near low-tide line and below in shallow water.

Range: North Carolina to Florida and Texas; Bahamas; West Indies.

Comments: This showy tropical file shell has a bright red mantle and long, white mantle tentacles which it extends when it is undisturbed.

345 **Common Jingle Shell**
(*Anomia simplex*)
Class Bivalvia

Description: 2¼″ (57 mm) long, 2¼″ (57 mm) wide. Irregularly circular. White to yellowish-orange, silvery, or brownish, with iridescent sheen. *Lower (right) valve fragile, with hole near top* for byssus attachment; upper (left) valve somewhat stouter, convex. *Smooth, or with irregular wrinkles.*

Habitat: Attached to rocks, shells, pilings, and other hard objects; from low-tide line to water 30′ (9 m) deep.

Range: Maine to Florida and Texas; Bahamas; West Indies.

Comments: When strung on a cord and suspended in the wind, these shells make a fine jingling sound. The Prickly Jingle Shell (*A. aculeata*) is smaller, ¾″ (19 mm) long and ¾″ (19 mm) wide, and whitish-tan, and its upper valve is covered with radiating rows of rough, prickly scales. It ranges from Maine to North Carolina.

---

344 **False Pacific Jingle Shell**
(*Pododesmus macrochisma*)
Class Bivalvia

Description: 4″ (102 mm) long, 4″ (102 mm) wide. Irregularly circular. Grayish-white tinged with green; soft parts bright orange. *Lower (right) valve with hole near top for byssus* attachment, moderately thin; *upper (left) valve translucent, with variable radiating ribs.*

Habitat: On rocks, shells, pilings, and floats, both in bays and harbors and on exposed rocky coasts; from low-tide line to water 200′ (61 m) deep.

Range: Alaska to Baja California.

Comments: The greenish color of this species is due partly to the animal's own pigment and partly to the 1-celled green algae living within the shell. The upper valve is commonly invaded by boring sponges and covered by bryozoans, tube-dwelling polychaete worms, and other sedentary creatures.

---

289 **Eastern Oyster**
(*Crassostrea virginica*)
Class Bivalvia

Description: 10″ (25 cm) long, 4″ (102 mm) wide. *Irregularly oval to elongate.* Grayish-white. *Upper (right) valve flattened,* smooth except for a few wrinkles or growth lines, sometimes with radial ridges; *lower (left) valve deeper, cemented to solid object.* Narrow at hinge, gradually widening. Umbo long and curved. Margins smooth, sharp.

Habitat: On soft and hard bottoms, in waters of reduced salinities; from low-tide line to water 40′ (12 m) deep.

Range: Gulf of St. Lawrence to Florida and Texas; Bahamas; West Indies. Introduced on the Pacific Coast.

Comments: This is the common edible oyster of the eastern seaboard, the basis of an extensive fishing industry. These oysters are prolific, each female routinely spawning 10–20 million eggs; very large oysters may spawn up to 100 million. An oyster may change its sex several times in successive seasons, but larger ones are generally functional females. A large oyster may spawn several times in one year.

---

290 **Giant Pacific Oyster**
"Japanese Oyster"
(*Crassostrea gigas*)
Class Bivalvia

Description: 12″ (30 cm) long, 5″ (127 mm) wide. Rough, irregular. Tan or grayish-white, sometimes with purple streaks. Shape varying from long and thin to round and deep. *Lower (left) valve deeper, cemented to solid object.* Surface rough, with *irregular, concentric, curved ridges.* Interior of valve white, with pale purple spot by muscle scar. Margin thin, irregular.

Habitat: Attached to rocks and shells, in bays and on mudflats; near low-tide line.

Range: British Columbia to s. California.

Comments: This species was introduced from Japan to the West Coast about the turn of the century. Populations have been established in Puget Sound and in several bays in northern California, where they are harvested for restaurants, or packaged and frozen. They reach marketable size in about 3 years, but can live about 20 years.

359 **Coon Oyster**
(*Dendrostrea frons*)
Class Bivalvia

Description: 2¾″ (70 mm) long, 2¼″ (57 mm) wide. Irregularly oval, moderately thin. Yellowish-white, rosy, or deep brown. *Upper (right) valve with coarse radiating folds; lower (left) valve often with several fingerlike, clasping spines* for attachment. Margin fluted.
Habitat: Attached to sea whips, corals, submerged brush, or wires; from low-tide line to water 15′ (5 m) deep.
Range: North Carolina to the West Indies.
Comments: The Coon Oyster has the interesting habit of producing clawlike spines near the hinge line for gripping a sea whip or other slender object. Not all coon oysters are so attached, however; they sometimes simply cling to each other in masses as big as a bushel basket. Raccoons like to feed on them—hence their common name.

291 **Native Pacific Oyster**
(*Ostrea lurida*)
Class Bivalvia

Description: 3½″ (89 mm) long, 3″ (76 mm) wide. Irregular, usually oval. Grayish-white, sometimes with brownish or purplish rays. *Surface rough, scaly,* sometimes with ridges that result in a scalloped margin. Upper (right) valve flattened; lower (left) valve deeper. *Margins usually saw-toothed.*
Habitat: Attached to rocks, pilings, and shells in quiet bays and estuaries; near low-tide line and below in shallow water.
Range: Alaska to Baja California.

Comments: *O. lurida* alternates yearly between being a male and being a female. This oyster is smaller than *Crassostrea gigas,* its chief competitor in West Coast markets, and is regarded by many as superior in taste.

---

340 **Wavy Astarte**
(*Astarte undata*)
Class Bivalvia

Description: 1¼″ (32 mm) long, 1″ (25 mm) high. Solid, roughly triangular. White, covered with thick, reddish-brown to dark brown periostracum. Valves equal, with about *15 strong concentric ridges and deep furrows,* strongest at middle of shell, decreasing at ends. *Umbones pointed, prominent, near center.*

Habitat: In sand, mud, or gravel bottoms; from low-tide line to water 625′ (191 m) deep.

Range: Labrador to Maryland.

Comments: This species is the most common Astarte in New England.

---

341 **Boreal Astarte**
(*Astarte borealis*)
Class Bivalvia

Description: 2″ (51 mm) long, 1⅝″ (41 mm) high. Solid, oval, compressed. White, covered with tough, *dark brown periostracum.* Valves equal, with regular concentric ridges near umbones, less prominent toward margins. *Umbones pointed near center, bent forward.*

Habitat: In sand, mud, or gravel bottoms; from below low-tide line to water 600′ (182 m) deep.

Range: Circumpolar: Arctic to Massachusetts Bay; Alaska.

Comments: This genus of northern clams is named after Astarte, the Phoenician goddess of fertility and sexual love. The Chestnut

Astarte (*A. castanea*) is about the same length, has more prominent umbones and many small concentric wrinkles, and is chestnut-brown. It ranges from Nova Scotia to New Jersey.

324 Broad-ribbed Cardita
(*Carditamera floridana*)
Class Bivalvia

Description: 1½" (38 mm) long, 1" (25 mm) high. Solid, thick, oval-oblong. Yellowish or grayish-white, with reddish-brown spots arranged concentrically on the ribs. Valves equal, with about *20 strong, rounded ribs* beaded with concentric growth lines. *Umbones large,* about ¼ way back from front end.
Habitat: In sand or mud bottoms; from below low-tide line to water 25′ (8 m) deep.
Range: S. Florida and Texas; Mexico.
Comments: This clam is common on the west coast of Florida, where it is washed ashore in great numbers. It is abundantly used in the shellcraft industry.

337 Carolina Marsh Clam
(*Polymesoda caroliniana*)
Class Bivalvia

Description: 1½" (38 mm) long, 1⅜" (35 mm) high. Oval, convex. White, with *dark brown to black, velvety periostracum.* Valves equal, smooth, with weak concentric growth lines. *Umbones elevated, usually eroded.*
Habitat: In mud of tidal marshes and river-fed estuaries; near low-tide line and below.
Range: Virginia to n. Florida and Texas.
Comments: This clam lives in brackish water usually more acidic than sea water, and this accounts for the erosion of its shell near the umbones. It very much resembles some of the freshwater clams found in rivers.

339 **Black Clam**
(*Arctica islandica*)
Class Bivalvia

Description: 5″ (127 mm) long, 4″ (102 mm) high. Stout, plump, almost circular. Limy white, with *thick, shining, brown to black periostracum.* Valves equal, with crowded, tiny concentric growth lines. *Umbones prominent and bent forward.*
Habitat: In sand and mud bottoms; from below low-tide line to water 500′ (152 m) deep.
Range: Arctic to Cape Hatteras.
Comments: Though by no means as abundantly available as the Quahog, this species is collected by dredge, processed commercially, and sold to restaurants for making clam chowder.

---

333 **Buttercup Lucine**
(*Anodontia alba*)
Class Bivalvia

Description: 2½″ (64 mm) long, 2¼″ (57 mm) high. Almost circular, broadly convex, thin. Dull white; interior of shell tinged yellow to orange. Valves almost smooth except for irregular concentric growth lines. *Front part of margin lower than rear part. Shallow furrow slanting away from both sides of umbones.*
Habitat: In bays and lagoons, in sand bottoms; from below low-tide line to water 300′ (91 m) deep.
Range: North Carolina to Florida and Texas; Bahamas; West Indies.
Comments: The Buttercup Lucine is much sought by the shellcraft industry because of its striking shell lining.

332 **Tiger Lucine**
"Great White Lucine"
(*Codakia orbicularis*)
Class Bivalvia

Description: 3¾″ (95 mm) long, 3″ (76 mm) high. Oval, almost circular, not very convex. White; inner surface pale yellow. Valves equal, with many small, curving, radiating ribs crossed by *fine, sharp concentric threads.* Umbones slightly forward of center, with clearly outlined, heart-shaped depression in front.

Habitat: In sand or mud bottoms; from low-tide line to water 100′ (30 m) deep.

Range: S. Florida and Texas; Bahamas; West Indies.

Comments: This clam, the largest lucine in our Atlantic waters, is abundant in the West Indies. In older references it is listed as *Lucina tigrina.* The name *Codakia* is taken from the French *le codok,* the name of a Senegalese shell in the same family of bivalves.

---

331 **Cross-hatched Lucine**
(*Divaricella quadrisulcata*)
Class Bivalvia

Description: 1″ (25 mm) long, ⅞″ (22 mm) wide. Almost circular, solid, plump. Glossy white. Umbones in middle. Valves equal, with sharp, *chevron-shaped parallel grooves* running obliquely downward in both directions from a line between umbones and lower border.

Habitat: In sand bottoms; from water 6–300′ (2–91 m) deep.

Range: Massachusetts to Florida; Bahamas; West Indies to Brazil.

Comments: This unusually sculptured shell is frequently found washed up on the beach, and is used extensively in shellcraft.

347 **Leafy Jewel Box**
(*Chama macerophylla*)
Class Bivalvia

Description: 3½″ (89 mm) long, 3″ (76 mm) wide. Irregularly rounded. Reddish-purple, reddish-brown, orange, yellow, or white. *Upper (right) valve covered with long, leafy scales,* smaller, nearly flat; *lower (left) valve larger, cuplike, attached to substrate.* Margins finely toothed.

Habitat: Attached to rocks, corals, wrecks, or other hard surfaces; from low-tide line to water 46′ (14 m) deep.

Range: North Carolina to Florida and Texas; Bahamas; West Indies to Brazil.

Comments: The Leafy Jewel Box is cemented to a solid surface by the left valve, like an oyster. It is so firmly attached that a hammer and chisel are needed to collect it. Shells washed ashore by storms have usually lost most of their leafy scales in the surf. The Clear Jewel Box (*C. arcana—C. pellucida* in some literature) ranges from Oregon to Baja California. It grows to 3″ (76 mm) long and 3″ (76 mm) wide and is attached to rocks and pilings in protected places, from above the low-tide line to shallow depths. It has an excellent flavor, and is eaten in southern California. *C. arcana* has white, pink, or orange valves covered with concentric rows of rough frills. Its beak is coiled to the right, distinguishing it from the similar Left-handed Jewel Box (*Pseudochama exogyra*), which often occurs with it along the same shores.

---

348 **Florida Spiny Jewel Box**
(*Arcinella cornuta*)
Class Bivalvia

Description: 1½″ (38 mm) long, 1¼″ (32 mm) wide. Roundish, irregular, solid, plump. White. Valves about equal, with 7–9 heavy radiating ribs, each with *erect, tubular spines.* Umbones bent forward, with heart-shaped depression in front.

Habitat: On rubble bottoms; below low-tide line in shallow water.

Range: North Carolina to Florida and Texas; Bahamas; West Indies.

Comments: This jewel box has an unusual life pattern. It begins as a swimming larva, settles and attaches to a solid object as the Leafy Jewel Box does, but later in life becomes free again. The Florida Spiny Jewel Box is commonly found along the beach, most of its spines worn off by the surf.

---

362 Atlantic Strawberry Cockle
(*Americardia media*)
Class Bivalvia

Description: 2″ (51 mm) long, 2″ (51 mm) high. Obliquely squarish, plump, angled from umbones to rear end. Yellowish-white, mottled or checkered with reddish-brown. Valves equal. *33–36 strong radiating ribs covered with curved scales.* Umbones high. Lower margin scalloped.

Habitat: In sand bottoms; from low-tide line to water 18′ (6 m) deep.

Range: North Carolina to Florida; Bahamas; West Indies.

Comments: Many cockles are somewhat heart-shaped when viewed from the end. This little cockle is common in Florida waters.

---

364 Nuttall's Cockle
(*Clinocardium nuttallii*)
Class Bivalvia

Description: 6″ (15 cm) long, 6″ (15 cm) high. Almost round, plump, grayish, with thin but tough, yellowish-orange to yellowish-brown periostracum. Valves equal, with 33–37 *coarse radiating ribs* crossed by concentric growth lines; margin scalloped. Siphons short, fringed; foot bright yellow.

Habitat: In sand or mud bottoms in protected bays and estuaries; from low-tide line to water 600′ (182 m) deep.

Range: Alaska to s. California.

Comments: These cockles are harvested in Puget Sound for the fish market. The Iceland Cockle (*C. ciliatum*), which ranges from Greenland to Massachusetts, and from Alaska to Puget Sound, grows to 2¾″ (70 mm) in length and 2½″ (64 mm) in width, has 32–38 radiating ribs, a white exterior with a fibrous, grayish periostracum, and fluted edges. It lives in sand or gravel bottoms in moderately shallow water.

---

366 Giant Atlantic Cockle
(*Dinocardium robustum*)
Class Bivalvia

Description: 5¼″ (133 mm) long, 5″ (127 mm) high. Very plump, roundish, rear part flattened. Yellowish, with irregular reddish-brown or purplish spots, rear part brownish; interior salmon-pink. Valves equal; *umbones high; 32–36 strong, rounded radiating ribs;* margin scalloped. Foot dark red; siphons short, frilled.

Habitat: In sand or mud bottoms; from low-tide line to water 100′ (30 m) deep.

Range: Virginia to n. Florida and Texas.

Comments: *D. robustum* is the largest American cockle.

---

330 Common Egg Cockle
(*Laevicardium laevigatum*)
Class Bivalvia

Description: 2¾″ (70 mm) long, 3″ (76 mm) high. Obliquely oval, thin, convex. Ivory-white, tinged or spotted with orange, rose, purple, or brown; thin, brown periostracum usually worn away. Valves equal, smooth, with about *60 fine, obscure ribs. Umbones prominent, with 2 curving lines at each end;* margin with fine teeth.

Habitat: In sand or mud bottoms; from low-tide line to water 65′ (20 m) deep.

Range: North Carolina to Florida and Texas; Mexico; West Indies.

Comments: This cockle, capable of leaping by means of powerful thrusts of its long foot, was observed by one collector to escape by jumping out of a boat.

---

342 Morton's Egg Cockle

(*Laevicardium mortoni*)
Class Bivalvia

Description: 1″ (25 mm) long, 1⅛″ (28 mm) high. Oval, convex. Yellowish-white with brown zigzag markings, *bright yellow interior. Valves equal, smooth, with fine, granular concentric ridges.*

Habitat: In sand or sandy mud in bays; from low-tide line to water 15′ (5 m) deep.

Range: Cape Cod to Florida and Texas; Mexico; West Indies.

Comments: This little cockle is a shallow-water species that is frequently washed ashore during storms. It is eaten by wild ducks.

---

363 Yellow Cockle

(*Trachycardium muricatum*)
Class Bivalvia

Description: 2¼″ (57 mm) long, 2½″ (64 mm) high. Almost circular, thick, plump. Pale yellow, with irregular red or orange spots. Valves equal, with *30–40 scaly radiating ribs;* umbones prominent; margins with *strong teeth.* Siphons short, fringed.

Habitat: In sand or mud bottoms; from low-tide line to water 30′ (9 m) deep.

Range: North Carolina to Florida and Texas; West Indies to Brazil.

Comments: Since their siphons are short, cockles lie buried just beneath the surface of the bottom. After sustained freezing weather, Yellow Cockles are reported to pop out of the bottom at low tide.

---

334 **Disk Dosinia**
(*Dosinia discus*)
Class Bivalvia

Description: 3″ (76 mm) long, 3″ (76 mm) high. Almost *circular, flattened.* Valves equal; umbones prominent, near front end of thick, strong hinge ligament. Surface almost smooth, with about 50 fine concentric lines.

Habitat: In sand and sandy-mud bottoms; near low-tide line and below to water 70′ (21 m) deep.

Range: Virginia to Florida and Texas; Mexico; Bahamas.

Comments: Beachcombers soon discover the Disk Dosinia's strong hinge ligament, which keeps its valves together when the shell is blown ashore during storms.

---

302 **Sunray Venus**
(*Macrocallista nimbosa*)
Class Bivalvia

Description: 5″ (127 mm) long, 2½″ (64 mm) high. Long, oval, thick, flattened. Pale salmon or pinkish-gray, with *broken, purplish-brown radiating lines;* thin, brown periostracum. Umbones flattened; ligament strong, set in groove, with *heart-shaped area in front.* Surface glossy, with faint concentric and radiating lines.

Habitat: In sand bottoms; below low-tide line in shallow water.

Range: North Carolina to Florida and Texas; Mexico.

Comments: This large and beautiful clam, frequently washed ashore during storms, was used by pre-Columbian Indians as a tool.

338 Quahog
(*Mercenaria mercenaria*)
Class Bivalvia

Description: 5″ (127 mm) long, 4¼″ (108 mm) high. Thick, solid, convex, oval. Dull gray, stained brownish or black when freshly dug; *interior white, with deep purple at siphonal end.* Valves equal; umbones prominent, forward; hinge ligament strong, with *heart-shaped area in front.* Surface with many closely-set concentric lines.

Habitat: In sand or mud; between high- and low-tide lines in shallow water.

Range: Gulf of St. Lawrence to Florida and Texas; introduced into California.

Comments: The name "Quahog" (pronounced Kwo-hog or Co-hog) is derived from two Narragansett Indian words meaning "dark" or "closed", and "shell." Among bivalves, this clam is second only to oysters in commercial importance in the United States—partly because, like an oyster, it can remain tightly closed and live for weeks out of water if refrigerated. Young Quahogs are known as "cherrystones" and are served on the half-shell. Larger ones are tough and are served steamed or in chowders. They are known as "hard-shell clams" or "littlenecks." The Quahog's purple shell lining gave it a special importance as wampum to prehistoric Native Americans, and as such it was circulated all over the continent.

336 **Southern Quahog**
(*Mercenaria campechiensis*)
Class Bivalvia

Description: 6″ (15 cm) long, 5″ (127 mm) high. Thick, solid, convex, oval. Dull gray, rarely with brown mottling; interior white. Valves equal, heavy; umbones prominent, forward; hinge ligament strong, with *broad, heart-shaped area in front.* Surface with *coarse concentric lines.*
Habitat: In sand or mud; from low-tide line to water 50′ (15 m) deep.
Range: Virginia to Florida and Texas; Cuba; Mexico.
Comments: This species is heavier and broader than its close relative, the Northern Quahog. Although their ranges overlap, the Southern Quahog has not been exploited for food nearly so much as its northern relative.

---

368 **Common Pacific Littleneck Clam**
(*Protothaca staminea*)
Class Bivalvia

Description: 2¾″ (70 mm) long, 2¼″ (57 mm) high. Oval, plump. Whitish to tan, with angular brown markings. Umbones smooth, moderately prominent, near center. Surface with *many strong radiating ribs, heaviest toward rear,* crossed by lower concentric lines, giving ribs a *beaded appearance.*
Habitat: In sand or sandy mud in bays and estuaries; sometimes in gravelly sand on open rocky coast; between high- and low-tide lines and in shallow water.
Range: Aleutian Islands to Baja California.
Comments: This clam, one of the most common on the Pacific Coast, is a favorite seafood item. It is easily dug at low tide from its burrow 1–3″ (25–76 mm) below the surface.

335 **Common Washington Clam**
"Butternut Clam"
(*Saxidomus nuttalli*)
Class Bivalvia

Description: 6″ (15 cm) long, 4½″ (114 mm) high. Oblong-oval, heavy, moderately convex. Tan or grayish to reddish-brown, sometimes with rusty stains. *Valves slightly gaping at rear,* equal; ligament large; umbones smooth, at front end of hinge line. *Strong, raised concentric ridges.* Siphons with hard plate at tips.

Habitat: Buried more than 12″ (30 cm) in sand or mud, in bays and protected places; from low-tide line to water 40′ (12 m) deep.

Range: N. California to Baja California.

Comments: This clam, highly regarded as a seafood item, is abundant in some bays in southern California. However, the depth of its burrow makes digging it very hard work indeed. Sea otters dig for these clams in the Monterey Harbor by diving repeatedly and excavating large holes. The Butter Clam (*S. giganteus*) measures 5″ (127 mm) long and 4″ (102 mm) wide, but lacks the concentric sculpture and rust-stain color of the Common Washington Clam. It is found in protected bays from Alaska to San Francisco Bay. The most important commercial clam in British Columbia, and prized by clam diggers, it is also common in many areas of Washington, Oregon, and Humboldt Bay in California.

---

322 **Pismo Clam**
(*Tivela stultorum*)
Class Bivalvia

Description: 6″ (15 cm) long, 4¼″ (108 mm) high. Oval, heavy, moderately convex. Creamy-tan to brown, usually with purplish-brown radiating bands;

periostracum varnishlike, greenish or brownish. Valves equal; umbones nearly central; ligament heavy, set in groove. *Surface smooth,* with fine concentric lines.

Habitat: In sandy beaches exposed to surf; from low-tide line to water 80′ (24 m) deep.

Range: C. California to Baja California.

Comments: The excellent flavor of this clam has led to its near disappearance from accessible habitats. When an area of Pismo Beach, California (for which the clam is named), was opened to clamming in 1949, after being closed for several years, over 50,000 clams were dug daily. In 1973 the same beach yielded 21 legal-sized clams.

---

297 **False Angel Wing**
(*Petricola pholadiformis*)
Class Bivalvia

Description: 2″ (51 mm) long, ¾″ (19 mm) high. Long, fragile, convex; *front end rounded, rear end long.* Chalky-white. Valves equal; umbones low, forward; slight gape for siphons at rear end. Surface with about *10 heavy, scaly, radial ribs at front end* ending as *saw-toothed margin;* many smaller ribs toward rear, crossed by strong growth lines. Siphons separate, large, tubular, gray.

Habitat: Boring into clay, peat, or soft rock in protected areas; near low-tide line and below in shallow water.

Range: Gulf of St. Lawrence to Florida, Texas and Mexico; introduced into San Francisco Bay and c. Washington.

Comments: This clam bores into a hard substrate by anchoring its foot in the end of the burrow, drawing the valves down, and alternately opening and closing them, rasping with their saw-toothed margins.

326 **Dwarf Tellin**
(*Tellina agilis*)
Class Bivalvia

Description: ½" (13 mm) long, ¼" (6 mm) high. *Moderately long,* delicate, flattened, *sloping toward rear,* front end rounded. Glossy white, yellowish, or pink. Valves equal; umbones low, nearer rear end; hinge ligament prominent. Surface with faint concentric lines. Siphons separate and long.

Habitat: In sand bottoms; near low-tide line and below in shallow water.

Range: Gulf of St. Lawrence to North Carolina.

Comments: Valves of this small clam are commonly washed up on the beach. It is a representative of a very large genus, most of which are found in warmer waters. The Alternate Tellin (*T. alternata*), which ranges from North Carolina to Florida and Texas, is much larger, 3" (76 mm) long and 1½" (38 mm) high, rounded in front, and sloping sharply to the rear. It is white, pink, or yellow, and its surface has many fine concentric lines. The Iris Tellin (*T. iris*) has the same range, but is the size and shape of *T. agilis,* with translucent white valves. The Candy Stick Tellin (*T. similis*) is found in south Florida and the West Indies. It is 1" (25 mm) long and ⅝" (16 mm) high, and is glossy white with radiating pink rays.

327 **Modest Tellin**
(*Tellina modesta*)
Class Bivalvia

Description: 1" (25 mm) long, ½" (13 mm) high. Oval, flattened. Iridescent white. Valves bluntly pointed at rear end, rounded at front end, *lower margin nearly straight.* Surface with *fine concentric threads* that almost disappear toward the rear ¼, then reappear coarsely at rear slope.

Habitat: In sand and silt bottoms in bays; below low-tide line.

Range: Alaska to Gulf of California.

Comments: This species is found close to shore after being washed up by heavy wave action, although it does not normally occur above low-tide line.

---

328 Carpenter's Tellin
(*Tellina carpenteri*)
Class Bivalvia

Description: ⅜″ (10 mm) long, ¼″ (6 mm) high. Oval, flattened. Iridescent, cream-colored, usually with pinkish tint. *Valves bluntly pointed at rear,* equal, rounded at front end; *lower margin curved. Surface smooth.*

Habitat: In sand and sandy-mud bottoms; below low-tide line to water 2200′ (671 m) deep.

Range: Alaska to Gulf of California.

Comments: This wide-ranging little bivalve burrows shallowly in soft bottoms, and is preyed upon by snails, crabs, and bottom-feeding fish.

---

343 Baltic Macoma
(*Macoma balthica*)
Class Bivalvia

Description: 1½″ (38 mm) long, 1¼″ (32 mm) high. Rounded, thin, *moderately flattened.* Chalky-white, with *ragged, dark brown periostracum,* usually worn off upper part of valves. Valves equal; umbones prominent, central. Surface with fine concentric lines. Separate tubular siphons extending more than 10 times length of valves.

Habitat: In mud or sandy mud in quiet bays and brackish estuaries; well above low-tide line and below in shallow water.

Range: Arctic to Georgia; Bering Sea to c. California.

Comments: Their remarkably long siphons enable macomas to live nearly 12″ (30 cm) below the surface of the bottom. At low tide one can sometimes see their whitish siphons in quiet tidepools, actively waving like worms. *M. balthica* is a principal winter food of the Black Duck (*Anas rubripes*).

---

320 **Bent-nosed Macoma**
(*Macoma nasuta*)
Class Bivalvia

Description: 4⅜″ (111 mm) long, 3″ (76 mm) high. Oval, flattened, bent. White, with gray or brown, much-eroded periostracum. *Valves with ridge down rear slope,* smooth, shallow, unequal, rounded at front end, bluntly pointed at rear; *both valves bent toward right;* umbones prominent. Siphons separate, long, white when extended, orange when contracted.

Habitat: In mud, clay, sand, or gravel bottoms; from low-tide line to water 165′ (50 m) deep.

Range: Alaska to Baja California.

Comments: This clam lies buried 4–8″ (10–20 cm) below the surface of the bottom, left side down, with its siphons extended to the surface. The incurrent siphon extends up and arches downward to suck up fine organic particles from the bottom. The excurrent siphon extends to just below the surface, its current making a little fountain of water and sand or mud.

317 **White Sand Macoma**
(*Macoma secta*)
Class Bivalvia

Description: 3⅝″ (92 mm) long, 3″ (76 mm) high. Oval, flattened. White, with thin gray, brown, or olive, much-eroded periostracum. *Valves with diagonal ridge* from umbones to rear; oval, smooth, thin, unequal; *left valve almost flat, right one deeper.* Siphons separate, long, white.

Habitat: In fine sand or sandy mud; from low-tide line to water 165′ (50 m) deep.

Range: British Columbia to Baja California.

Comments: The valves of this species, though unequal, are not bent as in *M. nasuta.* The White Sand Macoma prefers a fine-sand or sandy-mud habitat, and is not found in mud, clay, or gravel bottoms. It is a favorite prey of Lewis' Moon Snail (*Polinices lewisii*), which digs it out of the bottom and drills a hole through the shell to rasp out the soft parts.

321 **Coquina**
(*Donax variabilis*)
Class Bivalvia

Description: ¾″ (19 mm) long, ⅜″ (10 mm) high. Rounded, triangular, thin, somewhat flattened; *rear end short, slanting abruptly downward;* front end long. Solid pink, lavender, purple, bluish, yellow, orange, or white, or with radiating stripes. Valves equal; umbones low, smooth. Surface with many fine radiating and concentric lines.

Habitat: On sandy beaches; above low-tide line.

Range: Long Island to Florida, Texas and Mexico.

Comments: In parts of Florida, coquina rock, composed of huge accumulations of these shells, is used as a building material. It is soft and can easily be cut into blocks, which harden as they dry. The Spanish fort at St. Augustine, Florida, was made of this material nearly 400 years ago. The Bean Clam (*D. gouldii*), similar in size and shape but less colorful—usually buff or yellowish, sometimes with dark rays—ranges from Monterey Bay to Baja California.

---

305 Jackknife Clam
(*Tagelus plebeius*)
Class Bivalvia

Description: 4″ (102 mm) long, 1½″ (38 mm) high. Oblong, plump, *squarish at front, rounded at rear.* White, with thick, yellowish-brown or olive periostracum. Valves equal; *gape at both ends;* umbones low, central. Surface with faint growth lines; low ridge on rear slope.

Habitat: Deep in sandy mud in bays, lagoons, and estuaries; above low-tide line and below in shallow water.

Range: Cape Cod to Florida and Texas.

Comments: This clam, which tolerates brackish water, is a strong digger, using its muscular foot to position itself in a vertical burrow with an opening to the surface. Formerly listed as *T. gibbus.*

---

307 California Jackknife Clam
(*Tagelus californianus*)
Class Bivalvia

Description: 4″ (102 mm) long, 1½″ (38 mm) high. Long, narrow. Grayish-white, sometimes with yellow bands, yellowish-brown to brownish-gray periostracum partly eroded. Valves equal, *gape at both ends, upper and lower margins almost parallel,* umbones low, central. Surface with faint growth lines; low ridge on rear slope.

Habitat: Sandy mudflats; near low-tide line.

Range: N. California to Panama.
Comments: This clam lives in a vertical burrow 4–20″ (10–51 cm) deep, and migrates up and down as the tide rises and falls. Though an edible species, in California it is more commonly dug for fish bait.

303 **Pacific Razor Clam**
(*Siliqua patula*)
Class Bivalvia

Description: 6¾″ (17 cm) long, 2⅜″ (60 mm) high. *Oval-oblong, thin,* somewhat flattened. White, with strong, glossy yellow or yellowish-brown periostracum. Valves equal; umbones low, smooth, nearer front end; hinge ligament strong. Siphons fused except at tips.
Habitat: In sand on flat beaches with strong wave action; near low-tide line and below in shallow water.
Range: Alaska to c. California.
Comments: This active digger does not live in a permanent burrow, but moves about in sand pounded by surf. When extended, its foot forms a wide flange that acts as an anchor. If tossed free of the sand by a breaking wave, the clam quickly buries itself with rapid thrusts of its foot. Esteemed as food, the Pacific Razor Clam is taken by sport and commercial fishermen.

304 **Atlantic Razor Clam**
(*Siliqua costata*)
Class Bivalvia

Description: 2½″ (64 mm) long, 1¼″ (32 mm) high. *Oblong, oval, thin.* Purplish-white, with shiny, greenish-brown periostracum. Valves equal, umbones low, nearer front end, *strong internal white rib* extending downward from front end of hinge line, visible from outside.

Habitat: On sand bottoms; below low-tide line in shallow water.

Range: Gulf of St. Lawrence to Cape Hatteras.

Comments: The shells of this fragile species are commonly found at low tide on the sandy beaches of New England, but one has to dig below the low-tide line to find a living specimen.

---

308 **Common Razor Clam**
(*Ensis directus*)
Class Bivalvia

Description: 10″ (25 cm) long, 1⅝″ (41 mm) high. Long, thin, convex; ends squarish; *valves slightly curved downward.* White, with thin, glossy, greenish- or yellowish-brown periostracum. Valves equal, with *gape at each end* and sharp edges; umbones flat. Foot cream-colored to buff; siphons short, fringed at edges.

Habitat: In sand and sandy mud, in bays, estuaries, and protected places; above low-tide line and below in shallow water.

Range: Labrador to Florida.

Comments: A Common Razor Clam can sometimes be found at low tide on a mudflat by looking for an oval hole. The clam moves up and down in its burrow, and can descend quickly when disturbed. The author has often seen a 6″ (15 cm) clam, set upright in sand at the water's edge, bury itself within 15 seconds in 8 or 9 pulls.

---

319 **Surf Clam**
(*Spisula solidissima*)
Class Bivalvia

Description: 7″ (18 cm) long, 5½″ (140 mm) high. Heavy, nearly *triangular, with rounded lower margin.* Yellowish-white, with thin, brown periostracum. Valves equal; *gape at rear end* for siphons; umbones large, central; hinge ligament strong. Surface smooth, with fine concentric lines.

Habitat: In sand near shore on open, exposed beaches; at low-tide line and below in shallow water.

Range: Nova Scotia to South Carolina.

Comments: The Surf Clam lives in relatively clean sand and can be found in the zone of breaking surf. Winter storms wash many thousands ashore on beaches, where they provide feasts for gulls. This clam is edible, but tough. It is sold as fish bait.

---

318 Gaper Clam
(*Tresus nuttallii*)
Class Bivalvia

Description: 8″ (20 cm) long, 4¾″ (121 mm) high. Oval-oblong, large, strong. Yellowish, with shaggy, brown periostracum, worn off upper part. Valves equal; gape at rear end for siphons, not large enough to permit complete retraction of all soft parts; umbones low, smooth, nearer front end. Surface with fine concentric growth lines. Siphons fused into *long neck* covered with rough, dark skin; pair of *tough siphonal plates* at tip.

Habitat: In sandy mud in bays and protected places; from low-tide line to water 95′ (29 m) deep.

Range: British Columbia to Baja California.

Comments: The Gaper Clam lives in a burrow as much as 36″ (1 m) deep, and extends its long neck up to the surface of the bottom. Tufts of algae and sedentary invertebrates grow on the hard siphonal plates at either side of the neck. Gapers are used both as human food and as fish bait.

306 **Red Nose**
(*Hiatella arctica*)
Class Bivalvia

Description: 2″ (51 mm) long, ¾″ (19 mm) high. *Valves coarse, irregular,* oval-oblong. Chalky-white, with yellowish-gray periostracum. Valves equal; gape at rear end for siphons; *umbones close together,* nearer front end; hinge ligament prominent, oval. Surface irregular, with coarse concentric growth lines. *Young with lines of short spines* slanting backward from umbones. *Siphons with red tips* fused into a neck, not completely retractable, with rough, wrinkled skin.

Habitat: In cracks and holes in rocks, among coarse gravel, or growth on pilings; above low-tide line and below in deep water.

Range: The Arctic to the West Indies; Alaska to Panama.

Comments: The Red Nose is a nestler. Young clams settle in some crevice, attach themselves by byssal threads, and remain there for life.

315 **Geoduck**
(*Panopea generosa*)
Class Bivalvia

Description: 9″ (23 cm) long, 4¾″ (121 mm) high. Oblong, thick, rounded at front, squarish at rear. White to cream-colored, with thin, yellowish periostracum. Soft parts reddish. Valves equal; *gape at both ends;* umbones low, smooth; hinge ligament thick, large. Surface with coarse, wavy growth lines. Siphons united in *long, thick neck* with rough, brownish skin.

Habitat: In sandy mud in protected bays; near low-tide line and below.

Range: Alaska to Baja California.

Comments: The Geoduck (pronounced gooey-duck) is the largest American bivalve, reaching a weight of more than 13 lbs (6 kg), half of which is neck. It is esteemed as food, but is difficult to capture because its burrow may be more than 48″ (122 cm) deep, and is accessible to the clamdigger only for a limited time, at low spring tides. Its unusual name appears to be derived from the Nisqually Indians' phrase for "dig deep." The Geoduck's tough siphon is usually chopped and used for chowder, while the rest of the body is fried or otherwise cooked.

316 Soft-shelled Clam
(*Mya arenaria*)
Class Bivalvia

Description: 6″ (15 cm) long, 3½″ (89 mm) high. Thin, oval, rounded in front, somewhat pointed at rear. Chalky-white to pale gray, with thin, gray or tan periostracum. Valves equal, *gape at both ends;* umbones low, smooth. Surface with rough, wrinkled growth lines. Siphons fused into a *neck with tough, dark gray skin.*

Habitat: In both marine and brackish areas; burrowing to 10″ (25 cm) deep in sand and mud of protected bays, and below low-tide line to water 30′ (9 m) deep.

Range: Labrador to North Carolina; introduced into California and has spread north to Alaska.

Comments: This delicious clam is also known as the Long Neck Clam or Steamer Clam, an important commercial species. It lies in a vertical burrow with its siphons near the surface; at low tide, a spurt of water shooting out of a hole in the sand betrays the presence of a clam contracting its neck when disturbed.

296 **Angel Wing**
(*Cyrtopleura costata*)
Class Bivalvia

Description: 8″ (20 cm) long, 2⅜″ (60 mm) high. Fragile, brittle; rounded in front, long and narrowing to rear. White, with thin, gray periostracum. Valves equal; umbones prominent, forward. *30 strong radial ribs* crossed by coarse concentric growth lines; ribs at front end sharp and scaly, extending as *saw-toothed border.* Siphons fused into long, tannish neck.

Habitat: Boring into sandy mud, clay, or peat; above low-tide line and below in shallow water.

Range: Cape Cod to Florida, Texas, and Mexico; West Indies to Brazil.

Comments: This clam belongs to a family of borers, the pholads. When cleaned, the two delicate and graceful valves held together by the hinge ligament certainly suggest the wings of an angel.

313 **Striated Wood Piddock**
(*Martesia striata*)
Class Bivalvia

Description: 2″ (51 mm) long, 1″ (25 mm) high. Fragile, wedge-shaped; narrow at rear, round at both ends. Grayish-white. Valves equal; gape at both ends; umbones low. *Groove from umbones to lower margin.* Surface in front of groove with *toothed concentric ridges,* continuing to rear only as weak growth lines. Siphons fused into moderately long, grayish neck.

Habitat: Boring into submerged wood.

Range: North Carolina to Florida and Texas; West Indies to Brazil.

Comments: This wood piddock can be found in waterlogged driftwood. The clam bores into the wood, usually across the grain, by scraping with the rough, filelike forepart of its valves. A large population of such clams can completely demolish a submerged wooden structure.

310 **Flat-tipped Piddock**
(*Penitella penita*)
Class Bivalvia

Description: 3″ (76 mm) long, 1⅜″ (35 mm) high. Thin, globular at front end, narrower and flatter at rear. White, with brown *periostracum that continues beyond valves as flaps* at rear end. Valves equal; front end with *20 rasplike radiating ridges;* rear end with low concentric growth lines; both *front gape and hinge covered by plate* in adult; gape at rear open. Siphons fused into smooth, white neck with brownish marks around tip.
Habitat: Boring into clay, sandstone, shale, concrete, or brick; from low-tide line to water 300′ (91 m) deep.
Range: Alaska to Baja California.
Comments: Like other pholads, the Flat-tipped Piddock, the most common boring clam along the West Coast, bores by pulling its rasplike valves close to the bottom of the burrow and alternately opening and closing them, scraping ever deeper. Since the burrow gets wider as the animal grows, there is no way it can get out through the narrow upper part of the burrow. The clam grows to size and then stops boring and develops sexually.

314 **Great Piddock**
(*Zirfaea crispata*)
Class Bivalvia

Description: 3″ (76 mm) long, 1⅜″ (35 mm) high. Thin, oblong; round at rear, somewhat pointed at front. White, with brownish-black periostracum, mostly worn away. Valves equal; gape at both ends; hinge line and umbones covered by *hard, shieldlike plate. Groove from umbones to lower margin* divides smoother rear part from front part with *scaly ribs* and *saw-toothed margin.* Siphons fused into long, cream-colored neck with meshwork of brown lines about tip. Foot white.

Habitat: Boring into clay, peat, soft rocks, and submerged wood; near low-tide line and below in shallow water.

Range: Labrador to Maryland; introduced into California.

Comments: This plump clam is common in submerged peat beds in New Hampshire. The Rough Piddock (*Z. pilsbryi*), which ranges from Alaska to Baja California, is larger, 5¼″ (133 mm) long and 2½″ (64 mm) wide, with stronger valves and a comparably rough front valve surface. It is known to burrow over 18″ (46 cm) into sandstone, shale, and clay.

312 Pacific Shipworm
(*Bankia setacea*)
Class Bivalvia

Description: Shell ½″ (13 mm) long, ¾″ (19 mm) high; entire animal 39″ (1 m) long. Shell short, *arc-shaped.* White. Valves with sharp teeth at front end, umbones rounded, *gapes wide, meeting each other at lower end.* Mantle greatly extended, narrow, wormlike; separate siphons and *2 featherlike appendages* at rear end; foot suckerlike.

Habitat: In wharf piles, wooden boat and ship hulls, and submerged but not buried wood; from low-tide line to water 230′ (70 m) deep.

Range: Alaska to Baja California.

Comments: Shipworm larvae, almost too small to be seen with the naked eye, settle on submerged wood, and bore in by scraping with their larval valves. Even though a piece of wood may be riddled with shipworms, the burrows do not intersect one another. This tunneling can utterly destroy any wooden structure in the sea. Gould's Shipworm (*B. gouldii*), which is about the same size as *B. setacea,* ranges from New Jersey to Florida and Texas, and to the West Indies.

311 **Common Shipworm**
(*Teredo navalis*)
Class Bivalvia

Description: Shell ¼″ (6 mm) long, ¼″ (6 mm) high; entire animal more than 12″ (30 cm) long. Shell short, *arc-shaped.* White. *Valves sharp and bladelike at front end;* wide gape at each end, meeting at lower end; umbones flattened. Mantle greatly extended, narrow, wormlike; separate siphons and *2 paddlelike appendages* at rear end; foot suckerlike.

Habitat: In wharf piles, wooden boat and ship hulls, and submerged but not buried wood; near low-tide line and below.

Range: Newfoundland to Florida and Texas; West Indies.

Comments: This shipworm was the bane of navies and merchant fleets from ancient times until the advent of metal-hulled vessels. Many a ship, structurally weakened by burrowing shipworms, has broken apart during a storm at sea.

486 **Atlantic Long-fin Squid**
(*Loligo pealei*)
Class Cephalopoda

Description: 17″ (43 cm) long, 3⅝″ (92 mm) wide. Cylindrical, tapered toward rear. White, with variable red, purplish, yellow, and brown speckles. Head with pair of large eyes, 4 pairs of *arms ½ mantle length* and 1 pair of *tentacles ⅔ mantle length;* siphon under neck; *triangular fin ½ mantle length* on each side of rear end.

Habitat: Ocean surface to water 300′ (91 m) deep over continental shelf.

Range: Bay of Fundy to the West Indies.

Comments: These fast-swimming squid are abundantly used as fish bait and, to a lesser degree, as human food. They occur in large schools and are eaten by many commercially important fish, including sea bass, bluefish, and mackerel.

485 **Opalescent Squid,**
"Market Squid", "Calamari"
(*Loligo opalescens*)
Class Cephalopoda

Description: Mantle 7⅝″ (19 cm) long, 1¾″ (44 mm) wide. Cylindrical, tapered to point at rear. White, mottled gold, brown, or red; color changeable. Head with pair of large eyes, 4 pairs of *arms about ½ mantle length,* 1 pair of *tentacles ⅔ mantle length;* siphon under neck; *triangular fin ½ mantle length* on each side of rear.

Habitat: Ocean surface to bottom in open coastal waters.

Range: British Columbia to Mexico.

Comments: Most frozen squid found in American fish markets and food stores are of this species. They are taken by net in large numbers when they enter shallow water to breed.

Brief Squid
(*Lolliguncula brevis*)
Class Cephalopoda

Description: 3″ (76 mm) long, 1⅜″ (35 mm) wide. Cylindrical, *bluntly pointed at rear.* Cream-colored with reddish-purple spots; underside white. Head with pair

of large eyes, 4 pairs of arms ½ mantle length, 1 pair of *tentacles equal to body length;* siphon under neck; *short, rounded fin ⅓ mantle length* on each side of rear end.

Habitat: In bays and inlets.

Range: Maryland to Florida and Texas; Mexico; West Indies to Uruguay.

Comments: This little squid is important both as food for commercially valuable fish and as fish bait, but people netting it for the latter should be warned that it can deliver a sharp nip if handled.

484 Short-fin Squid
(*Illex illecebrosus*)
Class Cephalopoda

Description: 12″ (30 cm) long, 2¼″ (57 mm) wide. Cylindrical, tapered toward rear. White with variable red, brown, and gold speckles. Head with pair of large eyes covered by skin with small hole over each eye. *4 pairs of arms ⅖ mantle length, 1 pair of tentacles ½ mantle length;* siphon under neck; *triangular fin ⅓ mantle length* on each side of rear end.

Habitat: Open seas; near shore during summer.

Range: Greenland to North Carolina.

Comments: This squid travels in large groups noted for the speed and ferocity with which they charge into schools of small fish, sometimes killing many more than they can eat. It is also known for jetting to the surface and "flying" for a few feet above the water. The author saw one, collected from a Maine tidepool and placed in a bucket of water, jet its way to freedom as the surprised collector was wading along the shoreline.

480 **Common Atlantic Octopus**
(*Octopus vulgaris*)
Class Cephalopoda

Description: Body plus longest arm 120″ (3 m) long. Globe-shaped, with *4 pairs of arms.* Usually reddish-brown; color highly variable. Thick *arms 4 times length of mantle,* with 2 alternating rows of suckers; tubular siphon under neck; head almost as broad as body, eyes high on sides of head. Skin mostly smooth, but can temporarily raise variously shaped bumps.

Habitat: Among rocks and coral reefs near shore; near low-tide line and below in shallow water.

Range: Connecticut to Florida and Texas; Mexico; West Indies.

Comments: This octopus is secretive, hiding during the day in crevices and caves and under rocks. Small specimens may be found above the low-tide line. The Briar Octopus (*O. briareus*) ranges from southern Florida throughout the West Indies, among coral rocks and reefs, and in turtle grass beds. Its length is 18″ (46 cm), its arms are thick at the base and over 5 times the mantle length. It is usually pinkish-brown, but changeable, and its skin is smooth or finely granular.

---

482 **Long-armed Octopus**
(*Octopus macropus*)
Class Cephalopoda

Description: Body plus longest arm 36″ (1 m) long. Globe-shaped, with 4 pairs of arms. Tan to reddish-brown, with whitish spots; color changeable. *Arms long, more than 7 times body length,* with 2 alternating rows of suckers; tubular siphon under neck. Head almost as broad as body; eyes high on sides of head, with bump above each. Skin with small warts.

Habitat: Among coral reefs and coral rock.
Range: Florida Keys; West Indies.
Comments: This octopus discharges a dark brownish ink when disturbed, thus confusing possible predators. *O. macropus* feeds principally on various kinds of crabs, which it captures with one or more of its arms.

---

481 **Joubin's Octopus**
(*Octopus joubini*)
Class Cephalopoda

Description: Body plus longest arm 7″ (18 cm) long. Globe-shaped, with 4 pairs of arms. Tan, brown, or gray; color changeable. *Arms short, 2–3 times body length,* with 2 alternating rows of suckers; tubular siphon under neck. Head as broad as body, with eyes high on sides. Skin smoothish, with tiny scattered bumps.
Habitat: In empty clam shells, among coral rock; near low-tide line and below in shallow water.
Range: S. Florida; Bahamas; West Indies.
Comments: This little octopus is commonly cast ashore by storms on the gulf coast of Florida. When handled, it can deliver a painful bite without hesitation. Though the smallest of American octopods, it lays the largest eggs, ⅜″ (10 mm) long.

---

483 **Giant Pacific Octopus**
(*Octopus dofleini*)
Class Cephalopoda

Description: Body plus longest arm 16′ (5 m) long. Globe-shaped, with 4 pairs of arms. Reddish or brownish, with fine black lines; color changeable. *Arms 3–5 times body length,* with 2 alternating rows of suckers; tubular siphon under neck. Head as broad as body, with eyes high on each side. *Skin wrinkled and folded.*

Habitat: On rocky shores, in tidepools; from low-tide line to water 1,650′ (503 m) deep.

Range: Alaska to s. California.

Comments: This species is one of the largest known octopods, the heaviest on record weighing nearly 600 pounds (272 kg). It is fished commercially from Alaska to northern California. The Giant Pacific Octopus feeds on shrimps, crabs, scallops, abalones, clams, various fishes, and smaller octopods, and is eaten by seals, sea otters, sharks, and other large fishes.

478, 479 **Two-spotted Octopus**
(*Octopus bimaculatus*)
Class Cephalopoda

Description: Body plus longest arm 30″ (76 cm) long. Pear-shaped, with 4 pairs of arms. Usually gray, brown, olive, or reddish; mottled with black, mantle paler underneath, variable. *Arms 4–5 times mantle length,* with 2 alternating rows of suckers; web between arms. Head narrow, eyes high on sides of head; conspicuous *black spot below each eye.* Mantle with many prickly bumps.

Habitat: In holes, under rocks, and among kelps; from low-tide line to water 160′ (49 m) deep.

Range: S. California to Baja California and Gulf of California.

Comments: In 1949 it was discovered that there were two closely related species of 2-spotted octopods instead of one. The second was named the Mud Flat Octopus (*O. bimaculoides*). This is the species most commonly found between high- and low-tide lines, on mudflats as well as among rocks. Its arms are shorter, 2½–3½ times mantle length, and its eggs are larger. However, the first clue that there were 2 species instead of one was offered by the discovery that there were 2 different populations of kidney parasites (mesozoa) in the Two-spotted Octopus. It is noted that the males and females at breeding time have no trouble distinguishing between the species.

ARTHROPODS
(Phylum Arthropoda)

Of all the major groups of invertebrate animals, by far the largest and most familiar are the arthropods. More than 1,000,000 species are known, of which all but about 85,000 are insects—invertebrates familiar to us all. Although there are more insects than all other kinds of animals combined, they are chiefly land-dwellers, and so will not be treated in this book. We shall direct our attention, instead, chiefly to the preponderantly marine Crustacea, a class of more than 31,000 species, the entirely marine Pycnogonida, or sea spiders (500 species), and the Merostomata, or horseshoe crabs (5 species).

An arthropod's most obvious characteristic is the tough encasement of armor, or *exoskeleton*, which gives the animal rigidity and protects its soft insides. This armor is made principally of a substance called *chitin*, secreted by the underlying epidermal cells. The exoskeleton has joints, regions where the chitin is thin and flexible, permitting movement. Such joints are particularly obvious on the legs, and give the phylum its name, Arthropoda, which means "jointed foot" in Greek. Movement is achieved by muscles attached inside the skeleton, rather than on the outside, as in human beings.

The presence of an exoskeleton prevents increase in body size. Growth can be achieved only by a series of molts, the periodic shedding of the exoskeleton. Before an arthropod molts, a new, soft exoskeleton is deposited beneath the old one. The old exoskeleton splits, in a manner characteristic for each species, and the soft animal slowly climbs out. The new exoskeleton soon hardens, and the animal's size is again fixed until the next molt. During this process, the animal is soft and defenseless and thus vulnerable to predators; it usually goes into hiding while it molts.

The arthropods share a number of characteristics with the annelids. Both are segmented and have a similarly organized nervous system, and a heart that lies above the gut. Unlike the annelids', however, the segments of arthropods are not all alike, but are usually grouped together in functional regions: head, thorax, and abdomen. In some groups the head and thorax are fused and covered with a single plate, the *carapace.* Arthropods are bristly, with nearly all their bristles sense organs—some sensitive to touch, currents, taste, odor, or sound. They have eyes that may be simple, with one lens and retina, or compound, composed of many lenses and retinas. Their circulatory system is said to be "open," that is, with blood coursing through open spaces or *sinuses* to bathe the tissues, rather than through fine capillaries among the tissues. All marine forms, except very small ones, have gills for respiration.

Nearly all arthropods have separate sexes, and fertilized females lay eggs with shells. Marine forms generally pass through a series of larval stages before the adult form is reached.

In considering the economic importance of marine arthropods, we are prone to think only of those we prize as food: shrimps, lobsters, and crabs. We tend to overlook the small planktonic species so important in the marine food chain as harvesters of 1-celled photosynthetic algae, and themselves the prey of many species of fish.

Class Merostomata:

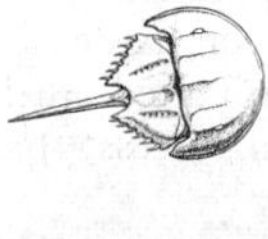

The class Merostomata is represented by 3 genera and 5 species, one of which lives on the coast of the eastern United States. That is the Horseshoe Crab (*Limulus polyphemus*). Its body consists of a convex forepart covered with a carapace (*cephalothorax* or *prosoma*), a rear part (*abdomen* or *opisthosoma*), and a long, spinelike tail (*telson*).

Class Pycnogonida: The pycnogonids, or sea spiders, are a strange group of small-bodied, long-legged marine arthropods. Though they walk on 8 legs, they are not spiders, which belong to a quite different group of arthropods. The pycnogonid body consists of a small *thorax* of 4 segments, each with a pair of side projections bearing the legs. The first segment has a necklike projection with a single 4-part eye on the top, and a sucking *proboscis* with a mouth at the tip. Beside the proboscis there may be paired accessory mouthparts in the form of *pinchers* and *feelers.* A given species may have one, both, or neither of these appendages. The rear end of the animal has a very small projection, the *abdomen,* the sole function of which is to bear the anus. In fact, there is so little room for organs in the body cavity of a pycnogonid that the sex organs are located in the long joints of the legs. In some pycnogonids there is an extra pair of slender legs curled under the first body segment, to which eggs are attached at breeding season. These are usually on the male, whose duty it is to carry the eggs until they hatch. In some species the female also has egg-carrying legs. The larvae, at hatching, have only 3 pairs of appendages and, in many species, become internal parasites in hydroids or corals while they develop into adults.

Pycnogonids feed by sucking the body fluids and soft tissues of hydroids, sea anemones, soft corals, sponges, or bryozoans. The sea spider mouth has 3 rasplike teeth, and the pharynx functions like a pump, drawing in the semiliquid food.

Class Crustacea: The crustaceans include a number of animals familiar because they are edible: shrimps, lobsters, and crabs. They also include a variety of other forms seen along the shore—isopods, beach fleas, sea roaches, water fleas, and barnacles—as well as many tiny marine and freshwater creatures too small to be seen with the naked eye.

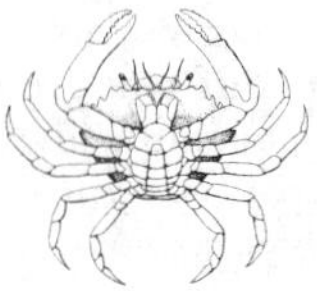

A crustacean is an arthropod with 5 pairs of appendages on 6 head segments: 2 pairs of *antennae,* a pair of jaws, or *mandibles,* 1 on each side of the mouth, and 2 pairs of manipulatory mouthparts, or *maxillae.* The number of segments in the body varies, depending on the group. In some forms the body may simply be a *trunk.* In more advanced types it may be divided into a thorax and an abdomen. The thorax has a maximum of 8 segments, and the abdomen, 6. Each segment bears a pair of appendages that have different forms and functions. The first 3 pairs on the thorax may be auxiliary mouthparts, or *maxillipeds.* The remaining 5 thoracic appendages may be *walking legs,* the first 2 or 3 pairs ending in *pincers.* The first 5 segments on the abdomen bear forked, flattened appendages called *swimmerets,* and the last segment ends in a flattened tailpiece, or *telson,* flanked by a pair of broad, flat appendages that together make up the tailfan. In crabs, the abdomen is folded forward and is recessed under the thorax.

Reproduction is almost entirely sexual, fertilization is usually internal, and the eggs are attached to the body of the female. The larvae usually float as plankton, and usually bear little resemblance to the adult.

666 **Horseshoe Crab**
(*Limulus polyphemus*)
Class Merostomata

Description: 24″ (61 cm) long, 12″ (30 cm) wide. Horseshoe-shaped carapace convex, with triangular abdomen and spikelike tail; older individuals usually covered with algae. Greenish-tan. Pair of compound eyes on each side of carapace, 2 simple eyes on forepart of midline. Sides of abdomen scalloped, with 6 spines. 1 pair of pinchers in front of mouth. Mouth surrounded by 5 pairs of walking legs, *each walking leg with a burrlike base; last pair of walking legs with circle of leaflets;* first pair on male heavy and rounded; others with pincher tips. Underside of abdomen has 6 pairs of overlapping flaps, the first covering openings of six ducts, the others covering 5 pairs of book gills comprised of many flat sheets.

Habitat: On mud or sand bottoms; from near low-tide line to water 75′ (23 m) deep.

Range: Gulf of Maine to Gulf of Mexico.

Comments: This animal is the only one of its kind in American waters, and cannot be confused with anything else. It feeds on clams, worms, and other invertebrates which it grinds with the burrlike bases of the walking legs that surround its mouth. In spring, horseshoe crabs congregate near the shore, males holding onto the abdomens of females with their heavy walking legs. When the tide is high, each female digs a hole above the low-tide line and lays 200–300 pale greenish eggs. As she does, the male spawns sperm to fertilize them. The eggs are then buried in sand, where they remain for several weeks, until hatching as miniature horseshoe crabs with tiny button tails.

577 Anemone Sea Spider
(*Pycnogonum littorale*)
Class Pycnogonida

Description: ½″ (13 mm) long, ⅛″ (3 mm) wide. Stubby, stout, flat, with tiny abdomen. Creamy-tan to brown. *Proboscis a blunt cone,* ⅓ total body length; *no accessory mouth parts.* 4 tiny eyes on projection behind proboscis. 4 pairs of stout legs with thick bases, and claws at tips. Males with pair of accessory legs for carrying eggs.

Habitat: Under rocks, associated with sea anemones; near low-tide line and below in shallow water.

Range: Gulf of St. Lawrence to Long Island Sound.

Comments: This sea spider feeds by approaching the base of a sea anemone, inserting its proboscis into the anemone's soft tissue, and sucking. Apparently the anemone is not disturbed, as it does not contract or take any kind of defensive action.

---

579 Stearns' Sea Spider
(*Pycnogonum stearnsi*)
Class Pycnogonida

Description: ½″ (13 mm) long, ⅛″ (3 mm) wide. Salmon-pink, ivory, or white. Surface smooth. Lacks 4 tiny eyes. Otherwise similar to the Anemone Sea Spider.

Habitat: Under rocks, associated with sea anemones, hydroids, and tunicates; from midtidal zone to low-tide line.

Range: British Columbia to c. California.

Comments: This species feeds chiefly on the Giant Green Anemone, but also on the Frilled Anemone, the Proliferating Anemone, and possibly the Ostrich-plume Hydroid.

---

578 Ringed Sea Spider
(*Tanystylum orbiculare*)
Class Pycnogonida

Description: 1⁄16″ (2 mm) long, 1⁄16″ (2 mm) wide. Circular, flat, with tiny abdomen. Creamy or grayish-white. *Proboscis broad, club-shaped,* ⅔ as long as body, with long, pointed, feelerlike appendage; *short pincher* on each side.

4 black eyes behind proboscis. 4 pairs of legs, each ⅛″ (3 mm) long, with claws at tips, 1 pair of accessory legs, larger in males. *End of abdomen cleft.*

Habitat: Among hydroids, tunicates, and other sedentary organisms, on rocks, pilings, and floats; near low-tide line and below in shallow water.

Range: Bay of Fundy to Brazil.

Comments: When *T. orbiculare* eggs, carried on the accessory legs of the male, hatch, the larvae invade the polyp of a hydroid, where they live parasitically until they develop adult features. The California Ringed Sea Spider (*T. californicum*), found on Feathery Hydroids, is about the same size as *T. orbiculare,* and is white with dark bands on the legs which match the hydroid. It ranges from Alaska to southern California.

---

576 **Lentil Sea Spider**
(*Anoplodactylus lentus*)
Class Pycnogonida

Description: ¼″ (6 mm) long, ¹⁄₁₆″ (2 mm) wide. Slender, flattened, long-legged, with long neck and tiny abdomen. Reddish-brown to reddish-purple. Proboscis rising under long neck, *cylindrical, rounded at tip,* ¼ as long as body, with *pinchers longer than proboscis* on each side, but no feelerlike appendages. 4 eyes on projection behind proboscis. Length of abdomen twice width. 4 pairs of slender legs ¾″ (19 mm) long. Only male has pair of accessory legs.

Habitat: On hydroids and tunicates on rock or shell bottoms; near low-tide line and below to water 900′ (274 m) deep.

Range: Bay of Fundy to Florida; West Indies.

Comments: The Lentil Sea Spider owes its color to a dark reddish pigment in its blood. It has longer legs than any other sea spider in the American Atlantic and looks like a tangled knot. The Spiny Sea Spider (*A. oculospinatus*) is a related species ranging from central California to Baja California. It is about the same size as *A. lentus,* and its body segments and the basal joints of its legs bear bands of 3–4 spines. Both species feed on hydroids.

575 **Clawed Sea Spider**
(*Phoxichilidium femoratum*)
Class Pycnogonida

Description: 1/8″ (3 mm) long, 1/16″ (2 mm) wide. Slender, flattened, with long neck, tiny abdomen. Pale tannish-gray. *Proboscis cylindrical, rounded at tip,* projecting from under long neck, about 1/4 length of body; *pinchers longer than proboscis* on each side, but no feelerlike appendages. 4 eyes on projection behind proboscis. 4 pairs of slender legs about 1/2″ (13 mm) long, with *extra clawlet* beside claw at tip. Only male has accessory legs.
Habitat: Among hydroids and other growth on rocks; near low-tide line and below to water 332′ (101 m) deep.
Range: Arctic to Long Island Sound; Alaska to c. California.
Comments: This sea spider is most commonly found among dense growths of tubularian hydroids on which it feeds. The extra spur on the claw helps to distinguish it from the Lentil Sea Spider, which it resembles in general proportions.

Splash Pool Copepod
(*Tigriopus californicus*)
Class Crustacea

Description: 1/16″ (2 mm) long, 3 times longer than wide. Segmented, oval, with narrow rear part comprising half the length. *Orange to brick-red. Single eye* in midline at front end. Pair of conspicuous

antennae at front, with hook at tip in males. Terminal segment has *2 appendages with long bristles.* Females frequently with egg sac underneath rear part.

Habitat: Splash pools near or above high-tide line.

Range: Alaska to Baja California.

Comments: Dense populations of this little crustacean sometimes occur in those tidepools replenished only by splashes of sea water at high tide, or during storms. Such pools typically are highly salty because of evaporation, except when diluted during the rainy season; and the copepod can tolerate salinities 3 times as great as that of sea water.

288 Common Goose Barnacle
(*Lepas anatifera*)
Class Crustacea

Description: 6″ (15 cm) long, 2¾″ (70 mm) wide. Flattened, lance-shaped, stalked. Enclosed in 5 strong, *limy, white, orange- or yellow-edged plates;* stalk purplish-brown. 6 pairs of feathery feeding appendages extendible through gape between plates. *Stalk ½ total length,* thick, rubbery. Surface of *plates almost smooth, with fine lines;* stalk smooth.

Habitat: Floating; attached to drifting objects, buoys, bottles, tar masses, and the Common Purple Sea Snail.

Range: Washed ashore on both coasts of the United States.

Comments: The Common Goose Barnacle is a creature of the high seas. Its swimming larvae are attracted by the shaded undersides of floating objects, where they settle gregariously. Anything long afloat, such as a navigational buoy, may be completely covered below with thousands of these barnacles.

Float Goose Barnacle
(*Lepas fascicularis*)
Class Crustacea

Description: 2⅜″ (60 mm) long, 1⅜″ (35 mm) wide. Flattened, triangular, stalked. Enclosed in 5 *delicate, bluish-white, limy plates;* stalk brownish in adults, white in young. 6 pairs of bristly feeding appendages extendible through gape in plates. *Stalk less then ⅓ total length,* thick, smooth, rubbery. Surface of plates *deeply grooved, edges thickened.*

Habitat: Floating; young attached to seaweeds, drifting debris, or gas-filled floats.
Range: Washed ashore on both coasts of the United States.
Comments: Young barnacles develop attached to some small, floating object, and are common inhabitants of Sargasso Weed. As each matures, it forms a gas-filled float at the end of the stalk, and becomes independent of other means of flotation. Once such a float is established, new larvae may settle and develop on it, giving rise to bunches of barnacles.

277, 287 **Leaf Barnacle**
(*Pollicipes polymerus*)
Class Crustacea

Description: 3¼″ (83 mm) long, 1⅛″ (28 mm) wide. Cylindrical, stout, stalked, scaly. Enclosed in 6 *major white plates surrounded by many smaller, overlapping plates.* Stalk grayish or brownish. 6 pairs of feathery appendages extend through gape between plates, forming a feeding net. Stalk thick, tough, *covered with fine spines.* Surface of plates smooth.
Habitat: Attached to wave-swept boulders; between high- and low-tide lines.
Range: British Columbia to Baja California.
Comments: The Leaf Barnacle frequently occurs in vast, dense populations, sometimes mixed with the California Mussel.

276 **Little Gray Barnacle**
(*Chthamalus fragilis*)
Class Crustacea

Description: ¼" (6 mm) high, ⅜" (10 mm) wide at base. Conical, low, top flattened. *Grayish-white.* Sides composed of *2 pairs of limy plates overlapping 2 unpaired end plates.* Top 2 pairs of plates with gape between pairs. 6 pairs of fine, feathery appendages extend through gape when open. Surface of plates smooth. Base membranous, without limy deposit.

Habitat: On rocks, singly or in small, uncrowded groups; near high-tide line.

Range: Cape Cod to Florida and Texas; West Indies.

Comments: This little barnacle lives high between the tide lines and above in the splash zone when it is wet; consequently, it can feed only at high tide, a short time in each tidal cycle. Dall's Barnacle (*C. dalli*) ranges from Alaska to southern California, and the Smooth Gray Barnacle (*C. fissus*) ranges from central California to Baja California. They are the same size as *C. fragilis,* and can be distinguished only by dissection and examination of the interior surfaces of their plates.

278, 286 **Northern Rock Barnacle**
(*Balanus balanoides*)
Class Crustacea

Description: 1" (25 mm) high, ½" (13 mm) wide at base. Conical if not crowded, flat at top. White. Sides composed of *2 pairs of limy plates overlapping only 1 of 2 unpaired plates.* 2 pairs of plates at top with gape between pairs. 6 pairs of feathery appendages extend through gape when feeding. Plates smooth, with few vertical grooves; scalloped at base. *Base membranous,* without limy deposit.

Habitat: On hard objects; from between high-

and low-tide lines to below low-tide line in shallow water.

Range: Arctic to Delaware.

Comments: This barnacle occurs in such dense populations in some places that it can grow only in length, and its shape becomes greatly distorted. The sharp edges of its plates pose a hazard to any bare skin that touches them.

285 **Rough Barnacle**
(*Balanus balanus*)
Class Crustacea

Description: 2″ (51 mm) high, 2″ (51 mm) wide. Conical, flat at top. Grayish-white. Sides composed of *2 pairs of limy plates overlapping only 1 of 2 unpaired plates.* 2 pairs of plates at top with gape between. *Side plates with strong ribs,* rough, craggy. *Base limy,* inside surface of plates grooved.

Habitat: On rocks; below low-tide line to water 544′ (165 m) deep.

Range: Arctic to Cape Cod.

Comments: The name *Balanus* means "acorn" in Latin. This species was named by Linnaeus in 1758, the first acorn barnacle to be given a scientific name.

275 **Ivory Barnacle**
(*Balanus eburneus*)
Class Crustacea

Description: 1″ (25 mm) high, 1″ (25 mm) wide. Conical, flat at top. Ivory-white. Sides composed of *2 pairs of limy plates overlapping only 1 of 2 unpaired plates.* 2 pairs of *grooved plates* at top with gape between. Side plates smooth. Base limy and full of hollow tubes.

Habitat: On rocks and pilings in bays and estuaries; near low-tide line and below in shallow water.

Range: Maine to South America.

Comments: This is the common estuarine barnacle. It occurs in brackish habitats almost into fresh water.

281 Thatched Barnacle
(*Semibalanus cariosus*)
Class Crustacea

Description: 2″ (51 mm) high, 2⅜″ (60 mm) wide. Conical, rough. Gray. Wall of 6 plates; *surface appears thatched, with many slender, tubular ribs.* 2 pairs of plates at top with gape between; *base membranous* (seen only by prying animal off rock).
Habitat: On rocks, along exposed shores; above low-tide line.
Range: Alaska to s. California.
Comments: This species grows in crowded colonies in the northern part of its range, but is more solitary in the southern part. *S. cariosus* broods its young in the winter, and in spring liberates larvae that then settle on a rocky surface. These barnacles may live as long as 15 years.

280, 283 Giant Acorn Barnacle
(*Balanus nubilis*)
Class Crustacea

Description: 3½″ (89 mm) high, 4⅜″ (111 mm) wide. Conical, flat at top. Whitish. Sides composed of *2 pairs of limy plates overlapping only 1 of 2 unpaired plates.* 2 pairs of plates at top with gape between. *Side plates heavy, rough, but without definite ribs;* top plates without grooves, *1 pair of top plates with long projection* extending upward.
Habitat: On rocks, pilings, and hard-shelled animals; from low-tide line to water 300′ (91 m) deep.
Range: Alaska to s. California.
Comments: This large barnacle is sometimes confused with the Eagle Barnacle (*B. aquila*), which gets even larger, 5⅜″

(136 mm) high and 5¼″ (133 mm) wide. *B. aquila* is also white and rough, but has grooved top plates. These large barnacles are roasted and eaten by Indians in the northwest.

279 Little Striped Barnacle

(*Balanus amphitrite*)
Class Crustacea

Description: ¾″ (19 mm) high, ¾″ (19 mm) wide. Conical, flat at top. White, with reddish-brown or purple stripes, cluster of 3–4 stripes in middle of each side plate. Sides composed of *2 pairs of limy plates overlapping 1 of 2 unpaired plates.* 2 pairs of plates at top with gape between. *Side plates smooth, bluntly pointed at top.* Top plates smooth. Base limy.

Habitat: On rocks, pilings, and shells in bays and estuaries; from low-tide line to water 60′ (18 m) deep.

Range: C. California to Panama; Cape Cod to Florida and Texas; Mexico; Bahamas; West Indies.

Comments: This species is commonly found on the bottoms of ships, which have carried it to many parts of the world. Though it can survive in colder waters, it requires a temperature of at least 68°F (20°C) to breed.

274 Bay Barnacle

(*Balanus improvisus*)
Class Crustacea

Description: ¼″ (6 mm) high, ½″ (13 mm) wide. Conical, flat at top. White. Sides composed of *2 pairs of limy plates overlapping only 1 of 2 unpaired plates.* 2 pairs of plates at top with gape between. *Side plates smooth.* Base limy.

Habitat: On rocks, pilings, oysters and other hard-shelled animals, in brackish estuaries; from low-tide line to water 120′ (37 m) deep.

Range: Oregon to Ecuador; Nova Scotia to Florida and Texas; Mexico; West Indies to Brazil.

Comments: This estuarine form was introduced from the East Coast to the West Coast before the middle of the 19th century along with the Eastern Oyster. It can tolerate fresh water at least part of the year, and has been found attached to freshwater crayfish.

284 Red-striped Acorn Barnacle
(*Megabalanus californicus*)
Class Crustacea

Description: 2″ (51 mm) high, 2⅜″ (60 mm) wide. Almost cylindrical, flat-topped. White, with *longitudinal red stripes.* Sides composed of *2 pairs of limy plates overlapping 1 of 2 unpaired plates.* 2 pairs of plates at top with gape between. Side plates with *12–15 ribs of different sizes converging to point at top.* Base limy.

Habitat: On rocks, pilings, kelps, and other hard-shelled animals; from low-tide line to water 30′ (9 m) deep.

Range: N. California to Mexico.

Comments: Because of their gregarious habits, these barnacles get much longer than wide. Splendid clusters are cleaned and sold in shell shops.

282 Volcano Barnacle
(*Tetraclita rubescens*)
Class Crustacea

Description: 2″ (51 mm) high, 2″ (51 mm) wide at base. Conical, volcano-shaped. Brownish or brick-red. Wall consists of *4 plates, almost indistinguishably fused.* 2 pairs of plates at top with gape between pairs. 6 pairs of bristly appendages extend through gape when feeding.

Plates highly ribbed, sometimes eroded. Young barnacles proportionally less high. Base consists of limy plate.

Habitat: On rocks, on open wave-swept shore; between high- and low-tide lines and sometimes below.

Range: C. California to Baja California.

Comments: This genus is chiefly tropical and is the only non-stalked barnacle with a wall of 4 plates. The West Indian Volcano Barnacle (*T. stalactifera*) is about the same size and generally has the same body form as *T. rubescens,* and is whitish or creamy-gray to dark gray. It is found in the Florida Keys and the West Indies.

---

597 Swollen-clawed Squilla
(*Gonodactylus oerstedii*)
Class Crustacea

Description: 1⅝" (41 mm) long, ¼" (6 mm) wide. Shrimplike. Cream-colored, mottled with light green or olive-green. Carapace flattened, short. Large, *jacknifelike appendage* near front end with *swollen claw* devoid of spines. 3 pairs of walking legs. *Tailpiece with ridge down middle,* border with strong *triangular tooth on each side of midline.*

Habitat: Burrowed under rocks in sand; at low-tide line and below in shallow water.

Range: North Carolina to Florida; Bermuda; West Indies to Brazil; Gulf of California to Ecuador.

Comments: This small mantis shrimp is the only American species whose claw is not armed with a row of sharp spines. The Pacific form is less mottled and more solidly colored than the Atlantic form.

595 **Common Mantis Shrimp**
(*Squilla empusa*)
Class Crustacea

Description: 10″ (25 cm) long, 2½″ (64 mm) wide. Shrimplike, somewhat flattened. Greenish or bluish-green, with darker green or blue margins to segments. *Large, jacknifelike appendage* near front with 6 *sharp spines* on claw. 3 pairs of walking legs. Tailpiece with blunt ridge down middle, and 6 *strong marginal spines.*
Habitat: Burrows in sand or mud; from low-tide line to water 500′ (152 m) deep.
Range: Cape Cod to Florida and Texas; south to Brazil.
Comments: This is the "shrimp snapper" well known and respected by shrimp trawlers. A quick slash of one of its large appendages can cut a shrimp or fish in two—or lacerate a finger. It is edible, and said to be delicious.

592 **Ciliated False Squilla**
(*Pseudosquilla ciliata*)
Class Crustacea

Description: 4″ (102 mm) long, ¾″ (19 mm) wide. Shrimplike, somewhat flattened. Yellowish-brown, greenish-brown, bright green, pale green, or whitish. Large, *jacknifelike appendage with 3 long spines* on claw near front end. 3 pairs of walking legs. *Tailpiece with several spines on top.* 3 pairs of spiny marginal teeth.
Habitat: In burrows around coral reefs, and on grass flats; below low-tide line.
Range: S. Florida; Bermuda; Bahamas; West Indies to Brazil.
Comments: The great color variation characteristic of this species shows how undependable color can be as a criterion for identification. Lesson's False Squilla (*Pseudosquillopsis marmorata*) is reported from southern California to the Galapagos Islands, and is found under

rocks at the low-tide line and below. It measures 4¾" (121 mm) long and ⅞" (22 mm) wide, and is brownish-gray or greenish.

---

596 Scaly-tailed Mantis Shrimp
(*Lysiosquilla scabricauda*)
Class Crustacea

Description: 12" (30 cm) long, 2½" (64 mm) wide. Shrimplike, somewhat flattened, stout. Cream-colored, with *dark, segmented cross-bands.* Large, jacknifelike appendages near front end with *8–11 sharp spines on claw.* 3 pairs of walking legs. Tailpiece covered with *scaly bumps and spines.*

Habitat: Burrowed in sand or mud bottoms; below low-tide line in shallow water.

Range: Florida to Texas; West Indies to Brazil.

Comments: This is the largest of the mantis shrimps, and a formidable creature to handle.

---

594 Harford's Greedy Isopod
(*Cirolana harfordi*)
Class Crustacea

Description: ¾" (19 mm) long, ¼" (6 mm) wide. Elongate, oval, flattened. Pale gray, tan, brown, blackish, 2-toned, variable. 2nd pair of antennae slender, ½ body length. Black eyes near side of head. 7 thoracic segments behind head, each bearing pair of *walking legs tucked under body,* all similar. Rounded, *triangular tailpiece* plus pair of flattened 2-part appendages form *tail fan.*

Habitat: Under stones on sandy bottom; in mussel beds between high- and low-tide lines, and kelp holdfasts below low-tide line.

Range: British Columbia to Baja California.

Comments: These scavengers occur in large numbers and quickly consume any dead creature in their vicinity. They are, in turn, preyed on by a number of inshore fishes. The Bay Greedy Isopod (*C. polita*) occurs in inshore waters from the Bay of Fundy to Cape Cod. It is ⅝" (16 mm) long and ⅛" (3 mm) wide, and has a rounded tailpiece.

585 **Baltic Isopod**
(*Idotea baltica*)
Class Crustacea

Description: 1" (25 mm) long, ¼" (6 mm) wide. Elongate, straight, somewhat widened in middle, flattened. Tannish-green, dark green, or mottled red, brown, or black, with white. 2nd pair of antennae directed forward and bent to the side. *Eyes at side of head.* 7 thoracic segments, each with pair of walking legs out to side, all similar. *Tailpiece ¼ body length, squarish at end with point in middle.* Last pair of appendages forms *doors enclosing gill-like abdominal appendages* under tailpiece.

Habitat: On seaweeds and rocks near shore, or swimming nearby in shallow water.

Range: Gulf of St. Lawrence to North Carolina.

Comments: This is the largest isopod along the northeastern coast. In early summer copulating pairs swim about, the larger female holding the smaller male beneath her. A closely related species, ranging from the Gulf of St. Lawrence to Cape Cod, is the Sharp-tailed Isopod (*I. phosphorea*), which grows to ½" (13 mm) long and ⅛" (3 mm) wide, has an arrowhead-shaped tailpiece, and occurs in a great variety of colors and patterns: solid, banded, or mottled brown, white, red, yellow, gray, or greenish. The Cut-tailed Isopod (*I. resecata*) ranges from Alaska to Mexico, is 1½" (38 mm) long and ⅝" (16 mm) wide, has a tailpiece with a concave border, and is green or brown. It clings to eelgrass and other seaweeds.

583 Vosnesensky's Isopod
(*Idotea wosnesenskii*)
Class Crustacea

Description: 1⅜″ (35 mm) long, ½″ (13 mm) wide. Elongate, straight, flattened. Almost black, also light brown, red, or green. Body sturdy, 2nd pair of antennae ¼ body length, straight. Eyes at side of head. 7 pairs of thoracic legs, all similar. *Tailpiece rounded, with small tooth at tip,* ¼ body length. Last pair of appendages form *doors enclosing gill-like appendages* under tailpiece.

Habitat: Under rocks, in mussel beds, and among seaweeds, in bays and exposed rocky shores; from between high- and low-tide lines to water 53′ (16 m) deep.

Range: Alaska to s. California.

Comments: This species is eaten by a number of shallow-water fishes. It was named after Russian zoologist I. G. Vosnesensky, who collected animals in Siberia, Alaska, and California in the mid-19th century.

584 Kirchansky's Isopod
(*Idotea kirchanskii*)
Class Crustacea

Description: ⅝″ (16 mm) long, ¼″ (6 mm) wide. Elongate, straight, flattened. Bright to dark green, sometimes mottled with pink. 2nd antennae stout, ¼ body length. Eyes at side of head. 7 pairs of thoracic legs, all similar. *Tailpiece rounded with slight point at tip.* Last pair of appendages form *doors enclosing gill-like abdominal appendages* under tailpiece.

Habitat: On surfgrass; above low-tide line on open rocky coast.

Range: C. California.

Comments: This animal clings tenaciously to surf-grass, and is well-adapted to life on a waveswept shore. The Monterey Isopod (*I. montereyensis*) is similar to *I.*

kirchanskii in size, and shares its habitat and range, but its coloration also varies to red and brown and often includes black and white.

580 Northern Sea Roach
(*Ligia oceanica*)
Class Crustacea

Description: 1″ (25 mm) long, ½″ (13 mm) wide. Oval, flattened, roachlike. Tannish-gray, mottled. *2nd pair of antennae ½ body length,* slender. Eyes at side of head. 7 clearly defined thoracic segments, each with pair of walking legs, all similar. Tailpiece short. *Last pair of appendages slender, ¼ body length, forked,* extending backward from sides of tailpiece.

Habitat: On rocks and pilings; near high-tide line and above.

Range: Cape Cod north, at least to Maine.

Comments: This animal's shape, speed, and habit of rushing into hiding when disturbed are reminiscent of a cockroach. The Exotic Sea Roach (*L. exotica*) ranges from Chesapeake Bay to Florida and the West Indies, and is extremely abundant along the Florida inland waterway. It is 1¼″ (32 mm) long and ⅝″ (16 mm) wide, and its last pair of appendages is about ½ body length. Its habits are similar to those of *L. oceanica.*

581, 582 Western Sea Roach
(*Ligia occidentalis*)
Class Crustacea

Description: 1″ (25 mm) long, ½″ (13 mm) wide. Oval, flattened, roachlike. Tannish-gray, mottled. Eyes at side of head. 7 clearly defined thoracic segments, each with pair of walking legs, all similar. Tailpiece short. *Last pair of appendages measuring more than ⅓ body length,* slender, extending backward from sides of tailpiece.

Habitat: On rocks and jetties; near high-tide line and above.

Range: C. California to Central America.

Comments: These creatures are more commonly hidden in crevices during the day. They become active late in the afternoon, and continue until after dawn. They undergo a daily color change, from darker during the day to paler at night.

589 **Red-eyed Amphipod**
(*Ampithoe rubricata*)
Class Crustacea

Description: ¾" (19 mm) long, ⅛" (3 mm) high. Arched, width equal to height. Grayish-green, pale below. *Antennae almost equal to body in length, 1st pair slender, 2nd pair heavier. Eyes oval, small, bright red.* 7 pairs of walking legs, first 2 pairs heavy, with large claw, last 2 pairs bent back along abdomen. *Tailpiece a rounded trapezoid.* Body smooth.

Habitat: Among seaweeds and in mussel beds on rocky coasts and in estuaries; above and just below low-tide line.

Range: Labrador to Long Island Sound.

Comments: These amphipods are early summer breeders, the female holding developing young in a pouch on her underside.

591, 598 **Scud**
(*Gammarus oceanicus*)
Class Crustacea

Description: 1" (25 mm) long, ¼" (6 mm) high. Arched, sides flattened, height greater than width. Olive-green or reddish. 2 pairs of *equally long antennae, ⅓ body length. Eyes kidney-shaped.* 7 pairs of walking legs, first 2 pairs heavy, with large claw, last 3 pairs bent back along abdomen. *Tailpiece longer than wide, with deep cleft* in middle. Back smooth, except last 2 segments, which are humped and have rows of 2–4 spines on each side of midline.

Habitat: Among rocks and seaweeds on open shores and in estuaries; above low-tide line and below to water 100′ (30 m) deep.

Range: Arctic to Chesapeake Bay.

Comments: A scud moves along the sea bottom lying on its side, flexing its body and kicking with the legs on the bottom side. It preys on worms, small crustaceans, and other invertebrates.

590 **Mottled Tube-maker**
(*Jassa falcata*)
Class Crustacea

Description: ⅜″ (10 mm) long, 1/16″ (2 mm) high. Arched, slender, nearly as wide as high. Reddish, mottled with paler spots. Antennae bristly, 2nd pair nearly twice as long as 1st. Oval eyes beside base of first antennae. 2nd walking leg with *huge hand nearly ⅓ body length, with large thumb and long, sharp claw* opposite it; last 2 pairs bent along abdomen.

Habitat: In tubes on pilings, wharves, buoys, eelgrass, and hydroid stems; near low-tide line and below to water 33′ (10 m) deep.

Range: Newfoundland to Florida and Texas; British Columbia to s. California.

Comments: This amphipod builds a tube out of mud, debris, and mucus, attaching it to almost any solid surface where there is good water flow. It feeds both by straining out suspended organic particles with its bristly antennae, and by preying on small invertebrates, grasping them with its huge second walking leg.

588 Noble Sand Amphipod
(*Psammonyx nobilis*)
Class Crustacea

Description: ¾″ (19 mm) long, ⅜″ (10 mm) high. Grayish-white. Arched, plump, sides flattened. Both pairs of antennae small, 1st pair shorter. Eyes kidney-shaped, black. 7 pairs of walking legs; *first 5 pairs of walking legs nearly covered with large plates;* 1st two pairs thicker at tips, *last 3 pairs short, broad, strong.* Tailpiece narrow, with deep cleft.

Habitat: In medium to coarse sand on exposed beaches, especially near stones; from between high- and low-tide lines to water 20′ (6 m) deep.

Range: Labrador to Long Island Sound.

Comments: This handsome sand-dweller can easily be found by shallow digging on New England beaches at low tide. It can bury itself with a few quick kicks of its stubby walking legs.

587 Big-eyed Beach Flea
(*Talorchestia megalophthalma*)
Class Crustacea

Description: 1″ (25 mm) long, ⅜″ (10 mm) high. Arched, broad, heavy. Gray to reddish-brown. 1st pair of antennae very short, *2nd pair of antennae less than ½ body length. Eyes large, bulging,* covering most of side of head. 7 bristly walking legs, 2nd enlarged at tip with claw; last 3 bent back along abdomen, last 2 long. Tailpiece triangular, with blunted end.

Habitat: On clean, sandy surf of exposed beaches, in burrows, or under logs and debris; near high-tide line and above.

Range: Newfoundland to Florida.

Comments: These lively creatures can leap a foot or more, much like fleas. Some bathers are unnecessarily apprehensive about being bitten by beach fleas, which feed only on organic debris. The Long-horned Beach Flea (*T. longicornis*) is the same size as *T. megalophthalma* and occurs in the same range. Its 2nd pair of antennae are almost body length and its eyes do not bulge.

586 California Beach Flea
(*Orchestoidea californiana*)
Class Crustacea

Description: 1⅛" (28 mm) long, ¼" (6 mm) high. Arched, stout, broad. Ivory. *Juveniles with dark "butterfly"-shaped marks on back.* 1st pair of antennae short, *2nd pair of antennae longer than body, orange in juveniles, rosy red in adults.* Eyes round, medium-sized, black. 7 pairs of walking legs, 2nd pair with thick hand and long claw, last 3 pairs heavy and strong, last 2 pairs bent back along abdomen.

Habitat: On wide, fine-sand beaches along open shore; near high-tide line and above.

Range: British Columbia to s. California.

Comments: At night these beach fleas can be seen in hordes on the sand above the breaking waves, leaping about and eating washed-up seaweed.

601 Linear Skeleton Shrimp
(*Caprella linearis*)
Class Crustacea

Description: ¾" (19 mm) long, 1/32" (1 mm) wide. Long, slender, arched, jointed. Pale buff. *Head rounded at front,* long. 1st pair of antennae ⅓ body length, twice as long as 2nd pair. 2nd pair of *antennae with row of long bristles* on underside. Eyes small, round, red. 1st and 2nd thoracic appendages enlarged, with grasping claw, *2nd thoracic appendage much larger.* Pair of saclike gills on next 2 segments. Last 3 segments with bristly grasping appendages directed backward.

Habitat: Among seaweeds, bushy hydroids and other growth; from low-tide line to water 660′ (201 m) deep.
Range: Arctic to Long Island Sound.
Comments: Males of this species are larger than females, each of which has a brood-pouch of leaflike plates on the underside.

600 Smooth Skeleton Shrimp
(*Caprella laeviuscula*)
Class Crustacea

Description: 2″ (51 mm) long, ¼″ (6 mm) wide. Long, slender, arched, jointed. Pale tan, greenish, or rosy. *Head rounded at front,* long, 1st antenna ⅓ body length, twice as long as 2nd antenna. Eyes small, round. Body smooth. 1st and 2nd thoracic appendages with grasping claw, *2nd thoracic appendage much larger.* Pair of saclike gills on next 2 segments. Last 3 segments with bristly grasping appendages directed backward.
Habitat: On hydroids, bushy algae, and other growth; near low-tide line and below in shallow water.
Range: British Columbia to s. California.
Comments: This species eats detritus, 1-celled plants, small invertebrates, and carrion, and is in turn eaten by various fishes, shrimps, and sea anemones. The California Skeleton Shrimp (*C. californica*) is larger, 1⅜″ (35 mm) long and 1/32″ (1 mm) high, and buffy; it lives on eelgrass in central and southern California.

599 Long-horn Skeleton Shrimp
(*Aeginella longicornis*)
Class Crustacea

Description: 2⅛″ (54 mm) long, 1/16″ (2 mm) wide. Long, slender, arched, jointed. Tannish, reddish, or almost colorless.

Head short, with spiny bumps. *First pair of antennae ⅔ body length,* 2nd pair of antennae shorter, less than ⅓ body length. Eyes small, round. Large *2nd thoracic appendage with 2–3 sharp teeth* on underside of palm, long claw. Pair of saclike gills on next 2 segments. Last 3 segments with non-bristly grasping appendages directed backward.

Habitat: On seaweeds, hydroids, and sponges; from low-tide line to water 7450′ (2271 m) deep.

Range: Labrador to North Carolina.

Comments: This is the largest skeleton shrimp in American Atlantic waters.

604, 606 **Opossum Shrimps**
(*Mysis* spp.)
Class Crustacea

Description: 1¼″ (32 mm) long, ⅛″ (3 mm) wide. Slender, shrimplike. Translucent, greenish, each segment with dark brownish spot. 1st pair of antennae with 1 short and 1 long slender branch; *2nd pair of antennae with 1 flattened, long, narrow, bladelike branch,* the scale, and 1 long, slender branch. Eyes large, stalked, black. Carapace covers most of thorax. *4th pair of abdominal appendages in male long and pointed.* Thoracic legs 2-branched, many-jointed, bristly. *Female with pouch for young under thorax.* Abdomen long, slender, with 5 pairs of forked, bristly appendages; last segment with pair of flattened, forked appendages and *notched tailpiece with bristles along entire margin,* making up tail fan.

Habitat: Among seaweed and eelgrass ranging into brackish estuaries; from low-tide line to water 600′ (183 m) deep.

Range: Gulf of St. Lawrence to New Jersey.

Comments: Opossum shrimps are so named because, like opossums, the females have a pouch. It is made up of flat, folding plates that close toward the midline, and hold the developing eggs.

603 **Red Opossum Shrimp**
(*Heteromysis formosa*)
Class Crustacea

Description: ⅜″ (10 mm) long, ¹⁄₁₆″ (2 mm) wide. Slender, shrimplike. *Red, pink,* or brownish. 1st pair of antennae with 1 short and 1 long slender branch; 2nd pair of antennae with 1 long, slender branch and *1 flattened, broad, oval, leaflike branch,* the scale. Eyes large, stalked, black. Carapace covers most of thorax. Thoracic appendages 2-branched, many-jointed, bristly. Female usually red, with pouch for young under thorax. Abdomen long, slender, with 5 pairs of forked appendages and *notched tailpiece with bristles along ½ the margin,* making up tail fan.

Habitat: In tidepools, under rocks and shells; from above low-tide line to water more than 800′ (243 m) deep.

Range: Bay of Fundy to New Jersey.

Comments: This species is more of a bottom-dweller than part of the plankton, and is usually found among rocks and shells, especially shells of the Surf Clam.

---

602 **Bent Opossum Shrimp**
(*Praunus flexuosus*)
Class Crustacea

Description: 1″ (25 mm) long, ¹⁄₁₆″ (2 mm) wide. Slender, shrimplike. Tannish, with branching, blackish spots. 1st pair of antennae with 1 short and *1 long slender antennal branch more than ½ body length; 2nd pair of antennae with 1 long, flattened, bladelike branch, or scale, with spine at tip, and shorter spines along outside margin,* and 1 short, slender branch. Eyes large, stalked, golden with black centers. Carapace covers most of thorax. Thoracic appendages 2-branched, many-jointed, bristly. Abdomen long, slender, with slight S-shaped bend; 5 pairs of forked, bristly abdominal appendages; last segment with flattened, forked appendages and *notched tailpiece* with bristles along margin, making up tail fan.

Habitat: Among seaweeds and around rocks; in shallow water.

Range: Nova Scotia to Cape Cod.

Comments: The Bent Opossum Shrimp was first found at Barnstable, Massachusetts, in 1960, presumably introduced from Europe. It is now common in the shallow waters of the Gulf of Maine as far north as Nova Scotia.

605 **Horned Krill**
(*Meganyctiphanes norvegica*)
Class Crustacea

Description: 1½″ (38 mm) long, ⅜″ (10 mm) high. Shrimplike. Transparent, with red spots on leg bases; stomach contents visible. 1st pair of antennae with spur on top of base, forked, slender, body-length; 2nd pair of antennae forked, one branch slender, ½ body length, the other flat, scalelike. Eyes large, stalked, black. Carapace covers thorax; thoracic appendages slender, bristly, bent forward, then downward and toward middle, forming *"thoracic basket"; gill not covered by carapace,* extending upward from base of each thoracic leg. Abdomen slender, with 5 pairs of forked swimming appendages; last segment with tailpiece and pair of forked, flat appendages, the tail fan.

Habitat: Floating in open sea; from surface to water more than 960′ (293 m) deep.

Range: Entire n. Atlantic Coast.

Comments: Krill is a collective name for several kinds of planktonic shrimps that occur in large schools. The Horned Krill is sometimes so abundant that its swarms cause the water to appear red. Krill is eaten by baleen whales, which charge, mouths open, into a school and then, mouths closed, strain out the krill.

609, 611 **Pink Shrimp**
(*Penaeus duorarum*)
Class Crustacea

Description: Males 6½″ (17 cm) long, 1″ (25 mm) high. Females 8⅜″ (21 cm) long, 1⅛″ (28 mm) high. Sides somewhat flattened. Gray, bluish-gray, or reddish-brown; dark spot between 3rd and 4th abdominal segments. Beak continuous with *ridge extending almost to rear end of carapace, with broad, rounded groove on each side.* 10 pointed teeth along ridge, extending onto beak; 2 teeth on underside of beak. *First 3 pairs of walking legs with pincers.* Surface of carapace and abdomen smooth.

Habitat: On various bottoms; from low-tide line to water 300′ (91 m) deep.

Range: Chesapeake Bay to Florida and Texas; Bermuda; Bahamas; West Indies to Brazil.

Comments: The Pink Shrimp is sought by fisheries from North Carolina through the Gulf Coast. The Brown Shrimp (*P. aztecus*) ranges from New Jersey to Florida and Texas, and from the West Indies to Uruguay; it is found from the shoreline to a depth of 300′ (91 m). It is similar to the Pink Shrimp, but has a groove down the middle of the rear half of the ridge, and it lacks a brown abdominal spot. It is brown or grayish-brown. The White Shrimp (*P. setiferus*) is bluish-white with dusky bands and scattered black dots, with the beak and sides tinged with pink, and with red swimmerets. Its beak is continuous with a ridge that extends only about ⅔ the distance to the rear of the carapace, does not have a groove in its midline, and is flanked by a pair of shallow grooves that extend only ½ the length of the carapace. Males measure 7¼″ (18 cm) long and 1″ (25 mm) high, and females 7⅞″ (20 cm) long and 1⅛″ (28 mm) high. These 3 species are the backbone of our Atlantic shrimping industry.

607 Common Shore Shrimp
(*Palaemonetes vulgaris*)
Class Crustacea

Description: Males 1½″ (38 mm) long, ¼″ (6 mm) high, females 1¾″ (44 mm) long, ¼″ (6 mm) high. Slender, elongate. Transparent, with a few red, yellow, white, and blue spots on back. Beak reaches beyond antennal scale, *beak tip directed upward;* 8–11 teeth along top of beak, *2 teeth behind eye socket. First 2 pairs of walking legs with pincers, 2nd pair larger.* Carapace and back of abdomen smooth.

Habitat: In bays and estuaries, usually among submerged seaweeds; from low-tide line to water 45′ (14 m) deep.

Range: Gaspé Peninsula to Yucatán Peninsula.

Comments: This shrimp has been used to study the hormonal control of color in crustaceans. It has cells—containing red, yellow, blue, or white pigment—whose independent expansion or contraction changes the animal's hue.

616 Pederson's Cleaning Shrimp
(*Periclimenes pedersoni*)
Class Crustacea

Description: 1″ (25 mm) long, ¼″ (6 mm) high. Long, tapered toward rear. Transparent, with *dark purple spots* on abdomen and tail fan, and *purple stripes on appendages. 3rd segment of abdomen enlarged, bent downward* almost at a right angle. 2nd pair of antennae longer than body length.

Habitat: Among tentacles of various sea anemones; around coral reefs in shallow tropical seas.

Range: Florida; Bahamas; West Indies; Yucatán.

Comments: This species lives among the tentacles of a sea anemone without being stung by the host.

617 **Spotted Cleaning Shrimp**
(*Periclimenes yucatanicus*)
Class Crustacea

Description: 1″ (25 mm) long, ¼″ (6 mm) high. Slender, elongate. Transparent, with brown speckles at front end; 3 large, round, white-bordered, tannish spots on the back; *3 white-bordered blue-black spots* on the side of the carapace, and similar spots on the flat paired appendages in the tail fan. *Legs and antennae banded with black and white. Third segment of abdomen enlarged and bent downward* almost at a right angle.

Habitat: Among the tentacles of the Pink-tipped Sea Anemone, Ringed Anemone and others; around coral reefs in shallow water.

Range: Florida; West Indies; Yucatán.

Comments: Like other cleaning shrimps, this species cleans parasites from the skin and mouth of fishes.

---

622 **Brown Pistol Shrimp**
(*Alpheus armatus*)
Class Crustacea

Description: 1½″ (38 mm) long, ⅜″ (10 mm) high. Brown, with light tan markings on side of body. *Antennae banded red and white.* Beak short, eyes small. Part of socket over eye with small tooth at front margin. *First pair of walking legs with pincers, one almost length of carapace.*

Habitat: Among rocks and coral rubble, and on reefs; near low-tide line and below in shallow water.

Range: Florida; West Indies.

Comments: The red-and-white-banded antennae make this species easy to distinguish from other species of pistol or snapping shrimps.

618 **Banded Coral Shrimp**
(*Stenopus hispidus*)
Class Crustacea

Description: 2″ (51 mm) long, ⅜″ (10 mm) high. Slender, elongate. *White, with red-edged purple bands* on body and pincers. Antennae twice body length. 1st pair of walking legs nearly body length, *hand of pincer long.* Pincers and *body surface spiny.*
Habitat: On coral reefs.
Range: Florida; West Indies.
Comments: This beautiful cleaning shrimp advertises its availability by standing in one place and waving its antennae to attract a fish in need of having its external and mouth parasites removed.

608 **Red Rock Shrimp**
(*Lysmata californica*)
Class Crustacea

Description: 2¾″ (70 mm) long, ½″ (13 mm) high. Slender, elongate. Cream-colored, with irregular longitudinal red stripes; red appendages. Beak prominent, ½ as long as antennal scale, ridged; *beak has 8 or more teeth above, 4 behind eyestalk level.* 2 longest joints of 2nd walking leg beaded, with *50 circular constrictions.*
Habitat: In rocky tidepools and crevices, and among seaweeds; from low-tide line to water 200′ (61 m) deep.
Range: S. California to Baja California.
Comments: These cleaning shrimp are also scavengers; scuba divers have even observed them picking dead skin from around divers' fingernails.

615 **Grabham's Cleaning Shrimp**
(*Lysmata grabhami*)
Class Crustacea

Description: 2⅛″ (54 mm) long, ⅜″ (10 mm) high. Long, tapered toward rear. Golden-yellow, with *white stripe* down middle of back flanked by *scarlet band on either side.* Beak with 6 teeth above, 2 behind eyestalk level. *2 longest joints of 2nd walking leg beaded.*

Habitat: On coral reefs; below low-tide line in shallow water.

Range: S. Florida; Bahamas; West Indies.

Comments: This handsome little shrimp is much sought after by underwater photographers because of its striking colors. The female with eggs adds yet another shade, as the egg mass is green.

---

613 **Red-lined Cleaning Shrimp**
(*Lysmata wurdemanni*)
Class Crustacea

Description: 2¾″ (70 mm) long, ½″ (13 mm) high. Long, tapered toward rear. Translucent white, with *longitudinal red stripes down back, transverse ones on side.* Beak prominent, ½ as long as antennal scale, ridged, *with 4–5 teeth above and 3–5 below.* 2 longest joints of 2nd walking leg beaded.

Habitat: On rocks, jetties, and coral reefs, among hydroids on pilings and buoys; from low-tide line to water 100′ (30 m) deep.

Range: Chesapeake Bay to Florida and Texas; Bahamas; West Indies to Brazil.

Comments: Specimens of this shrimp from South America differ from the North American ones in minor details, such as the number of teeth on the beak and the number of beads on the second walking leg. It is possible that they represent a distinct subspecies.

612 **Greenland Shrimp**
(*Lebbeus groenlandicus*)
Class Crustacea

Description: 2¼″ (57 mm) long, ⅝″ (16 mm) high. Slender, sturdy, elongate. Variable; gray, tan, brown, greenish, reddish, white, spotted or irregularly striped. Slender beak continues as *sharp ridge to end of carapace,* with 4 *strong teeth* behind eyestalk level. End joint of 2nd walking leg is constricted into 7 beadlike swellings. Lower margins of *abdominal side plates end in points.* Abdomen bent down at 3rd segment.
Habitat: In kelp beds and among rocks; near low-tide line and below to water 690′ (210 m) deep.
Range: Arctic to Massachusetts.
Comments: This cold-water species can be found at low tide in kelp beds in places of great tidal fluctuation, such as the Bay of Fundy.

614 **Montague's Shrimp**
(*Pandalus montagui*)
Class Crustacea

Description: 3¾″ (95 mm) long, ¾″ (19 mm) high. Long, tapered toward rear. Pink or red, young with red stripes. Beak long, slender, curved upward, *9–10 teeth above beak, 7 below.* 2nd walking legs unequal; end joint beaded with 20 segments.
Habitat: On various bottoms; young from low-tide line to shallow depths, adults in water 100′ (30 m) or more deep.
Range: Arctic to Rhode Island.
Comments: Though this species is smaller than the Maine Shrimp, it also is taken commercially. It is one of the edible prawns of northern Europe.

610 **Coon-stripe Shrimp**
(*Pandalus danae*)
Class Crustacea

Description: 5¾" (146 mm) long, 1¼" (32 mm) high. Large. Pale red, with irregular longitudinal blue stripes and occasional white spots on body; *legs striped brown and white. Beak curved upward,* long, slender, continuing as ridge on carapace; many teeth above and below on beak; *4 teeth on ridge behind level of eye socket.* 2nd walking legs unequal, end joint beaded with many constrictions. Tailpiece with 6 spines on each side.

Habitat: In bays, estuaries, and eelgrass beds, occasionally in tidepools; from low-tide line to water 600′ (182 m) deep.

Range: Alaska to c. California.

Comments: The Coon-stripe Shrimp is a functional male when young, subsequently changing to a fertile female. The Maine Shrimp (*P. borealis*) is fished commercially off New England. It is 7" (18 cm) long and 1½" (38 mm) high. Its beak is longer than its carapace, and it has a spine in the middle of the back of the 3rd and 4th abdominal segments. It ranges from the Arctic to Cape Cod.

593 **Sand Shrimp**
(*Crangon septemspinosa*)
Class Crustacea

Description: 2¾" (70 mm) long, ½" (13 mm) high. Somewhat flattened, top to bottom. Transparent, colorless, pale gray, buff, with many irregular, tiny, star-shaped, black spots; tail fan often blackish. *Beak short, without teeth,* reaching forward to level of eyestalks; spine in middle of back behind beak. 1st walking leg heavy, with *backward-bending claw* at tip; 2nd and 3rd walking legs very slender.

Habitat: On sandy bottoms and in eelgrass beds, on open shores and in bays and estuaries; from low-tide line to water more than 300′ (91 m) deep.

Range: Arctic to Florida.

Comments: Sand shrimps have a great tolerance for variations in salinity. The Franciscan Bay Shrimp (*C. franciscorum*) ranges from Alaska to southern California, living on sand or mud bottoms to a depth of 180′ (55 m). It is 2¼″ (57 mm) long and ¼″ (6 mm) high, and is similar to *C. septemspinosa* in color. It is trawled commercially in San Francisco Bay.

624 **Northern Lobster**
(*Homarus americanus*)
Class Crustacea

Description: 34″ (86 cm) long, 9″ (23 cm) high. Long, cylindrical, with large pincers. Greenish-black above, paler underneath; appendages and beak tipped with red. Rarely yellow or blue. Carapace cylindrical. Beak pointed, with 3 teeth on upper side, 1 behind level of eyestalk. 1st pair of antennae short, branch of *second pair of antennae longer than body.* First 3 pairs of walking legs with pincers, *1st pair of walking legs greatly enlarged, not alike.* One, usually left, heavier, blunt, with rounded teeth; other, usually right, less heavy, sharp, pointed; abdomen somewhat flattened. First 2 pairs of swimmerets of male modified for copulation.

Habitat: On rock bottoms, both in bays and open ocean; from near shoreline to continental shelf.

Range: Labrador to Virginia.

Comments: The lobster's two dissimilar pincers serve different purposes. The heavier one, or "crusher," is designed to crack hard objects like snails and bivalves. The sharper one, the "cutter," is used for tearing apart the prey, carrion, or plant material on which the animal feeds.

625 **West Indies Spiny Lobster**
(*Panulirus argas*)
Class Crustacea

Description: 24″ (61 cm) long, 6″ (15 cm) high. Long, cylindrical, spiny. Pale gray, tan, greenish, brownish, or mahogany, with large yellowish spots on carapace and tail, paler underneath; legs longitudinally striped with blue. Carapace has rows of strong spines, *largest pair of spines above eyestalks, bent forward.* 1st pair of antennae slender, branched, ⅔ body length; *2nd pair of antennae longer than body,* large, heavy, spiny, base with strong spines. *No pincers* on walking legs. Abdomen smooth, each segment with furrow across middle.

Habitat: Among rocks, sponges and other growth, and reefs; from low-tide line to water 300′ (91 m) deep.

Range: North Carolina to Florida; Gulf of Mexico; Bermuda; West Indies to Brazil.

Comments: Although called "crawfish" in many places, this lobster should not be confused with the freshwater crawfish or crayfish. It is heavily fished throughout its range, and in Florida is now abundant only in the Keys.

---

623 **California Rock Lobster**
(*Panulirus interruptus*)
Class Crustacea

Description: 16″ (41 cm) long, 4″ (102 mm) high. Long, cylindrical, spiny. Reddish-brown, spines red, underside lighter brown; legs with pale brown longitudinal stripe. Carapace with rows of strong spines, *largest pair of spines above eyestalks, directed forward.* 1st pair of antennae slender, branched, ⅔ body length; *2nd pair of antennae longer than body,* large, heavy, spiny, base with strong spines. No pincers on walking

legs. *Upper surface of abdomen pebbled.*
Habitat: Among rocks in tidepools; at low-tide line and below to moderately deep water.
Range: C. California to Baja California.
Comments: These lobsters are fished commercially in California and Baja California. Anyone handling them should wear gloves as protection against the spines.

Ridged Slipper Lobster
(*Scyllarides nodifer*)
Class Crustacea

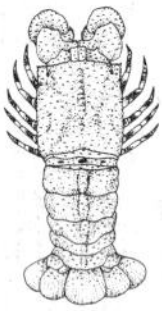

Description: 5″ (127 mm) long, 1¼″ (32 mm) wide. *Greatly flattened.* Yellowish to brownish; orange-red, low, round projections on edges; yellow underneath with brown spots; legs banded red and purple. *2nd pair of antennae strong, shovel-like;* walking legs simple, alike. Entire upper surface rough, with close-set, low, round projections.
Habitat: On mud, sand, shell, or coral bottoms, grassy beds and reefs; from low-tide line to water 240′ (73 m) deep.
Range: North Carolina to Florida and Texas; Bermuda; West Indies.
Comments: Like the spiny lobsters, this creature lacks pincers, and its antennae are modified into tools for shoveling and bulldozing. It preys on worms and other invertebrates.

626, 627 **Spanish Lobster**
(*Scyllarides aequinoctialis*)
Class Crustacea

Description: 12″ (30 cm) long, 3″ (76 mm) wide. Greatly flattened. Yellowish or reddish-brown, with brown spots, and *row of 4 reddish-purple spots* across top of 1st abdominal segment. Carapace thick, covered with low nodules and bumps. *2nd pair of antennae strong, shovel-like;*

walking legs simple, stout, similar.

Habitat: Around coral reefs and sand bottoms; in shallow to moderately deep water.

Range: S. Florida; Bahamas; West Indies.

Comments: This lobster is said to be delicious, but is not abundant enough to be commercially important.

---

621 **Flat-browed Mud Shrimp**
(*Upogebia affinis*)
Class Crustacea

Description: 2½" (64 mm) long, ⅜" (10 mm) high. Long, slender. Bluish or yellowish-gray. 1st pair of antennae small, *2nd pair of antennae ½ body length, bristly. Beak squarish, flattened,* extending beyond tiny eyes. *1st pair of walking legs coated with long bristles.* Abdomen large, twice length of carapace, separated from it by deep constriction. Tail fan large.

Habitat: Burrowing in sand and mudflats; in lower and midtidal zones.

Range: Cape Cod to Florida and Texas; West Indies to Brazil.

Comments: This crustacean is not a true shrimp, but is more closely related to the hermit crabs. It lives with a mate in a permanent burrow. The Blue Mud Shrimp (*U. pugettensis*) is larger, 6" (15 cm) long and 1" (25 mm) high. It is tannish-gray to blue-gray, with pale speckles on its upper surface, and is otherwise similar to the Flat-browed Mud Shrimp. It ranges from southern Alaska to Baja California.

---

619 **Beach Ghost Shrimp**
(*Callianassa affinis*)
Class Crustacea

Description: 2⅝" (67 mm) long, ½" (13 mm) high. Long, cylindrical. White to pink. Carapace ⅓ length of abdomen, *longitudinal groove* along each side. *Eyestalks short,* almost meeting in midline; beak small, rounded. 1st and 2nd walking legs with pincers, 1st pair unequal, *one walking leg huge in male.*

Habitat: Burrowing in gravel near large rocks on protected shores; in mid-tidal zone.

Range: S. California to Baja California.

Comments: These animals live in pairs in a permanent burrow which they share with a pair of blind Goby fish. The Short-browed Mud Shrimp (*C. atlantica*) ranges from Nova Scotia to Florida. It is 2½″ (64 mm) long, ½″ (13 mm) high, and whitish with yellow swimmerets and bristles, and burrows in sandy mud.

---

620 **Bay Ghost Shrimp**
(*Callianassa californiensis*)
Class Crustacea

Description: 4⅝″ (117 mm) long, ¾″ (19 mm) high. Long, slender. Whitish, with yellow swimmerets and yellow bristles. *Carapace ⅓ length of abdomen,* transparent on sides, gills visible. Base of antennae half as long as carapace, entire antennae as long as carapace. *1st pair of walking legs unequal, with pincers,* one huge in adult male.

Habitat: Burrowing in mud on flats, in bays and estuaries; between high- and low-tide lines.

Range: S. Alaska to Baja California.

Comments: This species burrows through loose sandy mud, and may dig down more than 24″ (61 cm) from the surface. Its burrow attracts numerous commensals.

---

647 **Say's Porcelain Crab**
(*Porcellana sayana*)
Class Crustacea

Description: ½″ (13 mm) long, ½″ (13 mm) wide. Round. Reddish or rust-brown, with yellowish-white, yellow, and bluish-white, round and elongate spots; variable. Carapace slightly convex. *Beak triangular, notched.* Strong *tooth above eye socket.* Pincers large, equal, hand longer than carapace. Last pair of walking legs folded up between carapace and abdomen. Abdomen loosely turned under body. Surface finely granular.

Habitat: Among rocks and clusters of oysters; as a commensal in snail shells occupied by large hermit crabs; from low-tide line to water 290′ (88 m) deep.

Range: Cape Hatteras to Florida and Texas; West Indies to Venezuela.

Comments: This little crab is sometimes commensal with large hermit crabs, which are rather messy eaters, shredding their food and dropping bits, which *P. sayana* then retrieves. This species may simply be opportunistic, adopting a host to find food for it.

---

641 Flat Porcelain Crab
(*Petrolisthes cinctipes*)
Class Crustacea

Description: 1″ (25 mm) long, ⅞″ (22 mm) wide. Carapace oval, flattened, almost circular. Reddish-brown, *legs banded near tips.* Pincers large, equal; *length of joint next to hand less than twice width,* margins not parallel, lobe on inner margin nearest body. 5th pair of walking legs folded up between carapace and abdomen. Abdomen turned under body.

Habitat: Under stones and in spaces between shells, in mussel beds, among sponges and tunicates; between high- and low-tide lines.

Range: British Columbia to s. California.

Comments: This is one of the most common crustaceans between high- and low-tide lines, especially in mussel beds.

672 **Thick-clawed Porcelain Crab**
(*Pachycheles rudis*)
Class Crustacea

Description: ¾″ (19 mm) long, ⅝″ (16 mm) wide. Carapace oval, flattened, nearly circular. Dull brown. *Granular pincers unequal, with scattered, long hairs.* 5th pair of walking legs folded up between carapace and abdomen; abdomen turned under body.

Habitat: Under rocks, in kelp holdfasts and holes left by boring clams, among mussels and on pilings; from low-tide line to water 60′ (18 m) deep.

Range: Alaska to Baja California.

Comments: These little crabs are usually found in pairs, living in a nook from which they have grown too large to escape.

679 **Fuzzy Crab**
(*Hapalogaster mertensii*)
Class Crustacea

Description: ¾″ (19 mm) long, ¾″ (19 mm) wide. Brown or reddish-brown. Carapace oval, wider toward rear, flattened; carapace and legs with *tufts of bristles.* Pincers not quite equal, right one usually larger; last pair of walking legs folded up between carapace and abdomen. *Abdomen thick and fleshy,* loosely turned under body.

Habitat: Under rocks in protected places; near low-tide line and below to moderate depth.

Range: Alaska to n. California.

Comments: This furry little crab strains scraps of algae out of the water with its hairy mouth parts, and crushes barnacles with its pincers. It is secretive, and is seldom found in the open during the day.

668, 674 **Butterfly Crab**
(*Cryptolithodes typicus*)
Class Crustacea

Description: 1½″ (38 mm) long, 2¾″ (70 mm) wide. Butterfly-shaped. Red, orange, yellow, brown, gray, variable. Carapace flattened, widest near rear, *prominent squarish beak; margin beyond eye sockets extending straight toward sides,* surface with row of low bumps toward each side. Pincers about equal, walking legs flattened, last pair small and tucked behind abdomen. Abdomen completely turned under body.

Habitat: Around rocks in quiet water; from low-tide line to water 50′ (15 m) deep.

Range: Alaska to s. California.,

Comments: This creature moves slowly, and is well camouflaged.

---

667 **Turtle Crab**
(*Crypotolithodes sitchensis*)
Class Crustacea

Description: 2″ (51 mm) long, 2¾″ (70 mm) wide. Oval. *Carapace margin concealing legs,* scalloped, wide; eye sockets notched into margin; *beak squarish, widest at front end.* Otherwise similar to the Butterfly Crab.

Habitat: Around rocks in quiet water; from low-tide line to water 50′ (15 m) deep.

Range: Alaska to s. California.

Comments: At rest, this little creature resembles an old shell more than a living animal.

---

685 **Land Hermit Crab**
(*Coenobita clypeatus*)
Class Crustacea

Description: 1½″ (38 mm) long, ½″ (13 mm) wide. Living in snail shell. Nearly cylindrical. Reddish or purplish-brown, *pincers purple or bluish with orange tips,* beak and mouth parts yellowish. Carapace widest at rear. Eyestalks long, not set in sockets. Antennae retractable. 1st pair of walking legs with rounded pincers, *left one much larger.* 2nd and 3rd pairs of walking legs reduced, turned upward. Abdomen long, soft, with only left appendages developed; reduced tail fan.

Habitat: Among plants; above high-tide line.

Range: S. Florida; West Indies.

Comments: This is one of the few species of terrestrial hermit crabs in the world.

---

682 **Giant Hermit Crab**
(*Petrochirus diogenes*)
Class Crustacea

Description: 4¾″ (121 mm) long, 2¼″ (57 mm) wide. Living in snail shell. Oblong. Reddish, with white spots on pincers, red-and-white-banded antennae, blue eyes. Rear end of carapace twice as wide as front. Antennae retractable. Eyestalks long, not in sockets. 1st pair of walking legs with pincers, *right one somewhat larger,* both covered with *heavy irregular scales.* 2nd and 3rd pairs of walking legs long, sturdy, scaly. 4th and 5th pairs of walking legs reduced, turned upward. Abdomen long, cylindrical, soft, with only left appendages developed; reduced tail fan.

Habitat: On sand bottoms and seagrass flats; from shallows to water 300′ (91 m) deep.

Range: North Carolina to Florida and Texas; West Indies.

Comments: This is the largest hermit crab in American waters.

---

684 **Striped Hermit Crab**
(*Clibanarius vittatus*)
Class Crustacea

Description: 1¼″ (32 mm) long, ½″ (13 mm) wide. Living in snail shell. Nearly cylindrical.

Greenish to dark brown; gray to white, occasionally tannish longitudinal *stripes on legs;* antennae orange, eyes yellow. Carapace widest at rear. Eyestalks longer than first 3 joints of base of antennae, not set in sockets. *1st pair of walking legs with pincers, equal in size, hand length twice breadth.* 2nd and 3rd pair 1½ times as long as pincers, 4th and 5th pair reduced, 5th pair turned upward. Abdomen long, cylindrical, soft; appendages reduced.

Habitat: Beaches, mudflats, jetties, harbors, and bays; from low-tide line to shallow depths.

Range: Virginia to Florida and Texas; West Indies to Brazil.

Comments: This is one of the most conspicuous hermit crabs along our southern Atlantic shores. It lives in very shallow water and sometimes appears above water level at low tide, but its inability to get oxygen from the air prevents it from wandering ashore.

---

687 **Star-eyed Hermit Crab**
(*Dardanus venosus*)
Class Crustacea

Description: 1¼" (32 mm) long, ⅞" (22 mm) wide. Living in snail shell. Oblong. Orange-red; broad, cream-colored and orange-red bands on legs; *fingers and undersides of pincers veined in red;* low, round projections on pincers blue to purple. Carapace smooth, widest at rear. Antennae retractable. Eyestalks stout, constricted in middle; *eyes with black center and 6 radiating black lines* (when seen in sunlight). 1st pair of walking legs with pincers, unequal, *left one much larger than right,* covered with scalelike low, round projections with flattened fringes of bristles. 2nd and 3rd pairs of walking legs long, *left 3rd leg larger, flatter, and ridged.* 4th and 5th pairs of walking legs reduced, 5th pair turned upward. Abdomen long, soft, with reduced appendages.

Habitat: On sand, grassflats, and mud bottoms; from low-tide line to water 300′ (91 m) deep.

Range: North Carolina to Florida; Bermuda; West Indies to Brazil.

Comments: The stars in the eyes of this hermit crab are light refractions, much the same as the lines in a star sapphire.

---

680 **Bar-eyed Hermit Crab**
(*Dardanus fucosus*)
Class Crustacea

Description: 1¼″ (32 mm) long, ⅞″ (22 mm) wide. Living in snail shell. Reddish-tan, broad, cream-colored and red-orange bands on legs; fingers and undersides of *pincers cream-colored, veined in red.* Carapace smooth, widest at rear, eyestalks stout, constricted in middle. *Black bar across blue eyes.* Antennae retractable, 1st pair of walking legs with pincers, unequal, *left first walking leg much larger,* covered with scaly, low, round projections and fringes of bristles. 5th pair of walking legs turned upward. Abdomen long, soft, with reduced appendages.

Habitat: On sand, grassflats and, mud bottoms; from low-tide line to moderately deep water.

Range: Florida to Texas; Bahamas; West Indies.

Comments: This hermit crab frequently occupies the shells of tulip snails and various conchs.

---

686 **Acadian Hermit Crab**
(*Pagurus acadianus*)
Class Crustacea

Description: 1¼″ (32 mm) long, 1″ (25 mm) wide. Living in snail shell. Oblong. Carapace brownish; legs orange or reddish-brown, white near bases; *hand of pincers with orange or reddish-orange stripe down middle,* almost white at borders; eyestalks and first pair of antennae blue; eyes yellow. Carapace widest at rear. Eyestalks moderately long, not set in sockets. 1st pair of walking legs pincers, *right walking leg much larger,* both covered with low, round projections. 2nd and 3rd pairs of walking legs longer than pincers. 4th and 5th pair reduced, 5th pair turned upward. Abdomen long, soft, cylindrical, with reduced appendages.

Habitat: In rocky tidepools and just below low-tide line in northern part of its range; to water 1600′ (488 m) in southern part.

Range: Labrador to Chesapeake Bay.

Comments: Hermit crabs, which must change shells as they grow, will readily do so when they find one that suits them better.

---

681 **Hairy Hermit Crab**
(*Pagurus arcuatus*)
Class Crustacea

Description: 1¼″ (32 mm) long, 1″ (25 mm) wide. Living in snail shell. Oblong. Rich brown to reddish-brown. Carapace widest at rear. Hairy-looking, *covered with long bristles.* 1st pair of walking legs with pincers, *right one much larger and covered with low, round projections.* 5th pair of walking legs turned upward. Abdomen soft, with reduced appendages.

Habitat: On rock bottoms, in tidepools where tidal fluctuation is great; from low-tide line to water 900′ (274 m) deep.

Range: Arctic to Long Island Sound.

Comments: Only smaller Hairy Hermit Crabs are found in tidepools; larger specimens prefer the shells of Stimpson's Whelk or the Waved Whelk.

677 Long-clawed Hermit Crab

(*Pagurus longicarpus*)
Class Crustacea

Description: ½″ (13 mm) long, ⅜″ (10 mm) wide. Living in snail shell. Oblong. Grayish or greenish-white, pincers with tannish-gray or tannish stripe down middle, edged with white. Carapace widest at rear. *Right pincer larger than left, hand nearly cylindrical, 3 times longer than wide,* smooth except for row of weak spines at edge and down middle. 5th pair of walking legs turned up. Abdomen long, soft, cylindrical, with reduced appendages.

Habitat: On sand, mud, rock, and weed bottoms, along open shores and in brackish estuaries; from low-tide line to water 150′ (45 m) deep.

Range: Nova Scotia to Florida and Texas.

Comments: This is the most common hermit crab in Atlantic waters. It usually uses periwinkle, mud snail, or oyster drill shells—frequently covered with Snail Fur hydroids—as a home.

676 Flat-clawed Hermit Crab

(*Pagurus pollicaris*)
Class Crustacea

Description: 1¼″ (32 mm) long, 1″ (25 mm) wide. Living in snail shell. Oblong. Whitish or pale tannish-gray, eyestalks brown. Carapace widest at rear. *Right pincer larger than left,* covered with low, rounded projections, especially at borders; *hand broad and rounded,* width ⅔ length, *finger with blunt angle* midway along its border. 5th pair of walking legs turned up. Abdomen long, soft, cylindrical, with reduced appendages.

Habitat: On sand bottoms along open shores, and in brackish estuaries; from low-tide line to water 150′ (46 m) deep.

Range: Cape Cod to Florida and Texas.

Comments: This large hermit crab is frequently

found living in shells of moon snails and the larger whelks. The Zebra Flatworm often shares the shell as a commensal.

683 **Grainy Hermit Crab**
(*Pagurus granosimanus*)
Class Crustacea

Description: ¾″ (19 mm) long, ⅝″ (16 mm) wide. Living in snail shell. Pear-shaped. Olive-brown, legs somewhat paler; pincers with white or blue speckles; antennae red. Carapace widest at rear, *beak rounded and toothlike.* Antennae retractable. 1st pair of walking legs with pincers, *right one larger; hand evenly covered with fine granules.* 5th pair of walking legs turned upward. Abdomen soft, cylindrical, long, with reduced appendages.

Habitat: On rocky shores in tidepools on gravel bottoms and among kelp holdfasts; between high- and low-tide lines, to water more than 50′ (15 m) deep.

Range: Alaska to Baja California.

Comments: Adults of this species show a marked preference for the shells of the Black Turban Snail. This crab is common in lower tidepools, but less able to withstand drying than the even more abundant Blue-handed Hermit Crab (*P. samuelis*), which is found in higher tidepools and under rocks drained at low tide. *P. samuelis* is the same size, also prefers the shell of the Black Turban Snail, and is easily identified by the bright blue bands near the tips of its walking legs, and its pale blue pincer tips. Another species similar in size and habitat preference is the Little Hairy Hermit Crab (*P. hirsutiusculus*), which has hairy legs with several white cross-bands, and white pincer tips against a tannish-gray background. All three species have the same range.

689 **Pacific Mole Crab**
(*Emerita analoga*)
Class Crustacea

Description: 1⅜″ (35 mm) long, 1″ (25 mm) wide. *Egg-shaped.* Pale grayish or tannish, appendages whitish. Carapace convex, with fine crosswise wrinkles in front, smooth in rear half. 1st pair of antennae hairy. *2nd pair of antennae long, feathery,* usually concealed under edge of carapace. Eyestalks long, slender. *1st pair of walking legs without pincers,* broad, sturdy; 2nd, 3rd, and 4th pairs less sturdy, hairy, leaflike; 5th pair very slender. Abdomen broad at front, tapering rapidly, with pair of forked, leaflike appendages and *long, spearhead-shaped tailpiece* on last segment, bent forward underneath body.

Habitat: On open sandy beaches; between high- and low-tide lines.

Range: Alaska to Peru and Chile.

Comments: These crabs often occur in dense populations on open beaches.

690 **Atlantic Mole Crab**
(*Emerita talpoida*)
Class Crustacea

Description: 1″ (25 mm) long, ¾″ (19 mm) wide. *Egg-shaped.* Pale grayish-tan. Carapace convex, crosswise-creased line immediately behind beak, another curved one farther back; rear end of carapace smooth. 1st pair of antennae hairy; *2nd pair of antennae long, feathery,* usually concealed under edge of carapace. Eyestalks long, slender. 1st pair of walking legs broad, sturdy, without pincers; 2nd, 3rd, and 4th pairs less sturdy, leaflike; 5th pair very slender. Abdomen broad at front, tapering rapidly, with pair of forked, leaflike appendages, and *long, spearhead-shaped tailpiece* on last segment, bent forward underneath body.

Habitat: On open sandy beaches; between high- and low-tide lines.

Range: Cape Cod to Florida and Texas; Mexico.

Comments: This species is commonly used for bait, and can be collected in large numbers by attaching a wire mesh net to a common garden rake, and then raking the wave-swept beach.

---

688 **Spiny Mole Crab**
(*Blepharipoda occidentalis*)
Class Crustacea

Description: 2⅜″ (60 mm) long, 1⅝″ (41 mm) wide. *Egg-shaped.* Carapace and pincers gray. Carapace has *3 spines along front midline and 4 on border* at each side and deep notch on rear margin. Front half has fine indentations; rear half smooth. 1st pair of antennae very slender, 2nd pair of antennae heavier, feathery. Eyestalks slender, eyes small. *1st pair of walking legs with pincers,* heavy, spiny, hairy. Remaining walking legs smaller, but strong, leaflike. Abdomen tucked under body.

Habitat: On open sandy beaches; from low-tide line to water 30′ (9 m) deep.

Range: C. California to Baja California.

Comments: Young Spiny Mole Crabs comb tiny organisms and organic particles from the water with their feathery antennae, but do not migrate up and down the beach with the tides. The adults are scavengers, feeding mostly on dead mole crabs (*Emerita*) that abound in the same habitat.

---

670 **Sponge Crab**
(*Dromia erythropus*)
Class Crustacea

Description: 5″ (127 mm) wide, 3¼″ (83 mm) long. Fan-shaped, hairy. Carapace whitish, densely covered with *stiff, brown to blackish hairs. Tips of legs light red.* Front margin of carapace with *5 teeth* between eye socket and side. Back convex, with several wide, low bumps. Pincers and walking legs hairy except at tips; last pair folded upward at rear of carapace.

Habitat: On rocks and reefs, among sponges and other growth; below low-tide line to fairly deep water.

Range: Florida; West Indies.

Comments: This crab camouflages itself with a piece of living sponge that it cuts to fit the top of its carapace, and holds there by the sharp tips of the upturned 5th pair of walking legs. The sponge continues to grow.

---

669 **Lesser Sponge Crab**
(*Dromidia antillensis*)
Class Crustacea

Description: 1¼″ (32 mm) wide, 1⅜″ (35 mm) long. Fan-shaped, hairy. Yellowish-green, buff, gray, orange-buff, or reddish, *hairs paler. Tips of pincers red.* Carapace densely hairy, convex, with *4–5 teeth* between eye socket and side. Pincers and walking legs hairy except at tips; last pair of walking legs folded upward at rear of carapace.

Habitat: On rocks and coral reefs, among sponges and other growth; from low-tide line to water 1020′ (310 m) deep.

Range: Cape Hatteras to Florida and Texas; Mexico; Bermuda; Bahamas; West Indies to Brazil.

Comments: The Lesser Sponge Crab usually camouflages itself with a piece of living sponge, but sometimes decorates itself with compound tunicates or a colony of mat anemones instead of a sponge. When the crab is quiet it almost completely escapes notice.

635 **Purse Crab**
(*Persephona punctata*)
Class Crustacea

Description: 2⅝″ (67 mm) wide, 2¾″ (70 mm) long. Globe-shaped. Grayish, bluish, or pinkish, with *large, round, reddish-brown spots.* Carapace highly arched, covered with whitish granules, with *3 sharp spines* at rear; front end projects from round outline of carapace. Pincers long, moderately stout, with short fingers.

Habitat: On sand or shelly mud bottoms; from below low-tide line to water 180′ (55 m) deep.

Range: New Jersey to Florida and Texas; south to Campeche.

Comments: The Purse Crab is so named because the recessed abdomen of the female forms a roomy sac that contains her eggs.

---

671 **Flame-streaked Box Crab**
(*Calappa flammea*)
Class Crustacea

Description: 5½″ (140 mm) wide, 4″ (102 mm) long. Semicircular, convex. Grayish, with flame-shaped, purplish-brown markings on carapace; pincers reddish-blue on outer surface, white on inside and at fingertips. *Carapace straight across at rear,* arched, round in front; sides at rear with *5 strong teeth.* Pincers heavy, unequal, fitting snugly against body; hands triangular, with finger at top; *large hand with 7 teeth on upper margin,* smaller hand with 6.

Habitat: On sand bottoms; from below low-tide line to water 240′ (73 m) deep.

Range: Cape Hatteras to Florida, Texas, and Mexico; Bermuda; Bahamas.

Comments: The powerful pincers of this crab are used to chip open snail shells occupied by hermit crabs. The action is like that of a hand-operated can opener.

673 **Yellow Box Crab**
(*Calappa gallus*)
Class Crustacea

Description: 4″ (102 mm) wide, 3⅛″ (79 mm) long. Semicircular, convex. Upper surface of carapace and pincers grayish-yellow with dark red or reddish-brown spots; underside yellow to orange; legs yellow with fine red lines. *Carapace almost straight across at rear,* arched, rounded in front, rough, with bumps and longitudinal ridges; sides edged with *20 teeth, smaller near front, increasingly large toward rear.* Pincers heavy, unequal, knobby, upper margin of *larger hand with 6 teeth, smaller hand with 5.*

Habitat: On sand bottoms; from low-tide line to moderately deep water.

Range: Florida; Bahamas; West Indies.

Comments: The Yellow Box Crab eats snails and hermit crabs, whose shells it can easily crush with its large pincer.

Calico Crab
"Dolly Varden"
(*Hepatus epheliticus*)
Class Crustacea

Description: 2⅝″ (67 mm) wide, 1⅞″ (48 mm) long. Fan-shaped. Dull yellowish, grayish, or brownish, with *many large, round or irregular, light red spots* with dark borders. Carapace convex, broadly rounded in front, granular, narrowing toward abdomen. Pincers strong, equal; hand with row of coarse, low, round projections on outer surface, and crest of 3–4 teeth at top.

Habitat: On sand bottoms in bays and open ocean; from below low-tide line to water 150′ (46 m) deep.

Range: Chesapeake Bay to Florida and Texas; West Indies to Campeche.

Comments: The Calico Crab is very active, and has

a greater gill area, and hence greater capacity to take in oxygen, than most crabs.

637 **Lady Crab**
(*Ovalipes ocellatus*)
Class Crustacea

Description: 3⅛″ (79 mm) wide, 2½″ (64 mm) long. Fan-shaped. Carapace yellowish-gray, with *closely set rings of reddish-purple spots,* metallic iridescence; pincers light brown, with bluish tips and purple spots on top. Carapace slightly wider than long; convex, granular; with 3 sharp teeth between eye sockets, *5 strong teeth* directed forward from eye socket to widest part. Pincers large, sharp; finger ½ length of hand; joint next to hand with spine on inner and outer side; *5th pair of walking legs paddlelike.*

Habitat: On sand, rock, or mud bottoms; from low-tide line to water 130′ (40 m) deep.

Range: Cape Cod to South Carolina. Isolated population on Prince Edward Island.

Comments: This handsome crab is known for its aggressive disposition and sharp pincers. It must be handled with great caution.

658 **Sargassum Crab**
(*Portunus sayi*)
Class Crustacea

Description: 2⅛″ (54 mm) wide, 1⅛″ (28 mm) long. Wide, spindle-shaped. Chocolate-brown, light brown, or olive, with irregular whitish spots, *orange spines on pincers.* Width of carapace twice length, with *sharp spine at side;* front margin nearly semicircular; 8 teeth between eye and spine on each side. Pincers large, long, equal; spine on outer margin of hand; *2 spines on next joint, 4 spines on long joint beyond;* 5th pair of walking legs broad, paddle-shaped.

Habitat: Among floating Sargasso Weed.

Range: Nova Scotia to Florida and Texas; Bermuda; West Indies to Brazil.

Comments: Although this crab is normally a creature of the high seas and a member of the Sargasso Weed community, it is frequently seen among weeds blown ashore by storms. Gibbes' Crab (*P. gibbesii*) is 2⅜" (60 mm) wide and 1⅛" (28 mm) long, and brownish-red with carmine-red pincers and spines. The front margin of the carapace has 7 teeth between the eye and the large spine at each side. Its pincers are twice body length, strong and sharp; the last pair of walking legs are paddle-shaped. It ranges from Massachusetts to Florida, Texas, and Venezuela. The Spiny-handed Crab (*P. spinimanus*) has a carapace less than twice as wide as its length, and a spine at each side the same size as the 8 teeth on the margin. It is 3½" (89 mm) wide and 2⅛" (54 mm) long. The pincers are more than twice as long as body width, the hand slender and ridged, the fingers straight and curved in at the tips. The joint next to the hand has 2 strong spines and the long joint next to that has 5. It ranges from New Jersey to Florida and Texas, to Bermuda, through the Gulf of Mexico, and the West Indies to Brazil.

---

657 **Blue Crab**
(*Callinectes sapidus*)
Class Crustacea

Description: 9¼" (23 cm) wide, 4" (102 mm) long. Spindle-shaped. Grayish or bluish-green, spines red; male with blue fingers on hand, female with red; underside white, with yellow and pink tints. *Carapace 2½ times as wide as long,* moderately convex, smooth; *4 triangular teeth between eye sockets, 8 sharp, strong teeth between eye socket and large spine at side.* Pincers powerful; hand ribbed, with spine at base; fingers nearly straight, toothed; long joint between wrist and body with *3 large teeth on front margin;* 5th pair of walking legs broad, paddle-shaped.

Habitat: In shallows and brackish estuaries; from low-tide line to water 120′ (37 m) deep.

Range: Nova Scotia to Florida and Texas; Bermuda; West Indies to Uruguay.

Comments: Because of its commercial importance, this species has been studied more extensively than any other crab. It supports a large seafood industry in Chesapeake Bay and along the entire southeastern and Gulf coasts.

---

633 **Flat-browed Crab**
(*Portunus depressifrons*)
Class Crustacea

Description: 1⅝″ (41 mm) wide, 1″ (25 mm) long. Spindle-shaped. Upper surface mottled light and dark gray, lower surface bluish-white; pincers and walking legs bright blue or purple on under surface, darker above. *Carapace with blunt spine at widest part,* width less than twice length, *8 triangular teeth between spine and eyesocket;* 6 short teeth between eyesockets. Pincers long, fingers toothed, outer surface of thumb hairy; *long joint, with 6 spines on front border, between wrist and body;* walking legs long, last pair broad, paddle-shaped.

Habitat: On sand bottoms of coves and inlets; in shallow water.

Range: North Carolina to Florida; Texas to Yucatan Peninsula; Bermuda; Bahamas; West Indies.

Comments: This crab is common from Florida south; in the Carolinas it is found in the stomach contents of bottom-feeding fish.

664 **Green Crab**
(*Carcinus maenas*)
Class Crustacea

Description: 3⅛" (79 mm) wide, 2⅝" (67 mm) long. Fan-shaped. Greenish, with blackish mottlings above, yellowish below, adult females orange-red underneath, young variable in color and pattern. Carapace slightly wider than long, with *3 teeth between eye sockets* and *5 strong teeth, forward,* curved toward the side of each eye socket. Pincers moderately large, equal. 5th pair of walking legs normal, not paddle-shaped.

Habitat: On rocks, jetties, and mud banks in wetlands; in tidepools; from open shore to brackish water.

Range: Nova Scotia to New Jersey. Introduced into Brazil, Panama.

Comments: Introduced from Europe, the Green Crab was unknown north of Cape Cod in the last century, but is now the most common crab along the shores of the Gulf of Maine.

644 **Pacific Rock Crab**
(*Cancer antennarius*)
Class Crustacea

Description: 4⅝" (117 mm) wide, 2¾" (70 mm) long. Fan-shaped. Upper side purplish-red; underside cream-colored, with many *red spots;* tips of fingers black. Carapace oval, finely granular; 11 teeth from eye socket to side; length ⅔ width, *widest at 8th–9th tooth.* Pincers stout, short, bent downward at tips; walking legs short.

Habitat: On rocky shores, kelp beds, gravel bottoms; from between high- and low-tide lines to water 130′ (39 m) deep.

Range: Oregon to Baja California.

Comments: This crab is both a scavenger and a predator, successful at devouring hermit crabs. The crab inserts the fingers of both pincers into the shell containing the hermit, chips away until the hermit can retreat no farther, and then eats it.

---

652 **Oregon Cancer Crab**
(*Cancer oregonensis*)
Class Crustacea

Description: 1⅞" (48 mm) wide, 1⅜" (35 mm) long. Oval, convex. Upper side dark red, underside paler; tips of pincers black. Carapace almost round, convex, coarsely granular; area between eye sockets with *5 unequal teeth; 10 teeth from eye socket to side, carapace widest at 7th and 8th tooth.* Pincers stout, short, bent downward at tips, walking legs short, hairy.

Habitat: On rocks and pilings on rocky shores; from low-tide line to water 1335′ (435 m) deep.

Range: Alaska to s. California.

Comments: This species lives among beds of mussels and barnacles. It feeds chiefly on these, but also eats worms, small crustaceans, and scraps of green algae.

---

653 **Jonah Crab**
(*Cancer borealis*)
Class Crustacea

Description: 6¼" (16 cm) wide, 4" (102 mm) long. Fan-shaped. Upper side dull rosy to brick-red, yellowish beneath, legs yellow mottled with reddish-purple. Carapace oval, granular, length ⅔ width, *front border rounded. 3 teeth between eye sockets,* middle one largest; *9 teeth to side of eye socket, each with several smaller points.* Pincers stout, short, fingers bent downward, black at tips; joint next to hand with sharp spine on inside upper border. Walking legs short, hairy, black-tipped.

Habitat: On rocky shores and bottoms; from

low-tide line to water 2620′ (799 m) deep.

Range: Nova Scotia to Florida; Bermuda.

Comments: This crab is a common species among seaweeds along the open rocky coast of northern New England, but it seldom moves into brackish estuaries. Though it has strong pincers, it is not at all aggressive and can be handled safely with moderate precautions.

650, 654 **Atlantic Rock Crab**
(*Cancer irroratus*)
Class Crustacea

Description: 5¼″ (133 mm) wide, 3½″ (89 mm) long. Fan-shaped. Upper side yellow, closely dotted with reddish or purplish spots, whitish to creamy yellow underneath. Carapace oval, fairly smooth, *front border rounded. 3 teeth between eye sockets,* middle one longest; *9 simple teeth* to side of eye socket. Pincers stout, short, fingers bent downward, black at tips. Walking legs short, hairy at edges.

Habitat: On rock, sand, or gravel bottoms, in estuaries and on open shores; from low-tide line to water 2600′ (780 m) deep.

Range: Labrador to South Carolina.

Comments: This crab enters lobster pots, and until a short time ago was regarded by lobstermen as a pest and a bait-stealer. But the rise in seafood prices has made it profitable to market the crabs as well as the lobsters.

655 **Dungeness Crab**
(*Cancer magister*)
Class Crustacea

Description: 9¼″ (23 cm) wide, 6⅜″ (16 cm) long. Fan-shaped. Upper side grayish-brown, tinged with purple, cream-colored underneath, *pincers not black-tipped.*

Carapace oval, surface granular; 5 unequal teeth between eye sockets; margin with 10 teeth from eye socket to side, last one largest. Pincers stout, fingers bent downward at tips; walking legs short.

Habitat: On sand bottoms; from low-tide line to water more than 300′ (91 m) deep.

Range: Alaska to s. California.

Comments: The Dungeness Crab is the chief crab species taken commercially on the Pacific Coast. While it occurs mainly in water more than 100′ (30 m) deep, it comes into shallow water to molt. The molted skeletons washed up on the beach have caused people to think the crabs were dying of some disease.

---

651 Red Crab
(*Cancer productus*)
Class Crustacea

Description: 6¼″ (16 cm) wide, 4¼″ (108 mm) long. Fan-shaped. Upper side brick-red, underside yellowish-white. Carapace oval, smooth, *area between eye sockets extended forward* beyond side margins, with 5 *equal teeth; 9 teeth from eye socket to side.* Pincers stout, short, bent downward at tips; walking legs short.

Habitat: Among rocks, in tidepools, bays and estuaries, and on open rocky shores; from low-tide line to water 260′ (79 m) deep.

Range: Alaska to s. California.

Comments: Although adults are uniformly brick-red, young Red Crabs are strikingly varied: white, brown, blue, red, or orange—either solid or patterned.

648 Coral Crab
(*Carpilius corallinus*)
Class Crustacea

Description: 6″ (15 cm) wide, 3½″ (89 mm) long. Oval. Brick-red, with large, scarlet to dark red spots and small, white or yellowish dots and irregular lines; pale yellow beneath; tips of pincers black. Carapace convex, smooth, oval, with *1 blunt spine at rear corners.* Pincers heavy, smooth.

Habitat: On reefs and coral rubble; in shallow water.

Range: Florida; West Indies.

Comments: This handsome crab—an unlikely relative of the generally unprepossessing mud crabs—is the largest in the West Indies and is highly prized as food.

Warty Crab
(*Eriphia gonagra*)
Class Crustacea

Description: 2″ (51 mm) wide, 1½″ (38 mm) long. Almost rectangular. Tannish-green, with purplish-brown markings on the back, sides pale yellow, white underneath; *hand with rows of large, flat, low, round projections;* purplish-brown on upper part of hand, bright yellow on lower part, finger dark brown. Carapace granular at front, smooth at rear, front wide; *eyes far apart; 5 strong spines* from eye socket to side. Pincers unequal, large, heavy.

Habitat: Under rocks and among seaweeds and sponges; in tidepools and brackish estuaries, and on coral reefs; near low-tide line and below in shallow water.

Range: North Carolina to Florida; West Indies to Argentina.

Comments: Although this handsome little crab has rather heavy pincers, it is not aggressive when handled. The author has found it among coral rocks at low tide on beaches crowded with vacationers.

645 **Flat Mud Crab**
(*Eurypanopeus depressus*)
Class Crustacea

Description: ¾″ (19 mm) wide, ⅝″ (16 mm) long. Small, fan-shaped. Grayish-olive to olive-brown, lighter underneath; fingers of pincers dark brown. Carapace rounded in front, sides slanting in toward rear border; *area between eye sockets almost straight, notched* in middle; *4 teeth on side margin, last 2 sharp-pointed.* Pincers stout, unequal, *larger pincer with fingers almost straight, tips hollowed out,* or "spoon-shaped."

Habitat: On mud bottoms, among oysters; in bays and brackish estuaries.

Range: Massachusetts Bay to Florida and Texas; Bermuda; West Indies.

Comments: This mud crab shows a preference for oyster shells. It preys on newly settled oysters, but is not considered a serious pest by oystermen.

646 **Black-clawed Mud Crab**
(*Lophopanopeus bellus*)
Class Crustacea

Description: 1⅜″ (35 mm) wide, ⅞″ (22 mm) long. Small, fan-shaped. Upper surface brown, tan, reddish-brown, white, or bluish, variously mottled; *fingers of pincers black with reddish-brown tips.* Carapace rounded across front, sides narrowing toward straight rear margin. *Area between eye sockets bowed,* notched in middle. *3 teeth on side margin, last 2 sharp-pointed.* Pincers stout.

Habitat: Under rocks on mud or sand bottoms, in quiet bays and estuaries, in tidepools and kelp holdfasts on rocky shores; from low-tide line to water 240′ (73 m) deep.

Range: Alaska to s. California.

Comments: Unlike many crabs, the female of this species mates while hard-shelled, and usually produces two broods per year.

642 **Stone Crab**
(*Menippe mercenaria*)
Class Crustacea

Description: 4⅝″ (117 mm) wide, 3⅛″ (79 mm) long. Oval. Mottled brownish-red with gray spots, tannish beneath, fingers of pincers black. Carapace oval, convex, smooth. Between eye socket and side, *margin divided into 4 lobes,* first 2 wide, last 2 toothlike. Pincers heavy, unequal, largest about ¼ longer than carapace width; *inner surface of hand rasplike.* Walking legs stout, hairy.

Habitat: Adults in burrows in sandy-mud shoals just below low-tide line; young in turtle grass beds and shell and rock bottoms of channels.

Range: North Carolina to Florida and Texas; Bahamas; West Indies to the Yucatán.

Comments: This is the largest of the mud crabs, and the source of the delectable seafood dish, crab claws. Its catch is carefully regulated in most states, and in Florida it is illegal to take females. When a male is taken, one may break off and keep the large pincer—which must be at least 4″ from tip to first joint—but must return the crab to the water so that it can regenerate a new pincer. Most crustaceans are able to regenerate an appendage in 2 molts.

636 **Say's Mud Crab**
(*Neopanope texana*)
Class Crustacea

Description: ⅞″ (22 mm) wide, ⅝″ (16 mm) long. Small, fan-shaped. Dark blue-green, slate-gray, buff, or brown, with reddish or purplish speckles. *Black color on fingers extends onto hand* of pincer. Carapace rounded across front, sides slanting toward straight rear border; *4 teeth* between eye sockets and side margin, *last 2 teeth sharp-pointed.* Pincers nearly equal, major one with large tooth at base of finger.

Habitat: On mud bottoms in bays and brackish estuaries, among oysters; from low-tide line to water 90′ (27 m) deep.

Range: Prince Edward Island to Florida.

Comments: This species shows variation in different parts of its range. From Virginia southward, these crabs lack the black color on the fingers of their pincers, and are slightly larger, 1⅛″ (28 mm) wide and ⅞″ (22 mm) long.

---

634 Commensal Crabs
(*Pinnotheres* spp.)
Class Crustacea

Description: ⅝″ (16 mm) wide, ½″ (13 mm) long. Oval. White, salmon, brown, or blue, with or without white spots. *Carapace round, smooth, soft, flexible.* Pincers small; hand flat on inside, swollen on outside; *fingers curve to meet;* walking legs slender, *last 2 joints of walking legs with fringe of hair. Female's abdomen wider than carapace.*

Habitat: In mantle cavity of various bivalve mollusks, tubes of parchment worms, or pharynx of tunicates, or on sea stars.

Range: Massachusetts to Florida and Texas; West Indies to Argentina; British Columbia to Peru.

Comments: These crabs live as commensals or parasites on the body or in the cavity of a host animal that feeds by capturing organic particles which the crab shares.

---

649 Sally Lightfoot
(*Grapsus grapsus*)
Class Crustacea

Description: 3⅝″ (92 mm) wide, 3¼″ (83 mm) long. Square-bodied. Mottled dark red and pale green, or solid red, with many white speckles. Pale greenish or bluish underneath. Pincers black with red patches and cream-colored fingertips. Carapace squarish, front vertical, slightly bulged at sides; *upper side creased with crosswise grooves* at sides and rear; bumpy behind eyes; *2 spines on margin* behind eyes. Pincers small, nearly equal, fingertips spoon-shaped; walking legs broad, flattened, *3rd pair of walking legs longest.*

Habitat: On rocky shores and jetties; at water's edge.

Range: Florida and West Indies to Brazil; Baja California to Chile.

Comments: Speed, agility, and keen eyesight make this large and colorful crab extremely hard to catch. It has been seen plucking ticks from the large marine iguanas of the Galapagos Islands in the Pacific.

639, 663 **Purple Shore Crab**
(*Hemigrapsus nudus*)
Class Crustacea

Description: 2¼" (57 mm) wide, 2" (51 mm) long. Round-bodied. Upper surface of carapace purplish-black, sometimes reddish-brown or greenish-yellow; white beneath. *Pincers covered with deep purple or red spots,* purple or reddish above, fading to white below. Carapace oval-oblong, arched in front, smooth; 3 short teeth on margin between eye socket and side. Eyes far apart. Pincers large in male, equal, both fingers toothed, *pincer tips bent toward each other,* cup-shaped. Walking legs sturdy, flattened, not very hairy.

Habitat: On open rocky shores, among seaweeds, and in bays and estuaries.

Range: Alaska to Baja California.

Comments: This crab feeds primarily on the film of small algae on rocks, but is also a scavenger of animal matter.

643 **Yellow Shore Crab**
(*Hemigrapsus oregonensis*)
Class Crustacea

Description: 1⅜" (35 mm) wide, 1¼" (32 mm) long. Round-bodied. Upper surface of carapace *yellow to gray, mottled with brownish-purple or black;* pincer tips yellow to white. Legs hairy. Otherwise similar to the Purple Shore Crab.

Habitat: On mudflats, in eelgrass beds, in bays and estuaries; between high- and low-tide lines.

Range: Alaska to Baja California.

Comments: This crab feeds most commonly at night, eating algae that it scrapes off rocks and shells. It can bury itself in the mud very quickly when pursued.

---

662 **Striped Shore Crab**
(*Pachygrapsus crassipes*)
Class Crustacea

Description: 1⅞" (48 mm) wide, 1½" (38 mm) long. Square-bodied. Upper surface of carapace brownish-purple or blackish, with *green cross-stripes,* fading to white beneath. Pincers mottled reddish-purple above, white below; upper joint of walking legs mottled green and purple; outer joints purplish-brown above, whitish below. Carapace with shallow crosswise grooves, *1 tooth on side.* Eyes almost at corners, far apart. Pincers heavy, equal; *fingers with spoon-shaped tips;* walking legs flattened, sturdy.

Habitat: On rocky and hard-mud shores; in tidepools, mussel beds, bays and estuaries; well above low-tide line.

Range: Oregon to Baja California.

Comments: This crab spends at least half its time on land, but submerges now and then to wet its gills and feed. The Mottled Shore Crab (*P. transversus*) ranges from North Carolina to Florida and Texas, and to Bermuda and the West Indies, and south to Uruguay. It is smaller, ½″ (13 mm) wide and ⅜″ (10 mm) long, has shallow crosswise grooves on the carapace, and other characteristics similar to *P. crassipes,* but it is variable in color: black, olive, yellowish-brown, or gray, mottled with dark brown, reddish, or dark olive. It inhabits rocks, pilings, mangrove roots, and sandy shores.

Flattened Crab
(*Plagusia depressa*)
Class Crustacea

Description: 1⅞″ (48 mm) wide, 1¾″ (44 mm) long. Rough, flat. Upper surface light reddish-brown, with red, low, round projections bordered with *short, black bristles;* yellowish-white underneath; upper surface of legs with large, red spots. Carapace nearly circular, flat; *4 teeth on side margin,* eyes far apart, 4 irregular teeth between eyes. Pincers of male large, as long as carapace; female's smaller. Movable finger of hand curved downward. Walking legs' *long upper joint with 3 spines on top.*

Habitat: On rocks, jetties, and pilings; below low-tide line.

Range: North Carolina to Florida; Bermuda; West Indies to Brazil.

Comments: The Flattened Crab can cling tenaciously to any rough surface, and is thought to have been introduced from Africa by clinging to the hulls of wooden sailing ships.

665 **Wharf Crab**
(*Sesarma cinereum*)
Class Crustacea

Description: ⅞″ (22 mm) wide, ¾″ (19 mm) long. Squarish. Upper surface brown to olive, gray beneath. *Carapace almost square,* nearly smooth, with few low, round projections toward front, *sides hairy;* eyes at front corners. Pincers stout, short, joints between hand and body with granular cross-bands; *immovable finger with large tooth.* 3rd pair of walking legs over twice length of carapace.

Habitat: Among rocks and drift logs, on wharf piles and boat hulls; in shallow burrows above high-tide line.

Range: Chesapeake Bay to Florida and Texas; Honduras; West Indies to Venezuela.

Comments: This crab leads a semi-terrestrial life, and is often called the "Friendly Crab" because of its habit of climbing into boats. The Marsh Crab (*S. reticulatum*) lives in salt marshes from Cape Cod to Florida and Texas. It is larger, 1⅛″ (28 mm) wide and ⅞″ (22 mm) long, is olive, blackish, or purplish, and has purplish pincers. It lives in burrows in salt marshes, frequently sharing the burrow with the Mud Fiddler Crab. It eats marsh grass, cutting swaths near its burrow.

---

632 Great Land Crab
(*Cardisoma guanhumi*)
Class Crustacea

Description: 5″ (127 mm) wide, 4″ (102 mm) long. Globular. Adults: bluish or ash-gray, or nearly white; young: violet. *Carapace globular,* smooth, with fine ridge along sides; eye sockets shallow, front surface high. Pincers unequal, huge one in male over 12″ (30 cm) long. Walking legs longer than body width, sturdy.

Habitat: Burrowing in sandy or muddy areas along backwaters, estuaries, and inland waterways, and among mangroves; above high-tide line.

Range: S. Florida to Texas; Bahamas; West Indies.

Comments: This is the largest land crab in the United States. It is well adapted to land life, and can go 3 days without water. Prized as food in the Bahamas and West Indies, this species is trapped and fed on soaked corn before being eaten.

---

638 **Mountain Crab**
(*Gecarcinus ruricola*)
Class Crustacea

Description: 4″ (102 mm) wide, 3″ (76 mm) long. Globe-shaped. Almost black, with purple tinge and small, yellow spot on each side near rear; *pincers and legs red with yellow tips.* Carapace oval, highly arched; eye sockets shallow, area between them narrow. *Pincers equal in both sexes.* 2 end joints of walking legs with 6 *longitudinal rows of spines.*
Habitat: In burrows in low-lying land near sea.
Range: S. Florida; West Indies to Brazil.
Comments: This crab is highly prized as food in the West Indies. A close relative, the Black Land Crab (*G. lateralis*), is smaller, 2″ (51 mm) wide and 1½″ (38 mm) long, is plum to reddish-brown, mottled with black, and has unequal red pincers with yellow palms. It ranges from southern Florida to southern Texas, Bermuda, and the West Indies.

---

631 **Ghost Crab**
(*Ocypode quadrata*)
Class Crustacea

Description: 2″ (51 mm) wide, 1¾″ (44 mm) long. Square-bodied. Upper surface gray, grayish-white, yellowish-white, or straw-colored; white underneath; pincers white or pale lavender; young mottled gray and brown. *Carapace rectangular,* sides nearly parallel and vertical, front corner acutely angled, *H-shaped depression on front half,* surface granulated. *Eyestalks large, club-shaped.* Pincers unequal, strong, rough, *margins saw-toothed,* both fingers toothed. Walking legs long, strong, hairy.

Habitat: On sandy beaches.

Range: Rhode Island to Florida and Texas; West Indies to Brazil.

Comments: If careful not to disturb them, one can watch these crabs walking along the beach, facing the moon when it is full. They are called Ghost Crabs with good reason; they blend closely with the sand on which they live, and are very swift. They seem to appear from nowhere, run, and suddenly disappear again.

---

630 **California Fiddler**
(*Uca crenulata*)
Class Crustacea

Description: ¾" (19 mm) wide, ½" (13 mm) long. Square, with 1 large pincer. Pale tan to brown, with white hands. Carapace nearly rectangular, somewhat narrowed at rear, convex, *tooth behind eye socket.* Eyestalks long, antennae small. Pincers of male unequal, one greatly enlarged, with *ridge on inner surface of hand;* pincers of female small, equal.

Habitat: In burrows in sandy mud in bays and estuaries; near high-tide line.

Range: S. California to Baja California.

Comments: This is the only fiddler crab found on the western coast of the United States, and its future is uncertain because of encroachment on its habitat by human construction.

---

628 **Sand Fiddler**
(*Uca pugilator*)
Class Crustacea

Description: 1½" (38 mm) wide, 1" (25 mm) long. Square-bodied, with 1 large pincer. Males: upper surface purplish or grayish-blue, with purple patch on front half and irregular markings of black, brown, or gray; brownish at side; hand of large pincer bluish, lavender, or reddish-brown; fingertips white. Females: similar but with more subdued colors, darker. Carapace nearly rectangular; convex, smooth, *H-shaped depression* near center; sides curved outward behind eye sockets; *eyestalks long, slender;* antennae small. Pincers of male greatly unequal, one very large, one small; those of female small, equal. Outer surface of hand with low, round projections on upper part, diminishing downward. *Inner surface without oblique ridge.* Fingers strong, movable one curved.

Habitat: On protected sand and sandy mud beaches, marshes, and tidal creeks.

Range: Boston Harbor to Florida and Texas; West Indies.

Comments: The forbidding-looking large pincer of the male Sand Fiddler is not at all dangerous. It is used in a courtship display prior to mating, when the crab rises on tiptoes, extending the pincer, then flexing it and bowing down. The Mud Fiddler (*U. pugnax*) is a little smaller, ⅞" (22 mm) wide and ⅝" (16 mm) long. It ranges from Cape Cod to northeastern Florida, and from northwestern Florida to Texas. The male has an oblique ridge on the inner surface of the hand of the large pincer. It is dark olive or almost black above, sometimes speckled white near the front, has a royal blue spot near the middle, and is grayish below. The large pincer is brownish-yellow at the base, becoming yellow on the hand; the fingertips are almost white. Females are similar in color, but lack the blue spot. This crab prefers a muddy habitat and frequently digs into mud banks along tidal marshes, sharing burrows with the Marsh Crab. Where sand and mud intergrade, populations of Sand Fiddlers and Mud Fiddlers may be intermixed, though they do not interbreed.

629 **Brackish-water Fiddler**
(*Uca minax*)
Class Crustacea

Description: 1½″ (38 mm) wide, 1″ (25 mm) long. Square, with 1 large pincer. Chestnut-brown above, gray at front; large *pincer red at movable joints, hand white;* walking legs olive or grayish-brown. Carapace nearly rectangular, somewhat narrowed at rear, convex, smooth, *H-shaped depression* near center. Eyestalks long, slender, antennae small. *Male's large pincer with an oblique ridge of low, round projections* on inner surface of hand; pincers of female equal, small.

Habitat: In muddy marshes of low salinity, sometimes at great distance from ocean; in burrows above high-tide line.

Range: Cape Cod to Florida and Texas; West Indies to Colombia.

Comments: This fiddler strongly prefers brackish-water mud habitats, and can survive at least three weeks in fresh water. It is omnivorous, eating organic matter sifted from mud, decaying marsh grass, and occasionally other fiddlers.

660, 661 **Toad Crab**
(*Hyas araneus*)
Class Crustacea

Description: 2½″ (64 mm) wide, 3¾″ (95 mm) long. Violin-shaped. Reddish to olive, legs banded red and orange. Older specimens less colorful. *Carapace widest at rear half,* narrow at middle and slightly wider from middle forward. Sharp corners behind eye sockets, continuing to rear as row of low, round projections; long, *triangular beak deeply cleft* along midline. Legs moderately long, pincers small, hands narrow, fingertips white.

Habitat: On rock and pebble bottoms, among kelps; from low-tide line to water 170′ (52 m) deep.

Range: Arctic to Rhode Island.
Comments: Older Toad Crabs bear coralline or green algae or bryozoans on their carapaces, making them difficult to see and hard to identify. The Lesser Toad Crab (*H. coarctatus*) ranges from the Arctic to Cape Hatteras. It is smaller, 1¼″ (32 mm) wide and 2″ (51 mm) long, and has a prominent toothed ridge behind the eye socket.

656 Common Spider Crab
(*Libinia emarginata*)
Class Crustacea

Description: 3¾″ (95 mm) wide, 4″ (102 mm) long. Round, spiny. Grayish-yellow or brown, tips of fingers white. Carapace globular; beak extended, with shallow, V-shaped notch; *midline with 9 spines* in a row, *border with 7 spines* behind eye socket, numerous smaller spines of different sizes, and hairs, on upper surface. Pincers equal; hand long, narrow; movable finger slightly curved; both fingers toothed; walking legs long, hairy. Legs and pincers of male almost twice as long as female's.
Habitat: Various kinds of bottoms; from low-tide line to water 410′ (125 m) deep.
Range: Nova Scotia to Florida and Texas.
Comments: Common Spider Crab males frequently have walking legs and pincers more than 6″ (15 cm) long, and are impressive in their slow, careful locomotion. They are often overgrown with fine algae which accumulates dirt and debris. The Doubtful Spider Crab (*L. dubia*) lives a similar life in comparable habitat, and ranges from Cape Cod to Florida, Texas, the Bahamas and Cuba. It differs in having a longer beak with a deeper central notch, 6 spines down the midline, and 6 spines along the margin, the last nearly 3 times as long as the others.

675 **Masking Crab**
(*Loxorhynchus crispatus*)
Class Crustacea

Description: 3½″ (89 mm) wide, 4″ (102 mm) long. Pear-shaped, covered with growth. Grayish-brown, covered with short, brownish hairs; fingertips white. Carapace widest in rear third; side margin without spines; *beak notched at tip,* thick, moderately long, bent down slightly; *sharp spine above and one beside eye.* Pincers long, slender; fingers short; 2nd pair of walking legs longest, 5th shortest.

Habitat: On rocks, pilings, and kelp holdfasts; below low-tide line to water 505′ (154 m) deep.

Range: N. California to Baja California.

Comments: The Masking Crab is a "decorator," camouflaging itself with seaweed, sponges, anemones, hydroids, or bryozoans, attaching them to the hooked hairs on its back.

---

640 **Spiny Spider Crab**
(*Mithrax spinosissimus*)
Class Crustacea

Description: 7⅜″ (19 cm) wide, 6¾″ (17 cm) long. Spiny, globular. Carmine, dark red, or orange-red; pincers rose-red, with yellow fingers; legs brick-red, with paler tips. *Carapace with blunt spines in middle and sharper ones toward sides,* somewhat pear-shaped, widest at rear half, rough. Side margin with 6 spines, first 2 double, *beak with U-shaped notch,* short. Pincers of male massive, longer than walking legs; those of female as long as walking legs; row of low, round projections on outer margin of hand; movable finger with single large tooth; walking legs spiny and hairy.

Habitat: On rocks and rubble bottoms; from low-tide line to water 590′ (180 m) deep.

Range: North Carolina to Florida; West Indies.
Comments: This is the largest spider crab on our Atlantic Coast. It is sometimes overgrown with encrusting organisms, but does not decorate itself as some other crabs do.

678 Sharp-nosed Crab
(*Scyra acutifrons*)
Class Crustacea

Description: 1½″ (38 mm) wide, 1¾″ (44 mm) long. Pear-shaped, encrusted. Reddish-brown, mottled. Carapace widest near rear, narrowing in front to a *beak of 2 flat horns; surface rough but not spiny.* Pincers long, equal; 2nd walking legs longest.
Habitat: Among seaweeds and sedentary animals, under rocks, on pilings; from low-tide line to water 300′ (91 m) deep.
Range: Alaska to Baja California.
Comments: The carapace of this crab is usually overgrown with sponges, tunicates, barnacles, bryozoans, or hydroids. When quiet, the crab usually sits with its beak down, making it very difficult to see.

659 Shield-backed Kelp Crab
(*Pugettia producta*)
Class Crustacea

Description: 3¾″ (95 mm) wide, 4¾″ (121 mm) long. Smooth, shield-shaped. Olive or reddish-brown with darker dots, paler below. Carapace smooth, squarish, extended forward to include V-notched beak and spine to inside of eye sockets, with distance between sockets ⅓ of total width of carapace; *prominent spine* on each side, ⅓ distance between widest point and eye socket; *2 more strong spines* toward rear end. Pincers equal, moderate size. Walking legs moderately long, sturdy; 2nd pair longest, 1½ times body length.

Habitat: Adults in kelp beds; young near low-tide line on rocks or inshore brown algae; from low-tide line to water 240′ (73 m) deep.

Range: Alaska to Baja California.

Comments: The color of this crab matches the color of the kelp on which it lives and which it eats. In Puget Sound, where kelps die back in winter, it is carnivorous, feeding on barnacles, bryozoans, and hydroids. This thin-shelled kelp crab does not tolerate brackish water, and is found only in areas of full salinity.

Atlantic Decorator Crab

(*Stenocianops furcata*)
Class Crustacea

Description: 4½″ (114 mm) wide, 5½″ (140 mm) long. Pear-shaped; covered with growth. Dark red; paler underneath; fingertips black. Carapace spiny. *4 strong spines in midline;* 2 pairs halfway to margin; *large spine in front of eye,* smaller one behind; *4 very large spines* at side. Beak of *2 long horns* curving slightly inward. Surface sparsely covered with *long, hooked hairs.* Pincer longer than other walking legs, with large, scattered, low, round projections; hand cylindrical, fingers short.

Habitat: On sand, shell, or coral-rubble bottoms; from low-tide line to water 360′ (110 m) deep.

Range: North Carolina to Florida and Alabama; Yucatán; West Indies; Cuba.

Comments: This is the largest decorator crab of our southeastern coast. Its tendency is to decorate itself not so much with growing plants and animals as with bits of algae, turtle grass blades, and other objects which it jams among its hooked hairs.

574 **Arrow Crab**
(*Stenorhynchus seticornis*)
Class Crustacea

Description: ½" (13 mm) wide, 2¼" (57 mm) long. Arrowhead-shaped. Pale gray, cream-colored, buff, or orange with *inverted, V-shaped, light-and-dark-brown or black stripes;* legs reddish, bright red at joints; fingers of pincers blue, eyes maroon. *Carapace triangular, smooth,* widest at rear, narrowing toward eyes, extending forward as *long, slender, spiny beak. Pincers twice body length, slender; legs slender, over 3 times body length,* with 2–3 rows of fine spines on longest joints.
Habitat: On rock, shell, sand, and coral-rubble bottoms, coral reefs, jetties, and wharf pilings; from low-tide line to water 4884' (1489 m) deep.
Range: North Carolina to Florida and Texas; Bermuda; Bahamas; West Indies to Brazil.
Comments: This dainty and elegant little crab is a delight to watch as it walks delicately among the rocks or on a reef. In recent years it has appeared in pet stores.

Pourtales' Long-armed Crab
(*Parthenope pourtalesii*)
Class Crustacea

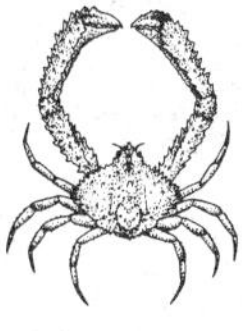

Description: 1⅞" (48 mm) wide, 1½" (38 mm) long. Spiny, long-armed. Purplish-red; buff cross-bands on pincers; palms of hand pinkish-brown. *Carapace triangular-oval, extremely spiny,* rounded at rear. Beak long, pointed, with strong tooth on each side; *pincers 2½ times body length, profusely spiny,* equal, heavy, flattened, rough; fingers bent downward, toothed; those on immovable finger heavy. 2nd–5th pairs of walking legs slender, smooth, equal to body width.
Habitat: On sand and sandy-mud bottoms; from

shallow water to water 800′ (244 m) deep.

Range: Cape Cod to Florida; West Indies.

Comments: The extraordinary size of this little crab's pincers and its spiny surface give it an unbalanced and grotesque appearance. The Saw-toothed Crab (*P. serrata*) ranges from North Carolina to Florida and Mexico, and from the West Indies to Brazil. It is smaller, 1⅛″ (28 mm) wide and ¾″ (19 mm) long. Its color varies from mottled gray-red with carmine fingers to grayish-tan with black fingers. The outer surface of its huge pincers has a ridge of large, sawlike teeth.

ECHINODERMS
(Phylum Echinodermata)

The Phylum Echinodermata is an entirely marine group of animals including sea stars, brittle stars, sea urchins, sand dollars, sea cucumbers, and sea lilies, whose most obvious feature is their *radial symmetry*. The echinoderm body is nearly always arranged in 5 parts, or multiples thereof. There is a body axis with the mouth at one end and anus at the other. In some forms the mouth faces up, in others, down. Echinoderms have an internal limy skeleton, covered by skin, and may also have spines, some movable, some fixed, variable in size and shape. The unique characteristic of the phylum is an internal hydraulic system, termed the *water vascular system*, that operates numerous *tube feet* (podia). These are slender, fingerlike appendages, arranged in rows, which the animal extends by pumping full of fluid, and retracts with muscles within the tube foot itself. Tube feet are used in locomotion and feeding. Many echinoderms have tube feet equipped with suction disks at their tips, enabling them to cling tenaciously to a surface. Inside the body is a complex system of canals, filled with sea water, (in sea cucumbers it is filled with body fluid) that operates the tube feet. Water passes back and forth between the canals and the sea through a *sieve plate*, a perforated disk through which the echinoderm can draw or expel sea water as needed.

Like other higher animals, an echinoderm has spacious body cavities containing a complex digestive system, a nervous system, and reproductive organs. There are no specific excretory organs; metabolic wastes diffuse either through the skin or the water vascular system.

Reproduction is sexual in all species, and sexes are separate in nearly all cases. Most species spawn eggs and sperm into the water but, in some, eggs are brooded by the female. Asexual reproduction is limited to a few sea stars and brittle stars that may divide and regenerate missing parts. The phylum is divided into 4 classes: the Class Stelleroidea, the starlike echinoderms, which is divided into the Subclass Asteroidea, the sea stars, and the Subclass Ophiuroidea, the brittle stars; the Class Echinoidea, the sea urchins, sand dollars, and sea hearts; the Class Holothuroidea, the sea cucumbers; and the Class Crinoidea, the sea lilies. Since the Class Crinoidea is poorly represented in waters adjacent to the continental United States and those species present occur in deep water, they will not be included in this book.

Class Stelleroidea: Subclass Asteroidea:

The asteroids include the sea stars, also called starfish, a term considered inappropriate by biologists, who would like to see the word "fish" reserved for finny vertebrates. The asteroid body has the form of a somewhat flattened star, with *arms* (rays) usually numbering 5 or a multiple of 5, rarely 6 or some other number, each in contact with adjacent arms where it joins the *central disk*. The surface of the central disk has the *anus* in the center, the sieve plate near the junction of 2 arms, and openings of sex ducts at each juncture of adjacent arms. The upper surface of each arm has the spines and other features of the species, and an eyespot, usually red, at the tip. The underside of a sea star has the *mouth* in the middle of the central disk, and an open *groove* from the mouth to the tip of each arm. 2 or 4 crowded rows of tube feet lie in each groove. In some sea stars there is a special skeletal structure for pinching small objects, a modification of 2 or 3 spines.

These *pinchers* (*pedicellariae*) may be shaped like small tweezers or pliers, or flattened like the jaws of a vise.

Sea stars travel on their tube feet. Tube feet with suckers reach ahead and attach, drawing the star in that direction while others detach, reach farther forward, attach, and so progress. Tube feet of burrowing sea stars lack suckers, and merely push the animal along.

The feeding of sea stars varies greatly with the species, but in many cases involves the extension of the forepart of the stomach out through the mouth to envelop the food.

Sea stars can regenerate lost arms. When an arm is damaged, it is shed at a point close to the central disk, even though the damage may be near the tip, a process called *autotomy*. In most species, after autotomy the cut surface heals over, regeneration of a new arm begins, and the autotomized limb dies. However, there are a few sea stars in which autotomy is spontaneous; not only does the star regenerate a limb, but the limb regenerates a star.

Subclass Ophiuroidea:

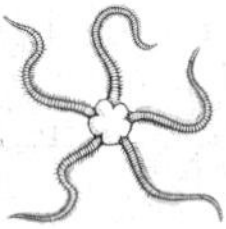

The brittle stars and basket stars make up the Subclass Ophiuroidea. In ophiuroids the base of an arm does not meet that of its neighbor as it does in a sea star; instead, a portion of the free border of the oral disk lies between them. The central disk may be round, pentagonal, or scalloped, its upper surface leathery or scaly. The mouth on the underside is shaped like a 5-pointed star, an arm joining the disk at each star point alternating with a triangular, pointed jaw with toothed margins. At the base of each jaw is a plate which may be perforated to form the sieve plate of the water vascular system, the number of sieve plates ranging from 1 to 5. Beside the base of each arm are 1 or 2 slits that open into a large *respiratory pouch*. Alternate swelling and contracting of the body wall moves water in and out of the pouch, whose lining takes up oxygen. Sex organs are adjacent to the walls of the pouches. The arms are long, jointed, and flexible, unbranched in brittle stars, extensively branched in basket stars. The segments commonly bear spines, a feature used in identifying species. While there is no groove on the underside of the arm, as there is in a sea star, there is a double row of active, suckerless tube feet that serve as sense organs, are used in feeding, and may be of some use in locomotion. Arms are used for locomotion and grasping food.

Class Echinoidea: The Class Echinoidea includes the sea urchins, cake urchins, sand dollars and heart urchins. Unlike sea stars and brittle stars, these creatures do not have arms, or rays. The skeleton, called a *test*, consists of rows of radially arranged plates immovably joined to each other. Movable spines, each with a concave base, fit on correspondingly convex bumps on each plate. Muscle fibers attached to each spine enable it to swing about in any direction.
Regular species, such as the sea urchins, are almost perfectly radially symmetrical, while *irregular species*, the cake urchins, heart urchins, and sand dollars, have a bilateral symmetry superimposed upon a radial pattern.

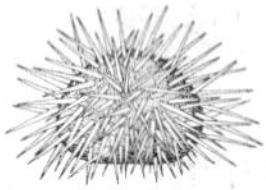

In sea urchins, the middle of the upper surface has a circular area, usually with scaly plates, bearing the anus. It is surrounded by 5 petal-shaped plates, each with a large pore, the opening of a sex duct. One of these plates is also full of small pores, and is the sieve plate. Alternating with these plates are 5 other plates which may or may not touch the area bearing the anus. Beyond these 10 plates are 20 longitudinal rows of firmly united plates extending toward the mouth, 5 pairs of rows perforated for tube feet alternating with 5 unperforated pairs. All plates bear spines.

Tube feet on an urchin are arranged in 5 pairs of rows that extend longitudinally around the test. They are tipped with suckers, and are long enough to reach beyond the spines. Urchins also have stalked pinchers, all of which have 3 jaws, and some have poison glands. These structures are defensive, protecting against predators and discouraging larval animals from settling on the urchins.

The body wall on the lower side extends beyond the border of the rows of plates as a flexible *lip* surrounding the *mouth.* Around the mouth are large tube feet which can attach to the substrate and pull the mouth against it for feeding. In all cases, feeding involves gnawing with a toothed organ called *Aristotle's lantern.* This remarkable structure consists of a set of skeletal rods and muscles arranged to open and close 5 teeth, like the jaws of a drill chuck. The lantern can be protruded out of and completely retracted back into the mouth. The area around the mouth is usually adorned with 10 frilly gills.

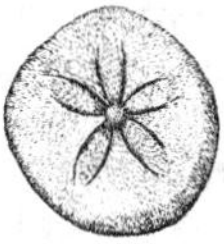

Cake urchins, heart urchins, and sand dollars are modified for burrowing in sand. They have shorter and more numerous spines than sea urchins do; tube feet are confined to the upper and lower surfaces, absent from the sides; and they have assumed a bilateral symmetry while retaining most of the general pattern of an echinoid. In heart urchins the mouth is well forward, and the anus at the rear end. In cake urchins and sand dollars the mouth remains central, but the anus is to the rear. The upper surface of the test shows the pattern of 5 sets of tube feet, one directed forward, 2 to the left, and 2 to the right. Though the anus is to the rear, the plates with reproductive pores and sieve plate remain at the upper center. Aristotle's lantern is not well developed in these irregular echinoids, and they feed on plankton and organic particles trapped on mucus.

Class Holothuroidea: Members of the Class Holothuroidea are generally called sea cucumbers, though some of them bear no particular resemblance to the vegetable. They are elongated, with the axis running from mouth to anal end. More primitive forms have 5 well-developed longitudinal rows of tube feet equally spaced around the circumference but, since such long animals must lie with one side down, many of them have well-developed tube feet on the 3 rows in contact with the substrate, and the other 2 rows reduced or missing. This imposes an almost bilaterally symmetrical pattern on these radially symmetrical animals.

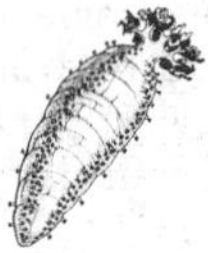

The mouth is surrounded by a row of *tentacles* which may be fingerlike, stalked with a buttonlike tip, or branched. Tentacles are actually modified tube feet, part of the water vascular system. They are used in feeding. The holothuroids differ from the echinoderms previously discussed in having a water vascular system full of body fluid rather than sea water; no sieve plate communicates with the sea. The body wall contains a large number of microscopic plates, scales, or hooks of varied shapes, whose patterns are important to the biologist in identification of the species, but of no use as field characteristics.

---

564 **Banded Luidia**
(*Luidia alternata*)
Class Stelleroidea

Description: Radius 6″ (15 cm). 5-armed. Upper surface gray, brown, greenish, purplish, or black, banded with creamy or yellow, paler below; tube feet frequently orange. Central disk small, arms narrow, upper surface covered with *mosaic of brushlike spines;* under surface paved by row of *narrow skeletal plates* at right angles to groove; each plate with *2 long, slender, movable spines* at margin.

Habitat: On sand or mud bottoms; below low-tide line to water 660′ (201 m) deep.

Range: North Carolina to Florida and Texas; Bahamas; West Indies to Brazil.

Comments: These sea stars can burrow into sand or mud using their tube feet and movable spines. They prey upon brittle stars and other bottom-dwelling invertebrates. The Striped Luidia (*L. clathrata*) is an even more common species that ranges from New Jersey to Brazil. It is about the same size, is bluish-gray, brown, salmon, or rose above, and has a dark stripe down each arm.

561 **Armored Sea Star**
(*Astropecten armatus*)
Class Stelleroidea

Description: Radius 6″ (15 cm). Flattened. Grey to beige. Upper surface covered with rosettes of small spines. Five arms edged with *row of plates bearing movable spines; tube feet pointed and lacking suckers.*

Habitat: In bays and outer coasts on sand bottoms; from low-tide line to water 200′ (61 m) deep.

Range: S. California to Ecuador.

Comments: This sea star feeds on the Purple Dwarf Olive and other snails, sand dollars, and sea pansies. It also scavenges. The Plated-margined Sea Star (*A. articulatus*) ranges from New Jersey to Uruguay, but is not common north of Cape Hatteras. Its radius is 5″ (127 mm), and it is usually purple above, with red-orange plates at the margin, or completely orange above; the underside is pale yellow.

563 **Spiny Sea Star**
(*Astrometis sertulifera*)
Class Stelleroidea

Description: Radius 5″ (127 mm). Star-shaped. Brown to green, spines purple, orange, or blue, with red tips. Small disk and five *spiny, flexible arms; spines surrounded by large pinchers.*
Habitat: On and under rocks and in sand channels, on rocky coasts; from low-tide line to water 130′ (39 m) deep.
Range: S. California to Baja California.
Comments: This is an active and mobile sea star; if turned over, it can right itself quickly. Prey, normally captured with the tube feet, includes a variety of mollusks, barnacles, sea urchins, and brittle stars.

---

535 **Mud Star**
(*Ctenodiscus crispatus*)
Class Stelleroidea

Description: Radius 2″ (51 mm). Almost pentagonal. Pale brownish-yellow; tube feet brown. Central disk large; *indentations between arms shallow, curved;* upper surface covered with brushlike spines; *furrow* from center to juncture of adjacent arms. Under surface paved with *narrow plates* at right angles to groove, each with flattened spines on surface; *3-pointed spine* extending over side of groove; short, *simple spine at outer margin.* 2 rows of thick tube feet.
Habitat: On mud bottoms; in water 20–6370′ (6–1942 m) deep.
Range: Arctic to North Carolina; Bering Sea to Chile.
Comments: This sea star not only lives in soft mud, creeping through it on thick, suckerless tube feet, but it eats mud as well, removing from it the digestible organic compounds.

541 **Cushion Star**
(*Oreaster reticulatus*)
Class Stelleroidea

Description: Radius 10″ (25 cm). 5-pointed, thick-bodied, highly arched. Larger specimens reddish-brown, orange-red, or yellowish, with spiny meshwork of contrasting lighter or darker color; young specimens olive-green or mottled purplish-brown; all ages cream-colored, yellow, or tan underneath. Arms not clearly demarked from oval disk; upper surface with knobby spines joined by raised ridges to form *network of squares and triangles;* margin with row of *equally spaced knobby spines,* under surface covered with knobby spines in neat rows, *sharper spines bordering grooves.* 2 rows of whitish tube feet. *Firm, hard skeleton.*

Habitat: On sand, coral, rubble, and growth-covered bottoms; below low-tide line in shallow water.

Range: North Carolina to Florida; West Indies to Brazil.

Comments: This is the largest sea star on our Atlantic Coast. Its remarkably hard skeleton maintains the animal's shape even when it is dried. One frequently sees garishly painted specimens of dried Cushion Stars in the windows of beachwear stores.

540 **Horse Star**
(*Hippasteria phrygiana*)
Class Stelleroidea

Description: Radius 8″ (20 cm). Nearly pentagonal. Scarlet above, creamy-white underneath. Central disk large, *indentations between arms shallow, curved.* Upper surface covered by large, blunt spines, each surrounded by ring of beadlike spines; row of upper and lower *marginal plates with beaded borders, and 1–3 large, and several small knobs* on each. *Viselike pinchers* on both surfaces. 2 rows of sturdy, sucker-tipped, white tube feet.

Habitat: On rock bottoms covered with growth; in water 66–2640′ (20–805 m) deep.

Range: Arctic to Cape Cod.

Comments: Its stomach contents reveal this handsome sea star to be an indiscriminate carnivore, eating mussels, worms and other echinoderms. Stormy weather sometimes beaches these stars.

---

546 Dawson's Sun Star
(*Solaster dawsoni*)
Class Stelleroidea

Description: Radius 10″ (25 cm). Radius of central disk ⅓ total radius. Many-armed. Upper surface gray, yellow, brown, or sometimes red, often with light patches. *8–13 arms.*

Habitat: On variety of bottom types; from low-tide line to water 1200′ (420 m) deep.

Range: Alaska to c. California.

Comments: *S. dawsoni* often preys on sea stars, even cannibalizing its own species. It also feeds on some sea cucumbers and the Diamondback Nudibranch (*Tritonia festiva*). It is seen washed up on the beach after storms. Stimpson's Sea Star (*S. stimpsoni*) is a closely related species with a similar range, a radius of 3¼″ (83 mm), and 9–11 arms which are more slender. Its color is red, orange, yellow, green, or blue, with a blue-gray spot on the central disk continuing as a stripe down each arm.

542, 543 Smooth Sun Star
(*Solaster endeca*)
Class Stelleroidea

Description: Radius 8″ (20 cm). Many-armed. Purplish, red, pink, or orange. *7–14*

	arms. Otherwise similar to Dawson's Sun Star.
Habitat:	On rock and gravel bottoms; from low-tide line to water 900′ (274 m) deep.
Range:	Arctic to Cape Cod; Alaska to Puget Sound.
Comments:	This sun star preys chiefly on small sea cucumbers and on other small sea stars. Its development is direct from egg to adult form, without a free-swimming stage.

545 **Spiny Sun Star**
(*Crossaster papposus*)
Class Stelleroidea

Description:	Radius 7″ (18 cm). Many-armed. Scarlet above, with concentric bands of white, pink, yellowish or dark red; white underneath. Central disk large, with *netlike pattern of raised ridges; 8–14 arms*, length ½ radius. Entire upper surface sparsely covered with *brushlike spines;* marginal spines larger. Mouth area bare. 2 rows of sucker-tipped tube feet in grooves.
Habitat:	On rock bottoms; from low-tide line to water 1080′ (329 m) deep.
Range:	Arctic to Gulf of Maine; Alaska to Puget Sound.
Comments:	Among the most beautiful of echinoderms, these sea stars seem to be sunbursts of color. They are predatory on smaller sea stars, swallowing them whole.

539 **Winged Sea Star**
(*Pteraster militaris*)
Class Stelleroidea

Description:	Radius 2¾″ (70 mm). Blunt-armed, thick bodied. Upper surface yellow, yellowish-orange, or red; white or tannish underneath. 5 broad arms; upper surface covered with *tentlike*

membrane supported by many *brushlike spines,* enclosing large space, *large hole* in center; *winglike membrane* on sides of arms supported by straight spines; bare area around large mouth, 2 rows of sucker-tipped tube feet in grooves.

Habitat: On rock bottoms covered with growth; in water 12–3600′ (4–1097 m) deep.

Range: Arctic to Cape Cod.

Comments: The tentlike membrane of this sea star covers a space used for brooding the young. A small number of large, yolky eggs are deposited there, hatch into miniatures of the adult, and then rupture through the membrane and creep away. This species can be found on the beach after storms.

---

538 **Badge Sea Star**
(*Porania insignis*)
Class Stelleroidea

Description: Radius 2¾″ (70 mm). Blunt-armed. Scarlet above, with white in creases between plates; white underneath. 5 arms broad, borders between them shallowly indented; upper surfaces with scattered, short, *whitish gills in creases in thick skin* covering skeletal plates, marginal plates with *3–4 short spines. Sieve plate white.* Plates on lower side long, oblique to grooves. 2 rows of sucker-tipped tube feet.

Habitat: On rock bottoms with heavy growth; in water 115–2240′ (35–683 m) deep.

Range: Cape Cod to Cape Hatteras.

Comments: *P. insignis* feeds on fine organic particles which it captures on its mucus-covered surface and moves by cilia into the mouth. It also directly ingests any organic sediment it encounters on the bottom. Beachcombers find it washed up on the sand after rough weather.

552 **Blood Star**
(*Henricia sanguinolenta*)
Class Stelleroidea

Description: Radius 4″ (102 mm). Slender-armed. Upper side usually blood-red, sometimes rose, purplish, orange, yellow, cream-colored, or white; underside whitish. Central disk small, *arms nearly cylindrical,* long; surface smooth, covered with closely set, equal-sized, blunt spines; red eyespot at tip of arm. Sieve plate whitish; 2 rows of sucker-tipped tube feet in grooves.

Habitat: On rock bottoms; from low-tide line to water 7920′ (2414 m) deep.

Range: Arctic to Cape Hatteras.

Comments: Besides trapping organic particles on its mucus and then moving them into its mouth by ciliary action, the Blood Star can also absorb dissolved nutrients through its skin.

551 **Pacific Henricia**
(*Henricia leviuscula*)
Class Stelleroidea

Description: Radius 3⅝″ (92 mm). Slender-armed. Tan, yellow, orange, red, or purplish, *usually spotted or mottled.* Otherwise similar to the Blood Star.

Habitat: On and under rocks with growth of sponges, bryozoans, or algae, in protected places; from low-tide line to water 1320′ (402 m) deep.

Range: Alaska to Baja California.

Comments: Breeding habits in this sea star vary with size. Smaller females brood their eggs in a depression around the mouth formed by arching the arms. Larger females discharge eggs directly into the water and do not brood them.

549 **Thorny Sea Star**
(*Echinaster sentus*)
Class Stelleroidea

Description: Radius 2¼" (57 mm). Small, with heavy spines. Red, reddish-brown, reddish-orange, or dark purple. Central disk small; 5 blunt, moderately long arms, upper surface of each having 5 *irregular rows of large spines, middle row largest,* each spine mounted on a rounded base; spines scattered over disk. Sieve plate raised, flat, rough. 2 rows of long tube feet in each groove.
Habitat: On rocky shores, grass beds, reefs; in shallow water.
Range: North Carolina to Florida; West Indies.
Comments: This is one of the commonest shallow-water sea stars in southern Florida. Certain species of *Echinaster* have been shown to be attracted to light, and *E. sentus* probably responds similarly, as it is found in the open on sunny days.

550 **Equal Sea Star**
(*Mediaster aequalis*)
Class Stelleroidea

Description: Radius 3½" (89 mm). Flattened. Red-orange. Symmetrical, with *5 arms edged with row of plates.* Upper surface covered with rosettes of small spines.
Habitat: On rock bottoms; near low-tide line and below.
Range: Alaska to Baja California.
Comments: *Mediaster* may, on casual inspection, be mistaken for the Bat Star, but its thinner arms edged with plates will readily distinguish it.

553 **Pacific Comet Star**
(*Linckia columbiae*)
Class Stelleroidea

Description: Radius 3⅝″ (92 mm). Long-armed. Gray, mottled with red or red-brown. 5 *long arms* usual, but may vary from 1 to 9; *arms rarely symmetrical.*
Habitat: On rocky shores; from low-tide line to water 240′ (73 m) deep.
Range: S. California to Colombia.
Comments: *Linckia's* powers of regeneration far exceed those of our other sea stars. A shed arm will regenerate a new disk and arms to complete a new star. The Common Comet Star (*L. guildingii*) ranges from Florida through the West Indies. Its radius is 3″ (76 mm), and it is dull red, brown, or purple.

536 **Leather Star**
(*Dermasterias imbricata*)
Class Stelleroidea

Description: Radius 4¾″ (121 mm). Blue-gray network with red or orange in the spaces between. *Disk large and high,* with 5 arms; upper surface *feels like wet leather;* body has a strong garlic odor.
Habitat: On rocky shores, in clean harbors on pilings and sea walls; from low-tide line to water 300′ (91 m) deep.
Range: Alaska to s. California.
Comments: *D. imbricata* feeds on anemones, sea cucumbers, the Purple Sea Urchin, and a variety of other invertebrates. The anemone Red Stomphia will release itself from its substrate and swim away when touched by the Leather Star.

537 **Bat Star**
(*Patiria miniata*)
Class Stelleroidea

Description: Radius 4″ (102 mm). Short-armed. Most commonly reddish-orange, but highly variable in color and pattern. Usually 5 but sometimes 4–9 *broad, short arms; lacks spines and pinchers.*
Habitat: On rocks, among surfgrass, and on rock and sand bottoms; from low-tide line to water 960′ (293 m) deep.
Range: Alaska to Baja California.
Comments: This is the most abundant sea star on the West Coast, where it is especially numerous in certain kelp forests.

547, 559 **Northern Sea Star**
(*Asterias vulgaris*)
Class Stelleroidea

Description: Radius 8″ (20 cm). Long-armed. Pink, rose, orange, tan, cream-colored, gray, greenish, bluish, lavender, or light purple. Body soft, flabby. Central disk moderately large; 5 *arms tapered to narrow tip,* somewhat flattened, widest just beyond juncture with disk; upper surface with abundant short, pointed spines, usually a *row of spines down middle* of arm. Red eyespot at each arm tip. *Sieve plate cream-colored to whitish,* prominent. Small, delicately tapered pinchers around spines at sides of arms. *4 rows of tube feet* in each groove.
Habitat: On rock or gravel bottoms; from between high- and low-tide lines to water 1145′ (349 m) deep.
Range: Labrador to Cape Hatteras.
Comments: In some coves in Maine this sea star is so abundant around mussel beds that one can see several hundred in a few minutes.

557, 558 **Forbes' Common Sea Star**
(*Asterias forbesi*)
Class Stelleroidea

Description: Radius 5⅛″ (130 mm). Long-armed. Tan, brown, olive, with tones of orange, red, or pink; *sieve plate orange.* Similar to the Northern Sea Star, but with *firmer skeleton,* arms blunter and thicker, and spines scattered, not usually in rows.

Habitat: On rock, gravel, or sand bottoms; from low-tide line to water 160′ (49 m) deep.

Range: Gulf of Maine to Texas.

Comments: Like the Northern Sea Star, this species feeds chiefly on bivalve mollusks. Experiments have shown that a star 3″ in radius can exert a 12-pound pull. A 2″ cherrystone clam exerts a 10-pound pull to keep its valves closed. This star needs only a tiny gape, 1/250″ (.1 mm), to gain its meal. It everts its stomach through its mouth, slips it between the mollusk's valves, and secretes digestive juices which begin to consume the clam's soft tissues. The clam soon dies, its valves gape, and the sea star finishes its meal.

554 **Troschel's Sea Star**
(*Evasterias troschelii*)
Class Stelleroidea

Description: Radius 8″ (20 cm). Slender. Orange to brown to blue-gray, sometimes mottled. *Small central disk and 5 slim, tapering arms.*

Habitat: On rock or sand bottoms, usually in quiet water; from low-tide line to water 230′ (70 m) deep.

Range: Alaska to c. California.

Comments: Like the Ochre Sea Star, from which it can be distinguished by its smaller disk and thin arms, Troschel's Sea Star does not do well in areas exposed to heavy wave action. The Fragile Scale Worm occurs as a commensal on this species, and may often match the color of its host.

560 **Slender Sea Star**
(*Leptasterias tenera*)
Class Stelleroidea

Description: Radius 1½″ (38 mm). Long-rayed. Red, rose, pink, lavender, or purplish above, whitish underneath. Usually 5, sometimes 6 slender *arms tapering to blunt tips;* moderately large central disk, upper surface rough with irregular rows of conspicuous slender spines, *ring of pinchers* around each; red eyespot at tip of arm; sieve plate white. *2 rows of long, sucker-tipped tube feet* in each groove.

Habitat: On rock or gravel bottoms; from low-tide line to water 495′ (151 m) deep.

Range: Nova Scotia to Cape Hatteras.

Comments: Like other species in this genus, *L. tenera* is a brooder, laying a small number of large, yolky eggs which she holds in the area around the mouth, assuming the humped-up position taken while feeding. The developing eggs, however, obstruct such feeding during their incubation, and she endures an enforced fast. The Green Slender Sea Star (*L. littoralis*) ranges from the Arctic to the Gulf of Maine, and has a radius of 1¼″ (32 mm). Its arms are not quite so long and slender as those of *L. tenera,* and it is olive-green, with a creamy-white sieve plate. It lives just above low-tide line in shallow water, on rocks with heavy growths of rockweed and knotted wrack.

555 **Broad Six-rayed Sea Star**
(*Leptasterias hexactis*)
Class Stelleroidea

Description: Radius 2″ (51 mm). 6-armed. Green, black, brown, or red, sometimes mottled. Disk moderate-sized with 6 *fairly broad arms; spines on upper surface dense and mushroom-shaped.*

Habitat: On rocky shores; well above low-tide line and below in shallow water.

Range: British Columbia to s. California.

Comments: *L. hexactis* eats small snails, limpets, mussels, chitons, barnacles, sea cucumbers, and other species, including dead animals. It produces yellow, yolky eggs that stick together in a mass after fertilization. These are brooded under the disk of the female until they hatch as miniature sea stars after 6 to 8 weeks. The small six-rayed sea stars of the West Coast are quite variable and have presented problems of identification. The only other species currently recognized is the Small Slender Sea Star (*L. pusilla*) which has sharp spines and longer, thinner arms than *L. hexactis,* and is a light gray-brown or reddish color. It also has a very limited range from San Francisco to Monterey Bay. It reaches a radius of 1″ (25 mm).

---

556 **Giant Sea Star**
(*Pisaster giganteus*)
Class Stelleroidea

Description: Radius 12″ (30 cm). Heavy. Red, brown, tan, or purple, with *blue rings around base of spines.* Body tough, firm; arms thick; *spines large and well-spaced.*

Habitat: On rocky shores and in shallow water on rock bottoms; near low-tide line and below.

Range: British Columbia to Baja California.

Comments: In southern California, Kellet's Whelk (*Kelletia kelletii*) has been observed sharing food with *P. giganteus*, but the whelk itself sometimes falls prey to the Giant Sea Star.

548 **Ochre Sea Star**
(*Pisaster ochraceus*)
Class Stelleroidea

Description: Radius 10″ (25 cm). Heavy. Yellow, orange, brown, reddish, or purple. Central disk moderately large, with 5 stout, tapering arms; upper surface with many short, white spines in *netlike or pentagonal pattern* on central disk.
Habitat: On wave-washed rocky shores; well above low-tide line, and sometimes below.
Range: Alaska to Baja California.
Comments: This is the commonest large, intertidal sea star, and it occurs in great numbers on mussel beds on the exposed coast. Many animals on which it preys can detect this star at a distance and escape.

544 **Sunflower Star**
(*Pycnopodia helianthoides*)
Class Stelleroidea

Description: Radius 26″ (66 cm). Many-armed. Purple, red, pink, brown, orange, or yellow. Disk broad, *24 arms;* surface soft and flexible; *spines, pinchers, and soft gills abundant.*
Habitat: On rocky shores and soft bottoms; from low-tide line to water 1435′ (437 m) deep.
Range: Alaska to s. California.
Comments: This is the largest, most active sea star on the Pacific Coast, and to see one in motion is an impressive experience. A large *Pycnopodia* has more than 15,000 tube feet that have to be coordinated in its stepping movements.

572 **Northern Basket Star**
(*Gorgonocephalus arcticus*)
Class Stelleroidea

Description: Disk diameter 4″ (102 mm), arm length 14″ (36 cm). Highly branching. Upper surface of disk yellowish-brown or darker brown; arms yellowish-tan, white at tips; white underneath. Disk pentagonal, naked, leathery, with 5 pairs of *spiny ridges* radiating from near center to sides of arms. *5 stout arms dividing into 2 equal branches* near disk, then *arms dividing equally again* 5 or more times to many *coiling, tendril-like branches;* arm joints with short, hooked spines, small tube feet. Mouth with 5 sawlike jaws.

Habitat: On rock bottoms; in water 18–4831′ (6–1472 m) deep.

Range: Arctic to Cape Cod.

Comments: This creature is a slowly coiling, squirming mass of branched arms, usually completely entangled in anything it can grasp. It is frequently seen when hauled up to the surface in lobster pots.

---

573 **Caribbean Basket Star**
(*Astrophyton muricatum*)
Class Stelleroidea

Description: Disk diameter 2″ (51 mm), arm length 15″ (38 cm). Highly branching. Light yellowish-tan to dark brown. Upper surface of central disk *smooth, leathery.* Arms jointed, very long; *arms without spines, with alternate branches rebranching several times to very fine twigs.* 1 pair slender tube feet per joint. Mouth 5-pointed, with 5 comblike jaws.

Habitat: On reefs and surrounding flats; on sea whips, sponges, and other growth; below low-tide line in shallow water.

Range: S. Florida; West Indies.

Comments: This animal is nocturnal. At night it positions itself on top of a sea whip or sponge, spreads its highly-branched arms out facing the current, and collects organic particles and planktonic animals as they float past. Later it contracts, retreats to a resting place, and consumes the collected food by running its tentacles across its comblike jaws. During the day it wraps itself up into a seemingly hopeless tangle around a sea whip or in a vase sponge.

---

566 **Panama Brittle Star**
(*Ophioderma panamense*)
Class Stelleroidea

Description: Disk diameter 1¾″ (44 mm), arm length 10″ (25 cm). Long-armed. Olive or grayish-brown, with buff bands on arms. Central disk pentagonal, flat, *granular above,* mouth 5-pointed, closable; *2 slits* into each respiratory pouch, 1 near mouth, 1 next to arm where it joins disk. *7–8 short spines* in vertical row on side of each joint, held close against arm.

Habitat: Among rocks and seaweeds on sand bottoms; near low-tide line and below in shallow water.

Range: S. California to Peru.

Comments: This predatory brittle star can live successfully in marine aquaria for several years, feeding on bits of fish or clam meat. When the star senses food, it advances, sweeping the lead arms back and forth. As soon as it touches the food, it coils an arm around it, brings it under the uplifted disk to the mouth, and quickly swallows it. The Short-spined Brittle Star (*O. brevispina*) ranges from Cape Cod to Florida, and from the West Indies to Brazil. It is greenish to black and vaguely mottled, with a disk diameter of ⅝″ (16 mm) and an arm length of 2⅝″ (67 mm). It is common in eelgrass beds near the low-tide line, but is found in water as deep as 390′ (119 m).

570 **Daisy Brittle Star**
(*Ophiopholis aculeata*)
Class Stelleroidea

Description: Disk diameter ¾″ (19 mm), arm length 3⅝″ (92 mm). Long-armed. Red, orange, pink, yellow, white, blue, green, tan, brown, gray, and black, in infinite variety of spots, lines, bands, and mottlings. *Central disk scalloped,* a lobe protruding between adjacent arms, covered with fine, *blunt spines and roundish plates.* Plates on top of arms surrounded by *row of small scales;* joints with 5–6 *bluntly-tapered spines* in vertical rows on side of arm.

Habitat: Under rocks in tidepools, among kelp holdfasts; from low-tide line to water 5435′ (1657 m) deep.

Range: Arctic to Cape Cod; Bering Sea to s. California.

Comments: These elegant brittle stars are an exotic sight in a tidepool, scrambling into hiding when one exposes them by lifting away their rock.

567 **Burrowing Brittle Star**
(*Amphiodia occidentalis*)
Class Stelleroidea

Description: Disk diameter ½″ (13 mm), arm length 6⅝″ (17 cm). Extremely long-armed. Gray to tannish-gray, mottled, dark spot on disk at base of each arm. *Disk almost circular, smooth,* lacking spines or plates; jaws with 4–5 teeth. *Arms 15 times width of disk,* slender; vertical row of 3 blunt spines at side of arm joint.

Habitat: In sand under rocks, in algal and kelp holdfasts, on mudflats, and among eelgrass roots; from low-tide line to water 1210′ (369 m) deep.

Range: Alaska to s. California.

Comments: This brittle star's long, slender arms are sometimes lost to predators, and it is common to find specimens in the process of regenerating ends of arms.

568 **Dwarf Brittle Star**
(*Axiognathus squamatus*)
Class Stelleroidea

Description: Disk diameter ¼" (6 mm), arm length 1" (25 mm). Tiny, long-armed. Tan, gray, or orange; white spot at margin near base of each arm. Disk round, plump, surface covered with fine scales, 2 large scales at base of each arm; jaws with 2 rounded teeth on each side. Arms with *oval plate* on top of each joint, vertical row of *3 short spines* on each side.

Habitat: Among rocks and gravel in tidepools, in crevices and algal holdfasts, on rocky shores; from between high- and low-tide lines to water 2716' (828 m) deep.

Range: Arctic to Florida; Alaska to s. California.

Comments: This little brittle star is bioluminescent, capable of emitting light. The Puget Dwarf Brittle Star (*A. pugetanus*), which ranges from British Columbia to southern California, has the same disk diameter, but longer arms, 1½" (38 mm), and is gray or banded gray-and-white.

---

571 **Esmark's Brittle Star**
(*Ophioplocus esmarki*)
Class Stelleroidea

Description: Disk diameter 1¼" (32 mm), arm length 3¾" (95 mm). *Brown or red-brown. Arm spines very short;* surface of disk smooth.

Habitat: Under rocks and in crevices; from low-tide line to water 220' (67 m) deep.

Range: N. California to s. California.

Comments: This slow-moving species captures slower-moving prey, or scavenges for particles of food. It is rugged and does not fragment when handled.

562 **Spiny Brittle Star**
(*Ophiothrix spiculata*)
Class Stelleroidea

Description: Disk diameter ¾″ (19 mm), arm length 6″ (15 cm). Spiny. Orange, yellow, tan, brown, green; variously patterned. *Long, thorny spines on margins of arms and disk.*

Habitat: Under rocks, in crevices and mats of algae or invertebrates; from low-tide line to water 660′ (2012 m) deep.

Range: C. California to Peru.

Comments: *O. spiculata* occurs in large concentrations in favorable habitats. Kelp holdfasts and clumps of bryozoans and worm tubes are often writhing masses of *Ophiothrix* arms. Individuals anchor themselves with the spines of one or more arms, and extend the others into the water for filter feeding. The Atlantic Long-spined Brittle Star (*O. angulata*), which ranges from Virginia to Florida, Texas, and Mexico, and from the West Indies to Brazil, is also highly variable in color and pattern. It is smaller, with a disk diameter of ½″ (13 mm) and an arm length of 2⅜″ (60 mm). It also has long, thorny spines along the margins of its arms and disk.

565 **Reticulate Brittle Star**
(*Ophionereis reticulata*)
Class Stelleroidea

Description: Disk diameter ⅜″ (10 mm), arm length 3″ (76 mm). Long-armed. Disk bluish-gray, with *network of dark lines,* arms white or pale yellow, with narrow brown or black bands. *Disk covered by tiny scales, 2 larger oval scales* at margin where arm joins; round. Jaws with 4–5 teeth on each side. Arm joints with *3 (4 near base of arm) tapered spines* per vertical row.

Habitat: In sandy areas, under rocks; near low-tide line and below in shallow water.

Range: S. Florida and West Indies.
Comments: This is a very common species in Florida. The Ringed Brittle Star (*O. annulata*) which ranges from southern California to Ecuador, has a disk diameter of ¾" (19 mm) and an arm length of 6½" (17 cm). Its color is variable, but it always has dark rings around the arms.

569 Spiny Brittle Star
(*Ophiocoma echinata*)
Class Stelleroidea

Description: Disk diameter ⅝" (16 mm), arm length 4" (102 mm). Spiny, long-armed. Black, often mottled with cream-color or white. Disk with *scalloped border,* deep crease midway between arms; *granular* above and over sides. Each arm joint with vertical row of *4 spines* on each side, *top spine longest;* length of spines decreasing toward tip of arm.
Habitat: On reefs and reef flats, in grass beds and under rocks and old coral heads; near low-tide line and below in shallow water.
Range: S. Florida; Bermuda; Bahamas; West Indies; Yucatán to Brazil.
Comments: One of the largest brittle stars in the American tropical Atlantic, *O. echinata* is common around reefs where its size and spines make it easy to identify.

517 Slate-pencil Urchin
(*Eucidaris tribuloides*)
Class Echinoidea

Description: 2½" (64 mm) wide, 1¼" (32 mm) high. Heavy-spined, oval. Tan with brown spots or stripes, sometimes mottled white, or tinged with green or red. Test red-orange to brown. *Primary spines thick, blunt, length of test width,* one per plate; *other spines tiny, flattened, abundant.* Area around anus with small spines. No gills.

Habitat: On coral reefs, rocks, and coral rubble; near low-tide line and below in shallow water.

Range: South Carolina to Florida; Bermuda; Bahamas; West Indies; Mexico to Brazil.

Comments: The large spines on this urchin are frequently coated with white or pink coralline algae, bryozoans, or sponges.

---

524 Long-spined Urchin
(*Diadema antillarum*)
Class Echinoidea

Description: 4″ (102 mm) wide, 1¾″ (44 mm) high. Long-spined, oval. Adults *dark, purple or black;* young urchins frequently with white bands or speckles on spines. Spines greatly variable in size, *longest spines on upper surface 16″ (41 cm) long.* Scaly plates on area around anus. 10 sets of gills.

Habitat: On coral reefs, rocks, coral rubble, reef flats, and tidepools; near low-tide line and below in shallow water.

Range: Florida; Bermuda; Bahamas; West Indies; Mexico to Suriname.

Comments: Bathers and divers in the American tropics know and respect the long, sharp spines of this urchin, which can puncture a wetsuit or tennis shoe and cause intense irritation if lodged in the skin.

---

518 Atlantic Purple Sea Urchin
(*Arbacia punctulata*)
Class Echinoidea

Description: 2″ (51 mm) wide, ¾″ (19 mm) high. Spiny, oval. Spines purplish-brown, reddish-gray, or sometimes nearly black; skin blackish. *Spines longitudinally grooved,* cylindrical, pointed, variable in size; *longest spines near top 1″ (25 mm) long;* top nearly free of spines. Area around anus with *4 plates,* each ¼ of a circle. 10 clusters of blue-black gills.

Habitat: On rock and shell bottoms, among seaweeds, in tidepools; from low-tide line to water 750′ (229 m) deep.

Range: Cape Cod to Florida; Texas; Yucatán; Cuba; Jamaica; West Indies; Trinidad.

Comments: This species is the most intensively studied of all sea urchins. It is omnivorous, and will eat various algae, sponges, coral polyps, mussels, sand dollars, and dead or dying urchins and other animals.

521, 528 **Variegated Urchin**
(*Lytechinus variegatus*)
Class Echinoidea

Description: 3″ (76 mm) wide, 1¼″ (32 mm) high. Spiny, oval. Green, white, pinkish-red, brownish-red, or reddish-purple; color variable. *Spines short, stout, abundant.* Area around anus with many scalelike plates.

Habitat: On turtle grass flats and sand and gravel bottoms with abundant plant growth, and among mangroves; from low-tide line to water 180′ (55 m) deep.

Range: North Carolina to Florida; Bahamas; West Indies.

Comments: This urchin is usually camouflaged with bits of shell, plant material, and other debris held over its body by tube feet. The Little Gray Sea Urchin (*L. anamesus*) is found from southern California to Baja California. It grows to 1⅝″ (41 mm) wide and ¾″ (19 mm) high. It is cream-colored to gray-purple with purple blotches, and its abundant spines grow to 1″ (25 mm) long. Its test is thin-walled, domed above and flattened below. It occurs on sand bottoms and in eelgrass beds in bays and on the outer coast from low-tide line to water 1000′ (305 m) deep.

525 Sea Egg
(*Tripneustes ventricosus*)
Class Echinoidea

Description: 6″ (15 cm) wide, 2½″ (64 mm) high. Short-spined, large, oval. *Spines white*, test brown to dark purple. Spines *very short*, about 1/15 test diameter, abundant, of nearly equal length. Area around anus with many scalelike plates of different sizes. 10 clusters of gills.

Habitat: In turtle grass flats; in shallow water around reefs.

Range: Florida; Bermuda; Bahamas; West Indies; Jamaica; Trinidad to Brazil.

Comments: This species has the largest test of any urchin in American waters. Its eggs are edible and, some say, delicious.

523 Green Sea Urchin
(*Strongylocentrotus droebachiensis*)
Class Echinoidea

Description: 3¼″ (83 mm) wide, 1½″ (38 mm) high. Spiny, oval. Test brownish-green, spines light green, gray-green, or, more rarely, brownish- or reddish-green; tube feet brownish. *Spines not greatly variable* in length, never over ⅓ diameter of test. Area around anus with *many scalelike plates* of different sizes. 10 clusters of gills.

Habitat: On rocky shores and in kelp beds; from low-tide line to water 3795′ (1157 m) deep.

Range: Arctic to New Jersey; Alaska to Puget Sound.

Comments: These urchins are so abundant in certain protected bays that it is impossible to walk through a bed of them without stepping on some.

522 **Purple Sea Urchin**
(*Strongylocentrotus purpuratus*)
Class Echinoidea

Description: 4″ (102 mm) wide, 1¾″ (44 mm) high. Spiny, oval. *Adults vivid purple;* juveniles greenish. Domed above and flattened beneath. Area around anus with *many scalelike plates.*
Habitat: On rocky shores with moderately strong surf; from low-tide line to water 525′ (160 m) deep.
Range: British Columbia to Baja California.
Comments: Above the low-tide line, these urchins often live in rounded depressions in the rock which they slowly erode with their teeth and spines.

520 **Red Sea Urchin**
(*Strongylocentrotus franciscanus*)
Class Echinoidea

Description: 5″ (127 mm) wide, 2″ (51 mm) high. Spiny, oval. Red, red-brown, or light to dark purple. *Spines abundant, variable, largest 3″ (76 mm) long.* Area around anus with *many scalelike plates.*
Habitat: On rocky shores of open coasts; from low-tide line to water 300′ (91 m) deep.
Range: Alaska to Baja California.
Comments: This large urchin is eaten by the California Sea Otter, and is in low abundance in the otter's range. A fishery in southern California processes the urchins' ovaries and ships them to Japan, where they are considered a delicacy.

519 **Rock-boring Urchin**
(*Echinometra lucunter*)
Class Echinoidea

Description: 2″ (51 mm) wide, ¾″ (19 mm) high. Spiny, oval. Reddish-brown or brownish-black, tinged with purple or green, spines light green, violet, light or dark reddish-brown. *Somewhat oval,* spine size variable, longest ¾″ (19 mm), pointed. Area around anus with *many small spines* and many plates of different sizes. 10 clusters of gills.

Habitat: In rocks with holes in areas of heavy wave action; near low-tide line and slightly below.

Range: Florida to Texas and Mexico; Bermuda; Jamaica; West Indies to Brazil.

Comments: This urchin bores its burrow during years of scraping with the teeth of its Aristotle's lantern. When the urchin braces its spines against the burrow, it is very difficult to remove.

Brown Sea Biscuit
(*Clypeaster rosaceus*)
Class Echinoidea

Description: 6″ (15 cm) long, 4½″ (114 mm) wide. Short-spined, oval. Reddish-, yellowish- or greenish-brown, or dark brown. *Test nearly oval,* narrowing toward front, widest near rear, top raised in middle, concave beneath; *mouth in middle, anus at rear,* margin beneath. Tube feet above in *5 petal-shaped loops.*

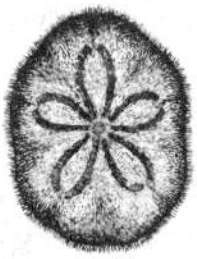

Habitat: In sandy areas, grass beds, and reef tracts, and sometimes on open beaches; below low-tide line in shallow water.

Range: North Carolina to Florida; Bahamas; West Indies to Brazil.

Comments: The Brown Sea Biscuit is very common around reef tracts in southern Florida, where it may be found burrowing just under the surface of the sand, or, if on the surface, with sand and bits of shell clinging to it.

530 **Common Sand Dollar**
(*Echinarachnius parma*)
Class Echinoidea

Description: 3⅛″ (79 mm) wide, ¼″ (6 mm) high. Flat, disklike. Reddish- or purplish-brown, darker above, paler below. Outline nearly circular, *test flattened,* covered with many close-set, short spines; *5 petal-shaped loops* of tube feet on upper surface. *Mouth central* on lower surface, *anus at rear margin. 5 branched furrows* leading to mouth.

Habitat: On sand bottoms; from low-tide line to water 5280′ (1613 m) deep.

Range: Labrador to Maryland; Alaska to Puget Sound.

Comments: Sand dollars feed on fine particles of organic matter, and are, in turn, eaten by flounder, cod, haddock, and other bottom-feeding fishes.

531 **Eccentric Sand Dollar**
(*Dendraster excentricus*)
Class Echinoidea

Description: 3″ (76 mm) wide, ¼″ (6 mm) high. Disk-shaped. Light lavender-gray, brown, reddish-brown, or dark purple-black. *Test almost circular, flattened, 5 petal-shaped loops* of tube feet on upper dorsal surface; both surfaces covered by short, fine spines. *Mouth central* on lower surface, *anus at rear margin.*

Habitat: On sand bottoms of sheltered bays and open coasts; from low-tide line to water 130′ (40 m) deep.

Range: Alaska to Baja California.

Comments: The sand dollar most familiar to beach-combers is the dead, clean test of this species found along sandy shores. The five petal-shaped areas on the upper surface are the slits from which the respiratory tube feet emerge.

533 **Michelin's Sand Dollar**
(*Encope michelini*)
Class Echinoidea

Description: 6″ (15 cm) long, 5½″ (140 mm) wide. Notched disk. Dark greenish to pale violet-brown. Test covered with fine spines, flattened, rounded in front, *squarish at rear,* with 2 pairs of *marginal notches, a single notch at front end,* and *slot* in middle between 2 rear notches. Mouth central on lower surface, anus at rear margin. 5 branched furrows leading to mouth.

Habitat: On sand bottoms, and reef flats; below low-tide line in shallow water.

Range: Florida to Texas; Yucatán.

Comments: This sand dollar differs from others in maintaining marginal notches that do not close over with age to form slots. The cleaned tests are snowy white, and are commonly sold as souvenirs in shell shops.

534 **Keyhole Urchin**
(*Mellita quinquiesperforata*)
Class Echinoidea

Description: 5½″ (140 mm) long, 6″ (15 cm) wide. 5-slotted disk. Tan, light brown, or grayish. Test covered with fine spines, almost *circular, flat,* with *2 pairs of slots* and a *single larger one* between the rear 2. Mouth central, anus at rear margin. 5 branched furrows leading toward mouth.

Habitat: On sand bottoms; below low-tide line in shallow water.

Range: Cape Cod to Florida; Bermuda; Jamaica; Puerto Rico; Mexico to Brazil.

Comments: This species is more closely related to other sand dollars than to sea urchins. Its 5 slots are notches when the animal is small, and are closed off by growth.

532 **Six-hole Urchin**
(*Mellita sexiesperforata*)
Class Echinoidea

Description: 4″ (102 mm) long, 4¼″ (108 mm) wide. 6-slotted disk. Silvery-gray to tan or yellowish-brown. *Test with 6 slots:* 2 pairs, 2 single slots in midline—1 toward front, 1 toward rear. Otherwise similar to the Keyhole Urchin.

Habitat: On sand bottoms; below low-tide line in shallow water.

Range: South Carolina to Florida; Bermuda; Bahamas; West Indies to Uruguay.

Comments: This urchin is similar in life habits to *M. quinquiesperforata.* Its species name, *sexiesperforata,* means "six holes" in Latin and is descriptive of its most distinctive characteristic.

Mud Heart Urchin
(*Moira atropos*)
Class Echinoidea

Description: 2″ (51 mm) long, 1¼″ (32 mm) wide. Spiny, egg-shaped. Test nearly as high as wide, covered with short spines above, longer ones below; upper surface with 5 *deep creases* for rows of long tube feet, *rear pair of tube feet short.* Mouth well toward front end of lower surface, with *prominent lower lip.* Anus at rear margin.

Habitat: On soft mud or sandy-mud bottoms; from low-tide line to water 480′ (146 m) deep.

Range: North Carolina to Florida and Texas; West Indies.

Comments: This irregular echinoid lies in its burrows beneath the mud, with a hole opening to the surface to admit water.

527 **Heart Urchin**
(*Lovenia cordiformis*)
Class Echinoidea

Description: 3″ (76 mm) long, 2⅜″ (60 mm) wide. White, gray, yellow, rose, or purple. Heart-shaped. Outline indented at front end, extending as groove containing front petal. Upper surface convex, covered with *numerous small spines* and *fewer long spines* more than 2¾″ (70 mm) long. Lower surface flat, *mouth forward of center, anus at rear margin.*

Habitat: On sand bottoms in bays and outer coasts; from low-tide line to water 460′ (140 m) deep.

Range: S. California to Panama.

Comments: *Lovenia* lives in a burrow just below the surface of the sand. The tube feet extend up a vertical shaft to the surface and sift surface deposits for fine organic particles.

526 **Long-spined Sea Biscuit**
(*Plagiobrissus grandis*)
Class Echinoidea

Description: 10″ (25 cm) long, 8¼″ (21 cm) wide. Spiny, oval. Yellow, tan, or reddish. *Test irregularly oval,* convex, indented at front, *sides nearly parallel;* covered with sharp, medium-sized spines; some *long spines at top* 4″ (102 mm) long. 5 rows of tube feet in deep furrows. Flat underneath, mouth near front end, anus at rear margin.

Habitat: In coral sand around reefs; to water more than 50′ (15 m) deep.

Range: Florida; West Indies.

Comments: *P. grandis* is the largest and one of the handsomest of the irregular echinoids. Its chief predators are large helmet snails, which grub it out of the sand and eat it.

529 **West Indian Sea Biscuit**
(*Meoma ventricosa*)
Class Echinoidea

Description: 7″ (18 cm) long, 5″ (127 mm) wide. Spiny, oval. Light to dark reddish-brown. Test oval, sides curved, somewhat narrower at rear, convex, indented at front, covered with *short, closely-set spines.* 5 rows of tube feet in furrows on upper surface. Lower side flat, *mouth area crescent-shaped,* near front end; anus at rear margin.

Habitat: In sand and grass beds around reefs; in shallow water.

Range: S. Florida; Bahamas; West Indies.

Comments: *M. ventricosa* lives on the surface rather than burrowing into the sand, but its tube feet gather and hold debris, with which it is usually covered.

---

238 **Four-sided Sea Cucumber**
(*Stichopus badionotus*)
Class Holothuroidea

Description: 14″ (36 cm) long, 2″ (51 mm) wide. Cucumber-shaped. Black, brown, yellowish, deep red, or purple, solid or mottled. Thick-skinned, smooth, somewhat bumpy. Flat-bottomed, narrower at top; *3 rows of tube feet on bottom, tube feet scattered elsewhere. 20 tentacles tipped with knobby disks about mouth.*

Habitat: In eelgrass and turtle grass beds, around coral reefs, and on reef flats; from low-tide line to water 6′ (2 m) deep.

Range: Florida; Bermuda; Bahamas; West Indies; Yucatán.

Comments: Adults of this species become slimy when disturbed, and few animals seem to want to eat them.

207 **California Stichopus**
(*Parastichopus californicus*)
Class Holothuroidea

Description: 16″ (41 cm) long, 2″ (51 mm) wide. Cucumber-shaped. Brown, red, or yellow, often mottled. Upper surface with *large and small, red-tipped, conical projections. Tube feet on lower surface only,* numerous.

Habitat: On protected rocky shores and pilings in clean, quiet water; from low-tide line to water 300′ (91 m) deep.

Range: British Columbia to Baja California.

Comments: In Puget Sound, natural populations of the California Stichopus normally eject their internal organs in late fall and then regenerate a new set for spring. A related species, the Parvima Stichopus (*P. parvimensis*) occurs from Monterey Bay to Baja California. It is similar, but has a chestnut-brown upper surface with short projections that are tipped with black.

237 **Fissured Sea Cucumber**
(*Astichopus multifidus*)
Class Holothuroidea

Description: 18″ (46 cm) long, 4″ (102 mm) wide. Rough, cucumber-shaped. Mottled yellowish to dark or light brown, paler underneath, tentacles pale. Upper surface with many *plump, conical projections,* larger along sides; under surface with many cylindrical, sucker-tipped tube feet not in rows. *21 tentacles tipped with knobby disks* about mouth, directed downward. Anus directed somewhat upward.

Habitat: On sand bottoms around reefs, reef flats, and turtle grass beds; from low-tide line to water 60′ (18 m) deep.

Range: S. Florida; West Indies; Mexico.

Comments: This plump sea cucumber, although not as common as Agassiz's Sea

Cucumber or the Four-sided Sea Cucumber, is a characteristic member of the reef-flat community.

236 **Florida Sea Cucumber**
(*Holothuria floridana*)
Class Holothuroidea

Description: 10″ (25 cm) long, 2″ (51 mm) wide. Cucumber-shaped. Dark brown, reddish-brown, brick-red, rarely paler, sometimes spotted, tentacles yellowish. Upper surface with *many conical projections,* 2 rows of slender tube feet; lower surface with many larger *tube feet not arranged in rows.* 20 larger tentacles with knobby tips, mouth pointing downward.

Habitat: In grass beds and on reef flats; in shallow water just below low-tide line.

Range: Florida; West Indies; Yucatán; Panama.

Comments: This sea cucumber sometimes occurs in great numbers just below the low-tide line on sandy shores. It produces sticky threads that are extruded from the anus when the animal is disturbed, doubtless a valuable deterrent to predators. This characteristic is locally known as "cotton spinning."

235 **Agassiz's Sea Cucumber**
(*Actinopyga agassizii*)
Class Holothuroidea

Description: 12″ (30 cm) long, 3″ (76 mm) wide. Thick, cucumber-shaped. Uniform brown to brown mottled with white, yellow, or other shades of brown, tentacles and tube feet yellow. Robust animal with thick, leathery skin, *3 bands of tube feet* on underside, small projections and 2 rows of tube feet on upper side. *25–29 tentacles* with knobby tips. Anus at tip of rear end, equipped with *5 white, limy teeth.*

Habitat: On reefs and reef flats, and in grass beds; in shallow water.
Range: Florida; West Indies.
Comments: *A. agassizii* has developed a remarkably intimate relationship with a small fish, *Carapus,* which normally lives in the cavity of the hind gut, just inside the anus, supplied with a constant flow of fresh sea water going into and out of the respiratory tree.

152 Orange-footed Sea Cucumber
(*Cucumaria frondosa*)
Class Holothuroidea

Description: 19″ (48 cm) long, 5″ (127 mm) wide. Cucumber-shaped. Dark reddish-brown above, paler beneath, paler in young specimens, tube feet orange-tinted. Round, long, tapered to rounded point at rear, somewhat tapered at head end; soft, smooth. Neck with mouth and tentacles retractable; *10 highly-branched tentacles* that can extend ½ body length or more. *5 crowded rows* of tube feet, 2 on upper side, 3 below.
Habitat: On rocky shores, under overhanging rocks, in crevices; from low-tide line to water 1208′ (368 m) deep.
Range: Arctic to Cape Cod.
Comments: This is the largest and most conspicuous sea cucumber in New England.

157 Red Sea Cucumber
(*Cucumaria miniata*)
Class Holothuroidea

Description: 10″ (25 cm) long, 1″ (25 mm) wide. Cucumber-shaped. Brick-red, bright orange, pinkish, or purple. Long, round, bluntly tapered at rear. Smooth, tough. *10 highly-branched, orange, retractable tentacles of equal size. 5 rows of tube feet,* 2 rows above, 3 below.

Habitat: Nestled in crevices and under rocks; near low-tide line and below to shallow depths.

Range: Alaska to c. California.

Comments: This animal's body is curved in the crevice it inhabits so that both the tentacular crown and the anus are exposed to moving water. The Peppered Sea Cucumber (*C. piperata*) is about half the size of *C. miniata*, and is yellowish-white and speckled with brown or black. It is found in similar habitats from British Columbia to Baja California.

---

151 **Stiff-footed Sea Cucumber**
(*Eupentacta quinquesemita*)
Class Holothuroidea

Description: 4″ (102 mm) long, ½″ (13 mm) wide. Cucumber-shaped. White or cream-colored with *yellow tentacles*. Body cylindrical, tapered at ends; 10 *short, bushy tentacles* around mouth; 5 rows of *stiff, non-retractable tube feet*.

Habitat: Under rocks and in crevices, in algal- and invertebrate-covered substrates; near low-tide line and below in shallow water.

Range: British Columbia to s. California.

Comments: This species, common in the shallow water of both outer coasts and clean harbors, is eaten by several species of sea stars.

---

150 **Hairy Sea Cucumber**
(*Sclerodactyla briareus*)
Class Holothuroidea

Description: 4¾″ (121 mm) long, 2″ (51 mm) wide. Hairy, sweet-potato-shaped. Blackish, brownish, greenish, or purplish. Wide in middle, strongly tapered, *mouth and tail ends bent upward;* nearly covered with *slender tube feet*, not in rows; soft.

10 large, bushy tentacles, 2 lowest ones smaller.

Habitat: On mud or sand bottoms; from low-tide line to water 20′ (6 m) deep.

Range: Cape Cod to Florida and Texas; West Indies.

Comments: The Hairy Sea Cucumber lies buried in mud or sand with only its tentacles, mouth, and anal end above the surface. It can regenerate lost tentacles in 3 weeks.

153, 155 Scarlet Psolus
(*Psolus fabricii*)
Class Holothuroidea

Description: 4″ (102 mm) long (contracted), 2″ (51 mm) wide. Scaly, flat-bottomed, oval. Scarlet above, sole tannish-orange. Covered with *granular scales;* front end extends as neck, with mouth surrounded by *10 long, red, highly-branched tentacles,* front ones longer, rear ones shorter. *Anus on cone, pointing upward,* with 5 anal teeth, surrounded by 1–2 rows of 5–6 scales. Lower side a flattened sole, with crowded rows of tube feet around margin, and partial row in center at front and rear, not meeting each other in middle.

Habitat: On boulders and rock walls; below low-tide line to water 1320′ (402 m) deep.

Range: Arctic to Cape Cod.

Comments: The Scarlet Psolus is a striking creature when fully extended, but appears as a mere red lump when contracted. It can be found on seashores at low tide, only in places of extreme tidal fluctuation, such as Passamaquoddy Bay.

154 **Slipper Sea Cucumber**
(*Psolus chitinoides*)
Class Holothuroidea

Description: 4¾" (121 mm) long, 2½" (64 mm) wide. Scaly, flat-bottomed, oval. Orange. Body covered with *large, granular plates.* Otherwise similar to the Scarlet Psolus.

Habitat: On rocks; from low-tide line to water 50′ (15 m) deep.

Range: Alaska to Baja California.

Comments: This animal adheres very firmly to a rock and is difficult to dislodge. Its chief predators are sea stars, including the Leather Sea Star, Sunflower Star, and Stimpson's Sun Star.

---

156 **Dwarf Sea Cucumber**
(*Lissothuria nutriens*)
Class Holothuroidea

Description: ¾" (19 mm) long, ⅜" (10 mm) wide. Oval, convex, soft. Upper surface red or orange; lower surface pink. *Lower side flat, with 3 rows of tube feet;* upper surface with many slender projections, mouth and anus directed upwards at opposite ends; *10 highly-branched tentacles* around mouth.

Habitat: On rocks, and among algal holdfasts, surfgrass roots, and encrusting invertebrates; from low-tide line to water 65′ (20 m) deep.

Range: C. California to s. California.

Comments: This interesting little sea cucumber broods its eggs in shallow depressions on its upper surface, and retains them until they develop into tiny juveniles with tube feet.

260 **Common White Synapta**
(*Leptosynapta inhaerens*)
Class Holothuroidea

Description: 6″ (15 cm) long, ⅜″ (10 mm) wide. Wormlike. Translucent white. Long, slender, smooth, *fragile;* 5 white longitudinal muscle bands visible through body wall. *No tube feet.* 12 retractable featherlike *tentacles, each with 5–7 branches* on opposite sides of stalk.
Habitat: On sand and sandy mud; from low-tide line to water 637′ (194 m) deep.
Range: Maine to South Carolina and Bermuda; British Columbia to s. California.
Comments: Its weight, when its digestive system is full of sand, and its delicate construction make this animal difficult to handle without breaking it. This species is listed as *L. albicans* by some authors. The Pink Synapta (*L. roseola*), which ranges from the Bay of Fundy to Long Island Sound, in shallow water in sand and under rocks, differs in being translucent rosy red, and in having tentacles with 2–3 branches on each side of the stalk.

267 **Silky Sea Cucumber**
(*Chiridota laevis*)
Class Holothuroidea

Description: 6″ (15 cm) long, ½″ (13 mm) wide. Wormlike. Translucent pinkish or pale peach. Cylindrical, bluntly rounded at ends, smooth, *no tube feet.* 12 tentacles, each tipped with *a tuft of 10 digits* arranged like fronds on a palm tree.
Habitat: In mud, among algal holdfasts, under rocks; from low-tide line to water 270′ (82 m) deep.
Range: Arctic to Cape Cod.
Comments: This creature is abundant almost anywhere one digs near the low-tide line in eastern Maine and New Brunswick. When collected and held in a bucket of sea water, it has a

disconcerting tendency to constrict its body into several segments, cast out its internal organs, and break into several messy pieces.

244 Sticky-skin Sea Cucumber
(*Euapta lappa*)
Class Holothuroidea

Description: 39″ (1 m) long, 2″ (51 mm) wide. Wormlike, large. Silvery-gray to light brown, with white patches, some with yellowish longitudinal stripe. Extensible, soft, flabby, with *fine protruding spines, no tube feet, 15 featherlike tentacles* with 10–35 simple branches on each side. Anus at tip of rear end.

Habitat: On old coral rock around reefs; in shallow water.

Range: S. Florida; West Indies.

Comments: This extremely long holothuroid is nocturnal, crawling about over the sea floor by night, to feed on organic matter, and hiding under an old coral boulder by day.

BRYOZOANS
(Phylum Bryozoa)

The Phylum Bryozoa, or Ectoprocta, includes more than 4000 species of colonial sedentary animals. Individuals (or *zooids*) in the colony are seldom as large as $\frac{1}{32}''$ (1 mm), though the colony itself may be several feet across. For some species a microscope is mandatory for positive identification, but this book will include only species with colonies that are large and distinctive.
The individual lies within a body-covering that is continuous with or fused to the body-covering of adjacent colony members. The covering may be gelatinous, membranous, rubbery, chitinous (made of the same tough material found in the exoskeleton of an insect or shrimp), or limy. The form of the colony may be branching, creeping, bushy, leafy, tubular, fleshy, or encrusting.
The case around the individual has an opening through which a crown of *tentacles* can be extended. The tentacles are ciliated and surround the *mouth*. The *anus* lies outside the crown of tentacles. The extended tentacles have the form of a funnel. Cilia drive water and food particles, mostly 1-celled plants and bacteria, into the funnel. Some species draw food into the mouth with their tentacles.
A colony increases in size asexually by budding, and new colonies are established by sexual reproduction. Most bryozoans are hermaphroditic, but some have separate sexes. Sperm are shed into the body fluid, and escape through pores at the tips of certain tentacles. Eggs that are small and numerous are shed into the sea; single large, yolky eggs are brooded, either inside the body cavity or in a separate brood chamber. The larvae are ciliated and are part of the plankton.
There are 3 classes of bryozoans: the

Classes Stenolaemata and Gymnolaemata: Stenolaemata, which are all marine, and are tubular and limy, with circular openings for the crown of tentacles; the Gymnolaemata, which are mostly marine, and are either tubular, encrusting, gelatinous, membranous, chitinous, or limy, with individuals specialized for specific functions; and the Phylactolaemata, which are all freshwater species, and hence not discussed here.

49 **Rubbery Bryozoan**
(*Alcyonidium hirsutum*)
Class Gymnolaemata

Description: Colony more than 3″ (76 mm) wide, ⅛″ (3 mm) high. Fleshy, encrusting. Yellowish-brown or reddish. *Colony gelatinous,* usually encrusting stems of rockweed and other plants; sometimes erect, branched. Surface covered with *many conical projections.* Closely packed individuals retract into *puckered openings.*

Habitat: On stems of seaweeds and on rocks; near low-tide line and below in shallow water.

Range: Arctic to Long Island Sound.

Comments: *A. hirsutum* is extremely variable in form, but is unlike all other species of *Alcyonidium* in having conical bumps on its surface. The Pussley Bryozoan (*A. verrilli*), which ranges from Cape Cod to Chesapeake Bay, grows in branching colonies 15″ (38 cm) high and 4″ (102 mm) wide, and is yellowish-brown.

34 **Porcupine Bryozoan**
(*Flustrellidra hispida*)
Class Gymnolaemata

Description: Colony 4″ (102 mm) long, 2″ (51 mm) wide. *Rubbery.* Reddish-brown. Encrusting, *bristling with small spines.* Individuals 1/16″ (2 mm) long, 1/32″

(1 mm) wide, *6-sided,* structures visible only at edge of colony where spines have not yet developed.

Habitat: On stems of rockweeds, on rocks and shells; from low-tide line to water 62′ (19 m) deep.

Range: Arctic to Long Island Sound.

Comments: This colony's expanded tentacles, when seen through the shallow water of a tidepool, look like a pale blue mist hovering over its surface. The Branched-spine Bryozoan (*F. corniculata*) is found from Alaska to southern California. It differs from the Atlantic species in having branching spines.

Bowerbank's Graceful Bryozoan
(*Bowerbankia gracilis*)
Class Gymnolaemata

Description: Colony more than ¾″ (19 mm) long, 1/16″ (2 mm) wide. Fuzzy, tangled masses. Straw-colored. *Individuals flask-shaped or tubular.* 1/16″ (2 mm) long, 1/32″ (1 mm) wide, arranged singly or in pairs or clumps along a connecting creeping stem. Tentacle crown with 8 tentacles.

Habitat: On almost any available hard surface; from low-tide line to water 165′ (50 m) deep.

Range: Maine to Florida; Alaska to Baja California.

Comments: The eggs of this species are brooded internally, and when many of the peach-colored embryos are present, the whole colony may have an orange tinge. A related species that occurs in similar habitats, but is not so tolerant of brackish water, is Bowerbank's Imbricated Bryozoan (*B. imbricata*). It is thicker, and has individuals with 10 tentacles. The clusters of individuals may remain creeping, or grow up from the surface as bushy tufts. *B. imbricata* is found from Massachusetts to Florida.

65 **Articulated Bryozoans**
(*Crisia* spp.)
Class Stenolaemata

Description: Colony 1″ (25 mm) long, 1″ (25 mm) wide. Bushy. Chalky-white to yellow. *Individuals tubular*, punctured by tiny pores. *Branches formed by alternating series of connected individuals. Teardrop-shaped embryo-brooding chambers* interspersed with other individuals along branches.

Habitat: On rocky overhangs, hydroid stems, algae, or dock pilings, or embedded among masses of sponges or clusters of mussels; from low-tide line to water 330′ (101 m) deep.

Range: Labrador to Florida and Texas; Bahamas; West Indies.

Comments: *Crisia* embryos developing in the brood chambers divide until a hundred or more greenish-white, genetically identical larvae are produced. These squeeze out of the brood chamber and swim away. Because they do not feed, they can swim for only a short time, and usually settle to the bottom within an hour and develop into the first individuals of a new colony.

---

51 **Coralline Bryozoan**
(*Diaperoecia californica*)
Class Stenolaemata

Description: Colony 4″ (102 mm) high, 10″ (25 cm) wide. Branched. White to tan. Colony a meshwork of branches, ⅛″ (3 mm) wide, 4″ (102 mm) high, made up of *successive rows of tubular individuals*, each with round opening at end of tube through which tentacles protrude. *Brood chamber broad, perforated,* usually occurring below forking branches, and surrounding parts of several tubes.

Habitat: On rocks, shells, giant kelps, and other hard substrates; from just below low-tide line to water more than 600′ (183 m) deep.

Range: British Columbia to Costa Rica.

Comments: The form of this species varies with its environment. Colonies from deep, quiet water have tall, thin branches; those from exposed, shallow water have thicker and shorter branches.

76 Bushy Twinned Bryozoan
(*Eucratea loricata*)
Class Gymnolaemata

Description: Colony 10″ (25 cm) high, 6″ (15 cm) wide. Densely bushy, stiff. Whitish. Highly branched colonies of individuals *joined in pairs, back to back* in tapered, limy cases. Opening oval, without spines.

Habitat: On rocks; from low-tide line to water 4485′ (1367 m) deep.

Range: Arctic to Cape Cod.

Comments: This bryozoan can readily be mistaken for a seaweed until closely examined. Its bushy nature makes it the habitat of many small invertebrates, including flatworms, snails, and various crustaceans.

112 Lacy-crust Bryozoan
(*Membranipora membranacea*)
Class Gymnolaemata

Description: Colony 1/32″ (1 mm) high, more than 3″ (76 mm) wide. Lacy crust. White. Thin crust of *radially arranged individuals* in long, rectangular, limy cases with a *knob at each front corner;* large frontal membrane over much of surface.

Habitat: On kelps and other seaweeds; near low-tide line and below in shallow water.

Range: Alaska to Baja California.

Comments: Members of this genus are common in shallow water, but their simple structure makes identification to species difficult, even for the expert. The

Gulfweed Bryozoan (*M. tuberculata*), a colony the same size as *membranacea,* is found on gulfweed and rockweed in the Atlantic, and on kelps and other seaweeds in the Pacific.

113 Hairy Bryozoan
(*Electra pilosa*)
Class Gymnolaemata

Description: Colony 2" (51 mm) wide, paper-thin. Lacy network, circular, lobed, or irregularly star-shaped. White, with *long, gold spines.* Colony shape variable; patches on stones or broad-bladed seaweeds; may take a cylindrical shape around narrow stems or blades of seaweeds. Individuals elongated, oval, 1⁄32" (1 mm) long, with *oval, membranous frontal area bordered by 4–12 spines,* the *middle bottom spine usually longer* than the rest. No brood chambers.

Habitat: On almost any substrate, but particularly common on seaweeds, especially the Irish Moss *(Chondrus crispus);* from low-tide line to water 100′ (30 m) deep.

Range: Arctic to Long Island Sound.

Comments: This is one of the most common bryozoans in the New England region. Its long golden spines give it a delicately hairy appearance. Both the colony and the individual zooids are extremely variable in form.

48 Sea Lichen Bryozoan
(*Dendrobeania murrayana*)
Class Gymnolaemata

Description: Colony 1½" (38 mm) long, 1" (25 mm) wide. Flexible, bushy tuft. Straw-colored. *Branches broad, flat, ribbon-shaped,* 3–12 individuals wide, with tangled rootlets anchoring it to substrate. Individuals elongate-rectangular, 1/32″ (1 mm) long, with most of the frontal surface membranous, but with *2–5 curving spines at sides,* as well as *2 spines at upper margins.* Brood chambers globular and beadlike. *Bird's-beak pinchers* on lower part of front surface of individuals.

Habitat: On stones, shells and other bryozoans; from low-tide line to water 514′ (157 m) deep.

Range: Arctic to Martha's Vineyard.

Comments: This bryozoan is found near the low-water mark only in northern New England. A related species, the Loose Sea Lichen Bryozoan (*D. laxa*), has broader, more leaflike branches, and ranges from British Columbia to southern California.

73 Spiral-tufted Bryozoan
(*Bugula turrita*)
Class Gymnolaemata

Description: Colony 12″ (30 cm) long, 4″ (102 mm) wide. Erect, flexible, branching. Orange-brown to tan. *Individuals arranged in 2 parallel rows,* main branches having *secondary branches arranged in a spiral pattern* about the long axis. Individuals 1/32″ (1 mm) long and half as wide, tapering downward, with several spines at their upper ends and with *stalked bird's-beak pinchers* located halfway down the front margins. Mature colonies with globular brood chambers.

Habitat: On seagrasses, algae, pilings, and rocks; from low-tide line to water 90′ (27 m) deep.

Range: Massachusetts to Florida.

Comments: Species of the genus *Bugula* may easily be confused with seaweeds. The California Spiral-tufted Bryozoan (*B. californica*) is a West Coast species that occurs as a similar colony consisting of spiral whorls of branches. The pinchers of *B. californica* have been observed to pull apart small crustaceans which were then captured by the tentacles.

47 **Ellis' Bryozoan**
(*Caberea ellisii*)
Class Gymnolaemata

Description: Colony 1″ (25 mm) high, 2″ (51 mm) wide. Fan-shaped, branching. Pale tan. Colony of *individuals all in one plane, closely set, in rows,* rising free of attachment. Individuals at edge of colony with *2 toothlike spines. Long, whiplike structures* scattered over colony.
Habitat: On rocks and shells; from low-tide line to water 360′ (110 m) deep.
Range: Arctic to Cape Cod.
Comments: The whiplike structures on the colony of Ellis' Bryozoan are modified individuals whose function is to lash about, keeping other organisms from settling on and overgrowing the colony.

116 **Black-speckled Bryozoan**
(*Celleporaria brunnea*)
Class Gymnolaemata

Description: Colony 3/8″ (10 mm) high, more than 2½″ (64 mm) wide. Rough, encrusting mass. Gray to orange-brown, black-speckled. Colony rough-textured, bristly, *multilayered, in bumpy mounds or short branches.* Individuals irregular in shape, 1/32″ (1 mm) long and half as wide, each with a sharp point just below opening. With brood chambers; jaws of bird's-beak pinchers large, dark brown or black, giving colony salt-and-pepper appearance.
Habitat: On solid substrates; from low-tide line to water 600′ (183 m) deep.
Range: British Columbia to Colombia and Galapagos Islands.
Comments: In this and related species of *Celleporaria,* individuals appear to

coordinate their feeding activities. Groups of them retract and writhe their tentacles, an action which may aid in pumping feeding currents through the colony.

117 Single-horn Bryozoan
(*Schizoporella unicornis*)
Class Gymnolaemata

Description: Colony ⅓₂" (1 mm) high, more than 4" (102 mm) wide. Limy, encrusting, variable in form. White, pale reddish-orange to dull red. Usually several layers thick, new layers growing over old dead ones. *Individuals squarish,* less than ⅓₂" (1 mm) wide and high, *aperture round, with indentation at rear, single spine below aperture,* surface perforated with *many pores.* Brood chambers rounded, perforated; stalkless bird's-beak pinchers on some individuals, sometimes scarce.

Habitat: On rocks, shells, pilings, worm tubes, and various seaweed stalks, on open shores and in estuaries; near low-tide line and below in shallow water.

Range: Arctic to Florida.

Comments: This bryozoan's habit of growing new layers over older ones sometimes results in the formation of thick, round nodules around a pebble or shell. The Staghorn Bryozoan (*S. floridana*), which ranges from North Carolina through Florida, the Gulf of Mexico, and the West Indies, forms branching colonies several layers thick, 4" (102 mm) high, 2" (51 mm) wide.

114 Common Red Crust Bryozoan
(*Cryptosula pallasiana*)
Class Gymnolaemata

Description: Colony ⅓₂" (1 mm) high, more than 3" (76 mm) wide. Hard, encrusting, thin, roundish. Red to orange, or pink. Individuals in elongate, limy, boxlike cases arranged radially and spirally, with keyhole-shaped *opening, rounded at head end, widest at rear end,* with a pair of toothlike points, separating fore from rear parts, through which animal's crown of feeding tentacles extends; *16–20 smaller pores* irregularly scattered over case surface.

Habitat: On rocks, shells, and other hard objects; near low-tide line and below in shallow water.

Range: Nova Scotia to Florida.

Comments: The crown of tentacles used in filtering microscopic organisms from the water as food can be seen only if a rock or shell encrusted with this bryozoan is placed in a bowl of sea water and kept at ocean temperature. The extended tentacles will appear as white fuzz on the surface of the colony.

---

50 **Lattice-work Bryozoan**
(*Phidolopora pacifica*)
Class Gymnolaemata

Description: Colony 2½" (64 mm) high, 8½" (22 cm) wide. Lacy, erect. Pink to orange. *Brittle, ruffled meshwork.* Individuals 1/32" (1 mm) long and half as wide, *all opening on one side of lacelike sheets* that make up colony. Brood chambers and small bird's-beak pinchers present.

Habitat: Attached to rocks, in tidepools, in bays and along the open coast; from low-tide line to water 650′ (200 m) deep.

Range: British Columbia to Peru.

Comments: When individuals expand for feeding, their tentacles are arranged around the holes in the branches. In this way a rapid feeding current is created, and the water passes efficiently in one side and out the other.

ENTOPROCTS

(Phylum Entoprocta)

The Phylum Entoprocta includes about 60 species of small sedentary animals, most of which are marine and colonial. An entoproct has an oval body mounted on a stalk. Its upper surface is surrounded by a crown of 8–30 ciliated tentacles. Within the crown lies the mouth at one end and the anus at the other; in this, entoprocts differ from bryozoans, whose anus is outside the crown of tentacles. The ciliated tentacles create a current of water, and organic particles are trapped on a coating of mucus on the tentacles and moved by the cilia into the mouth and the U-shaped digestive tract consisting of *esophagus, stomach,* and *intestine.* When the feeding animal is disturbed, its tentacles retract by shortening and curling to the center.

Colonial entoprocts have an attached creeping stem from which arise a number of stalks, sometimes branched, with individuals at the tips. Within the stems are muscle fibers which cause the stalk to bend or "bow" at times, and then, as quickly, to straighten up again.

The phylum includes both hermaphroditic species and those with separate sexes. In either case fertilized eggs are attached in a depression between the mouth and anus of the parent, and develop into ciliated, swimming larvae. Asexual reproduction by budding occurs both in colonial and in solitary species, and some regularly shed the bodies from the ends of the stalks and regenerate new ones.

Colonial entoprocts might be mistaken for hydroids at first glance, but the complete digestive tract and non-stinging tentacles are among the characteristics that distinguish between them. Measurements will be given for the colony size in the two genera described here.

85 Thick-based Entoprocts
(*Barentsia* spp.)

Description: ⅜″ (10 mm) high, 1¼″ (32 mm) wide. Dense. Whitish or yellowish; visible digestive contents yellow to brown. Slender stalks rising from creeping stem, *each stalk with thickened, muscular base and globular individual at tip.* Some species branched. Individual with ring of 10–20 ciliated *tentacles that curl toward center when disturbed.*

Habitat: On rocks, shells, pilings, seaweeds, and other solid objects; near low-tide line and below in shallow water.

Range: Atlantic and Pacific coasts.

Comments: Colonies of these small animals may at first be mistaken for hydroids. However, their habits of curling rather than contracting their tentacles, and of bowing at the stem when disturbed, quickly mark them as entoprocts.

Bowing Entoproct
(*Pedicellina cernua*)

Description: ¼″ (6 mm) high, 1″ (25 mm) wide. Colony of stalked individuals. Whitish, with visible digestive contents yellow to brown. Stalk stout, usually *spiny, without muscular swellings,* rising from creeping stem. Individual with 12–16 tentacles.

Habitat: On rocks, pilings, shells, and marine growth; above low-tide line and below in shallow water.

Range: Gulf of St. Lawrence to Florida; California.

Comments: This animal thrives in brackish-water estuaries as well as in seawater.

PHORONIDS
(Phylum Phoronida)

The Phylum Phoronida includes only 2 genera and 10 species of marine wormlike animals that live in tubes either buried in sand or attached to rocks, shells, or other solid surfaces. The phoronid body is long and cylindrical, without paired appendages or specialized regions. At the front end is a large double row of tentacles arranged in a *crescent,* with the 2 ends spiralled. In the middle of the crescent is the *mouth,* which leads into a U-shaped digestive system consisting of an *esophagus, stomach,* and *intestine* which terminates in an *anus* lying on the opposite side of the crescent from the mouth. Feeding involves a current of water moved by cilia on the tentacles. Most phoronids are hermaphroditic. Eggs and sperm are shed in the body cavity and escape through the kidney duct. Larvae develop in the seawater in most species, but some brood the young in the crescent of tentacles. Asexual reproduction occurs in some, either by transverse fission and regeneration of missing parts, or by budding.

43 Green Phoronid Worm
(*Phoronopsis viridis*)

Description: 5″ (127 mm) long, 1/16″ (2 mm) wide. In *straight sand tubes* 7″ (18 cm) long, 1/8″ (3 mm) wide. Long, slender. Reddish-brown, crown of *tentacles light green.* Body without segmentation; front end with *distinct collar* bearing *crescentic row of numerous tentacles, ends of row spiralled;* mouth on one side of crescent, anus on opposite side. Blood vessels with red blood visible in body and tentacles.

Habitat: In sandy mudflats; in shallow water.
Range: Oregon to c. California.
Comments: When high tide covers the flat, the Green Phoronid Worm probes its head end up out of its vertical tube and expands its tentacular crown, which serves as both a respiratory organ and a food collector.

BRACHIOPODS
(Phylum Brachiopoda)

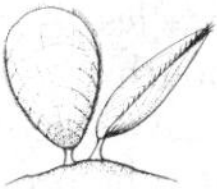

The Phylum Brachiopoda includes about 280 living species of shelled animals, but over 30,000 fossil species have been described from as far back as 600 million years ago. The genus *Lingula* is the oldest genus of animal life of which there are still living species, and dates back over 425 million years.

The brachiopod shell consists of 2 valves and superficially resembles that of the bivalve mollusks. Unlike those of mollusks, however, the valves are upper and lower instead of left and right. Brachiopods are sizeable animals, with shells usually 1–3″ (25–76 mm) long. They have a stalk that anchors them to the substrate. The phylum contains 2 classes: Inarticulata and Articulata. These names refer to the nature of articulation, or joining, of the 2 valves.

Classes Inarticulata and Articulata: The valves of inarticulates are the same size and are joined to each other only by muscles, with the stalk emerging from between them at that juncture. The articulates have a larger lower valve to which the upper is hinged. The stalk emerges through a hole in the lower valve to the rear of the hinge line. The bowl-like lower shell with its hole looks like an ancient Roman oil lamp, giving the group its common name of "lampshells."

The interior of the valves is lined with a mantle that secretes shell material. As the valves gape, they expose a large crescentic structure with a coiled arm at either side bearing a double row of long tentacles directed toward the gape. Cilia on the tentacles drive water over them, trapping fine organic particles and moving them to the *mouth* in the middle of the crescent. Inarticulates have a digestive system that ends at an

anus; articulates have an intestine that ends as a blind pouch, and undigested matter bound by mucus into small pellets that do not foul the tentacles, is expelled through the mouth.
Articulate brachiopods have a short, muscular stalk that is attached to rocks or other solid objects. They are capable of twisting about on the stalk.
Inarticulate brachiopods have a long stalk with a tuft of fibers at the tip by which the animal is anchored in a mud or sand bottom. The stalk can contract or extend, permitting the animal to gape its valves at the surface of the bottom, or retreat under the surface at low tide or when disturbed. In some Asiatic countries these stalks are cooked and eaten.
Nearly all brachiopods have separate sexes, with ovaries or testes in the rear part of the body cavity. Eggs or sperm are discharged through the kidney ducts. Most species are spawners, with development to swimming larvae taking place in the sea, but a few brood the developing eggs. None is capable of asexual reproduction.

309 Tongue-shell Brachiopod
(*Glottidea albida*)
Class Inarticulata

Description: Shell 1⅜″ (35 mm) long, ½″ (13 mm) wide. Stalk 1¾″ (44 mm) long, ⅛″ (3 mm) wide. Tongue-shaped. White, frequently with reddish to brownish markings. Shell of *2 equal valves*, sides nearly parallel, top end squarish, rear end tapered; edges bristly. *Stalk emerging between rear ends of valves.* Double-coiled tentacular arms protrude from between valves while feeding.

Habitat: Sandy mudflats, in quiet water; from low-tide line to water 530′ (162 m) deep.

Range: C. California to s. Mexico.

Comments: This brachiopod can be located by its slitlike hole in the surface of a sandy mudflat. At low tide it lies in its burrow, but at high tide the tips of its valves rise above the surface, and it expands its tentacular arms for feeding and respiration. The Green-banded Brachiopod (*G. pyramidata*) is similar in size, appearance, and life habits, and has a white shell with green cross-bands. It ranges from North Carolina to Florida and Texas.

---

360 **Common Pacific Brachiopod**
(*Terebratalia transversa*)
Class Articulata

Description: 1¾″ (44 mm) long, 2¼″ (57 mm) wide. Bivalved, short stalked. Gray, yellowish, or reddish. Valves unequal, fan-shaped, *lower valve larger,* narrow at rear. *Hole in beak of lower valve for attachment stalk* to pass through. *Valves smooth, or with 20–25 strong, radiating ribs;* concentric growth lines fine to coarse. Coiled tentacular arms inside.

Habitat: Attached by stalk to sides or undersides of rocks; from low-tide line to water 6000′ (1829 m) deep.

Range: Alaska to Baja California.

Comments: These brachiopods become sexually mature at 2 years, but require about 10 years to reach maximum size. Their chief predators in shallow water are crabs, which chip away at the edge of the valves until they can get at the soft parts.

---

358 **Northern Lampshell**
(*Terebratulina septentrionalis*)
Class Articulata

Description: 1¼″ (32 mm) long, ⅞″ (22 mm) wide. Bivalved, short stalked. Yellowish-white. Oval to teardrop-shaped, with many fine riblets. *Lower shell larger* than upper shell, narrowing to beak at rear. *Hole in beak permits attachment stalk to pass through.*

Habitat: Attached to rocks; from low-tide line to water 12500′ (3810 m) deep.

Range: Labrador to New Jersey.

Comments: This is the only species of lampshell found at low-tide line on the northern Atlantic Coast. It occurs in shallow water only in northern Maine.

ACORN WORMS
(Phylum Hemichordata)

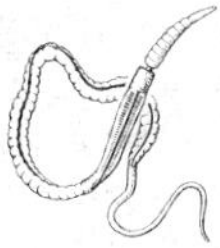

The acorn worms are burrowing forms with a 3-part body consisting of a muscular *proboscis*, usually short and cone-shaped, attached by a *stalk* to a short *collar* which bears the mouth just below the proboscis stalk, and a long *trunk*, the first part of which has many paired *gill slits* on the upper surface. These slits permit the escape of water taken in through the mouth and passed over *gills* in the walls of the foregut, or *pharynx*.

The pharynx continues into the midgut, where digestion takes place, and subsequently into the hindgut, which terminates in the anus at the rear tip of the worm. The animal has well-developed muscular, circulatory, and nervous systems, and simple sex organs along the sides to the rear of the pharynx, each with its own pore to the outside.

Probably all species are suspension feeders, trapping organic particles from the water on the mucus-coated proboscis, and moving the mucus into the mouth by means of cilia. Many species also eat great quantities of mud or sand which they eventually deposit in stringlike strands, having digested out the organic matter.

All acorn worms have separate sexes. The female sheds eggs outside the burrow in a mucus mass; nearby males shed sperm, and fertilization occurs. The developing eggs, dispersed by currents, develop into ciliated, swimming larvae which live in the plankton for a period of time before developing adult body form and burrowing into the bottom. These animals are very fragile. Most species can regenerate at least parts of the trunk, and some can reproduce asexually by division and regeneration of missing parts.

272 Golden Acorn Worm
(*Balanoglossus aurantiacus*)

Description: 6″ (15 cm) long, ⅜″ (10 mm) wide. Wormlike. Proboscis cream-colored or yellowish; collar reddish-orange; trunk purplish or greenish, with golden bands. *Proboscis rounded,* nearly same length as collar; first ⅓ of trunk with *flattened, winglike extensions* at the side, curling up over back, containing gonads; next ⅓ with *rounded papillae,* region of digestive glands; last ⅓ cylindrical, tapered.

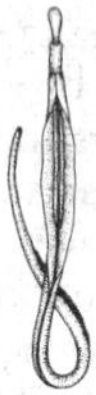

Habitat: On mudflats; near low-tide line and below in shallow water.

Range: North Carolina to Florida.

Comments: The winglike extensions of *Balanoglossus* containing the reproductive organs can be folded over the worm's back to adapt to life in a burrow.

Kowalewsky's Acorn Worm
(*Saccoglossus kowalewskii*)

Description: 6″ (15 cm) long, ¼″ (6 mm) wide. Slender. Proboscis whitish to pale peach; collar orange; trunk brownish. Tapered toward rear, *fragile.* Proboscis ½″ (13 mm) or more long, tapered to blunt point, attached by narrow stalk to *short collar* bearing mouth. Trunk long, first part with *many pairs of gill slits* lying on each side of a *ridge* extending down back.

Habitat: On mud bottoms; between high- and low-tide lines and just below in very shallow water.

Range: Maine to North Carolina.

Comments: This species is a mud feeder, and can be spotted by the pile of long, stringlike mud castings deposited about the opening of its burrow. A similar species, the Little Acorn Worm (*S. pusillus*), found in southern California and Baja California, is about the same size and lives in similar habitat.

CHORDATES
(Phylum Chordata)

If it were not for the existence of sea squirts and lancelets, the Phylum Chordata would consist only of vertebrate animals—those with a vertebral skeleton or backbone; but sea squirts and lancelets necessitate a broader view of the Phylum Chordata. This phylum takes its name from the *notochord*, a stiffened rod consisting of a fibrous sheath around translucent cells whose turgid condition provides both firmness and flexibility. No member of any other phylum has a notochord. Possession of a notochord prevents a chordate's body from telescoping when its longitudinal muscles contract. Instead, it bends from side to side. Lancelets retain the notochord throughout life, whereas sea squirts and vertebrates possess one only during larval or embryonic stages of development. Above its notochord, a chordate has a tubular dorsal nerve cord. Chordates have a pharynx perforated with a number of paired gill slits in aquatic species.

Subphylum Urochordata:

The Subphylum Urochordata, which includes tunicates, salps, and larvaceans, is a subgroup of the Phylum Chordata, along with the Subphylum Cephalochordata and Subphylum Vertebrata.

Adult urochordates bear little resemblance to other chordates. Most are sedentary animals whose body is enclosed in a jacket or tunic. Urochordates have a large pharynx with slits in its walls and a food groove in its floor. The pharynx functions both in respiration and in filtering food. The adult shows no evidence of notochord or tubular nerve cord, which are found only in the tadpole-shaped larva and which the larva loses, along with its muscular tail, as it matures.

Class Thaliacea: Salps (Class Thaliacea) are large and planktonic. They are barrel-shaped and translucent, and swim by taking water in the front end and forcing it out the rear both by ciliary action and muscular contraction.

Class Ascidiacea: The tunicates or sea squirts (Class Ascidiacea) are all attached forms, either solitary or colonial, the latter with many individuals produced by budding. They have a continuous tunic covering the body which is attached to a solid. At the end opposite attachment there is one opening through which water enters the animal and another nearby through which water escapes. These are called the *incurrent* and *excurrent siphons*, respectively. The incurrent siphon opens into a large pharynx with slitted walls, surrounded by a cavity, the *atrium*, opening to the outside through the excurrent siphon. Water is thus moved by cilia into the pharynx through the incurrent siphon and the slits, into the atrium, and out of the atrium through the excurrent siphon. The pharynx is continuous with the rest of the digestive tract, which loops about and terminates in an *anus* situated just inside the excurrent siphon.

Asexual reproduction by budding produces numerous individuals in the compound tunicates. Nearly all tunicates are hermaphroditic, with both ovary and testis. Some species spawn eggs and sperm, and fertilization and development occur in the sea. Others produce yolky eggs that remain in the atrium, where development to the tadpole stage occurs.

Subphylum Cephalochordata: Of all the invertebrates, those most similar to the vertebrate animals are the cephalochordates, or lancelets. Adult lancelets clearly show the 3 chordate characteristics: pharyngeal slits, tubular nerve cord above the body cavity, and

notochord, the latter extending from the tip of the head to the tip of the tail. Lancelets differ in having, like urochordates, a chamber around the pharynx to receive and remove water coming through the slits, and they lack the brain, eyes, and internal ears common to vertebrates.
The subphylum has only a small number of species, most of which belong to the genus *Branchiostoma*, 1 of which is cited below.

102, 103, 104 **Light Bulb Tunicate**
(*Clavelina huntsmani*)
Class Ascidiacea

Description: Individuals 2″ (51 mm) long, ⅜″ (10 mm) wide. Colony 3″ (76 mm) high, 20″ (51 cm) wide. *Tunic transparent; bright pink to pale yellow line* curves around edge of pharynx.
Habitat: On shaded, vertical rock faces, and under ledges; from low-tide line to water 100′ (30 m) deep.
Range: British Columbia to s. California.
Comments: This species buds asexually to form colonies of separate individuals. Bright orange embryos are brooded in the cavity next to the pharynx. Small swimming tadpole larvae are released and settle, developing into miniature adults within a day.

100 **Painted Tunicate**
(*Clavelina picta*)
Class Ascidiacea

Description: Individuals ¾″ (19 mm) long, ¼″ (6 mm) wide. Colony a cluster 8″ (20 cm) high, 12″ (30 cm) wide. *Tunic translucent white to yellowish*, with red, purple, or cream-colored band around upper part, and red or purple internal organs. Otherwise similar to the Light

Bulb Tunicate.

Habitat: On mangrove roots, soft corals, stony corals, and other hard objects; near low-tide line and below in shallow water.

Range: S. Florida; West Indies.

Comments: A colony 12″ wide includes over 1,000 individuals formed by budding from the rootlike base of older individuals.

---

98 **Northern Sea Pork**
(*Aplidium constellatum*)
Class Ascidiacea

Description: ⅜″ (10 mm) high, 1/32″ (1 mm) wide. Colony 1″ (25 mm) high, 3″ (76 mm) wide. Colony of *stalked lobes.* Cream-colored, with red or orange individuals. *Colony fleshy, soft,* with long, slender individuals embedded vertically in the soft substance.

Habitat: On rocks, pilings, and other hard objects; from low-tide line to water 20′ (6 m) deep.

Range: Maine to Gulf of Mexico.

Comments: A large colony of this compound tunicate resembles a piece of salt pork. A related species, the Common Sea Pork (*A. stellatum*), is similar in color and in range, but forms colonies 12″ (30 cm) wide and 1″ (25 mm) high, and is more firm and rubbery. The Pacific Sea Pork (*A. solidum*), which ranges from British Columbia to southern California, forms colonies 8″ (20 cm) wide and 2″ (51 mm) high, with individuals arranged on raised ridges. These fleshy colonies are sometimes hooked and brought up by fishermen.

125 **Northern White Crust**
(*Didemnum albidum*)
Class Ascidiacea

Description: ⅛″ (3 mm) high, more than 4″ (102 mm) wide. Rubbery crust. White, less commonly yellowish or salmon. Colony with many *tiny holes,* the siphons of individuals, made opaque by *abundant spherical granules* covered with blunt points, obscuring individuals.
Habitat: On rocks, wharf piles, and other hard objects; from low-tide line to water 1350′ (411 m) deep.
Range: Arctic to Cape Cod.
Comments: Without the aid of a microscope it is impossible to see that these commonly occurring white crusts contain individual tunicate bodies. The Glossy White Crust (*D. candidum*) ranges from the Bay of Fundy to Florida, the West Indies, and Brazil. Its colony form and color resemble those of *D. albidum,* but its spherical granules are much smaller and have sharp points.

118 **Pacific White Crust**
(*Didemnum carnulentum*)
Class Ascidiacea

Description: ⅛″ (3 mm) high, 4¾″ (121 mm) wide. *Rubbery crust.* White, gray, pink, or orange. Otherwise similar to the Northern White Crust.
Habitat: On rocks, shells, pilings, algae, and other solid objects; from low-tide line to water 100′ (30 m) deep.
Range: Oregon to Panama.
Comments: This species can be distinguished from its Atlantic relatives only by microscopic detail. It is eaten by the Leather Star and the Winged Star.

27 **Elephant Ear Tunicate**
(*Polyclinum planum*)
Class Ascidiacea

Description: Colony 1⅝" (41 mm) high, 8" (20 cm) or more wide. *Flattened lobe.* Yellow, tan to brown. *Short stalk* on one margin attaches colony to substrate. *Rings of 8 or more individuals on both sides of lobe with common excurrent opening.*

Habitat: On rocky shores with moderate wave action; from low-tide line to water 100′ (30 m) deep.

Range: N. California to Baja California.

Comments: The characteristic growth form makes this an easily recognized species. Smaller colonies are hemispherical, but become lobed as they grow.

---

91 **Mushroom Tunicate**
(*Distaplia stylifera*)
Class Ascidiacea

Description: Colony ⅝" (16 mm) high, ¾" (19 mm) wide. Stalked. Yellowish-brown, with darker brown individuals. Colony rubbery, *mushroom-shaped,* convex, with stalk ½ its height; tiny *individuals embedded in irregular radial rows* in common tunic.

Habitat: On rocks, shells, and other solids; from low-tide line to water more than 300′ (91 m) deep.

Range: South Carolina to Florida; Bahamas; West Indies to Colombia.

Comments: This tunicate broods its young in a special brood pouch, about 12 at a time. They emerge as tadpole-shaped, swimming larvae.

122 **Disk-top Tunicate**
(*Chelyosoma productum*)
Class Ascidiacea

Description: 2⅜″ (60 mm) high, 2″ (51 mm) wide. Globular, solitary. Translucent, almost colorless, whitish, or cream-colored. *Tunic thin, firm, tough;* upper surface a *flat, oval disk* of thickened plates bearing siphons. Siphons with 6 *plates of thickened tunic.*

Habitat: On rocks, pilings, and hard surfaces, on open shores and in bays; from low-tide line to water 165′ (50 m) deep.

Range: Alaska to s. California.

Comments: The thickened plates of tunic on the disk and around the siphons of this tunicate have a form distinctive in the species. Groups of individuals sometimes live close together, but the species is not colonial.

105 **Sea Vase**
(*Ciona intestinalis*)
Class Ascidiacea

Description: 6″ (15 cm) high, 1″ (25 mm) wide. Slender, vaselike. Pale yellow or greenish, translucent, *openings ringed with yellow. Pharynx long;* longitudinal muscles and digestive system visible through *translucent body wall.* Openings close together near tip.

Habitat: On rocks, pilings, floats, and boat hulls in harbors and protected bays; from low-tide line to water 1650′ (500 m) deep.

Range: Alaska to s. California; Arctic to Rhode Island.

Comments: The ubiquity, abundance, size, and easy availability of this species have made it the best-studied tunicate in the world. A large *Ciona* pumps 4 or 5 gallons of seawater each day to filter its food, obtain oxygen, and excrete waste products.

101 **Creeping Tunicate**
(*Perophora viridis*)
Class Ascidiacea

Description: ⅛″ (3 mm) long and nearly as wide. Colony 3″ (76 mm) or more long, 3″ (76 mm) wide. Creeping, vinelike. Translucent greenish-yellow. *Oval individuals* rising on short stalks from *creeping stems.* 2 siphons clearly visible with aid of hand lens.
Habitat: On rocks, pilings, bases of algae, and other solid substrates, in bays and estuaries; from low-tide line to water 90′ (27 m) deep.
Range: Cape Cod to Florida and Texas.
Comments: Both the vinelike character and green color of this tunicate might readily lead one to confuse it with a plant. But a pair of sharp eyes, especially if aided by a hand lens, can make out tiny individual sea squirts with their siphons, and even the pharynx is apparent when illumination is right. The Yellow-green Creeping Tunicate (*P. annectens*), which ranges from British Columbia to southern California, grows to 4″ (102 mm), wide and 2″ (51 mm) long, and has yellow-green or even orange-tinted individuals about the same size as those of *P. viridis.* This tunicate is transparent enough to reveal inner activity in the living state.

95, 96 **Mangrove Tunicate**
(*Ecteinascidia turbinata*)
Class Ascidiacea

Description: 1″ (25 mm) long, ⅜″ (10 mm) wide. Colony 10″ (25 cm) high, 8″ (20 cm) wide. Cluster of globules. *Transparent,* colorless tunic, clearly visible internal organs yellow, orange, or pink. Long, *globe-shaped* individuals, spiralling in *dense clusters* of several hundred growing from creeping stem.

Habitat: On roots of mangroves and stems of turtle grass; near low-tide line and below in shallow water.
Range: S. Florida to Texas; West Indies.
Comments: This attractive compound tunicate looks like a cluster of long, fluid-filled glass bulbs growing around a plant root or stem.

Black Tunicate
(*Ascidia nigra*)
Class Ascidiacea

Description: 4″ (102 mm) long, 2⅜″ (60 mm) wide. Leathery oval. *Blue-black or black.* Solitary; tunic thick, tough; *incurrent siphon higher* than excurrent siphon, both opposite attached end.

Habitat: On rocks, pilings, sea walls, and other hard objects; near low-tide line and below in shallow water.
Range: Florida to Texas; West Indies.
Comments: *A. nigra* is easily distinguished from all other tropical American species by its black color, and occurs by the hundreds on sea walls of marinas in the Florida Keys. When touched, it quickly closes its siphons and shortens, assuming the smallest possible volume. The Callused Tunicate (*A. callosa*) ranges from the Arctic to Cape Cod. It varies in color from tan to brownish, and is translucent when young, turning opaque with age. It is somewhat flattened against its substrate, and attached by its left side. Its incurrent siphon is 8-lobed and the excurrent siphon 6-lobed. It is 3½″ (89 mm) long and half as wide. The Horned Tunicate (*A. ceratodes*) which ranges from British Columbia to northern Chile, is 2¾″ (70 mm) long, and 1⅝″ (41 mm) wide. It is compressed and grayish or yellowish-green, with prominent hornlike siphons.

93 **Blood Drop Tunicate**
(*Dendrodoa carnea*)
Class Ascidiacea

Description: ¼″ (6 mm) high, ½″ (13 mm) wide. Low, dome-shaped. Pinkish to red, *siphons scarlet.* Round, *wider than high,* surface finely wrinkled; siphons low, squarish, 4-lobed.
Habitat: On rocks and shells in protected places; from low-tide line to water 234′ (71 m) deep.
Range: Newfoundland to Long Island Sound.
Comments: The Blood Drop Tunicate is a brooder, and late in summer liberates scarlet tadpole larvae.

108 **Club Tunicate**
(*Styela clava*)
Class Ascidiacea

Description: 6″ (15 cm) high, 2″ (51 mm) wide. Club-shaped. Yellowish-gray to reddish-brown. Tunic thick, leathery, with *conspicuous bumps;* no clear distinction between body and rooted stalk. Siphons near top, *both siphons pointed upward.*
Habitat: On rocks, pilings, and other solid objects; from low-tide line to water 80′ (24 m) deep.
Range: C. and s. California.
Comments: This species is a native of Asia and was carried on ships' bottoms into harbors in California, Europe, and Australia at the beginning of the century. It is used as seafood in Korea.

42 **Monterey Stalked Tunicate**
(*Styela montereyensis*)
Class Ascidiacea

Description: 10″ (25 cm) high, 2″ (51 mm) wide. Shaped like long bowling pin. Yellow to reddish-brown. *Thick, leathery tunic, ridged longitudinally;* openings close together, near tip, one pointed down.

Habitat: On rocks in both exposed and protected habitats; from low-tide line to water 100′ (30 m) deep.

Range: British Columbia to Baja California.

Comments: This is a common, easily recognized, and broadly distributed solitary tunicate. An individual followed in Monterey grew to 2″ (51 mm) high in less than 3 months and attained a height of over 9″ (23 cm) in 3 years.

106, 107 **Striped Tunicate**
(*Styela plicata*)
Class Ascidiacea

Description: 3⅝″ (92 mm) long, 1¾″ (44 mm) wide. Oval. Grayish or tannish-white, with *red or purple stripes on siphons. Tunic unstalked,* thick, leathery. Siphons with 4 lobes. Otherwise similar to the Monterey Stalked Tunicate.

Habitat: On rocks, pilings, shells, and other solid objects; from low-tide line to water 100′ (30 m) deep.

Range: North Carolina to Florida and the West Indies; s. California.

Comments: These solitary tunicates may be found in clumps attached to the same substrate, but their numbers do not result from budding. Each is the product of a sexually produced larva.

---

120 **Pacific Star Tunicate**
(*Botryllus tuberatus*)
Class Ascidiacea

Description: Colony ⅛″ (3 mm) high, more than 2″ (51 mm) wide. Thin, rubbery crust. Usually orange, but may be yellow, purple, green, or brown. Individuals 1/16″ (2 mm) or more long, 1/64″ (0.5 mm) wide, arranged in *circular or oval systems* with a *common excurrent opening.*

Habitat: On rocks, pilings, floats, and other solid substrates in bays, harbors, and protected places; near low-tide line and below in shallow water.

Range: N. California to s. California.

Comments: Colonies of *Botryllus* may persist for more than a year, although the individuals making it up live only 7–10 days. Each individual produces buds that can both replace the parent and add to colony growth.

---

121 Golden Star Tunicate
(*Botryllus schlosseri*)
Class Ascidiacea

Description: Colony ⅛″ (3 mm) high, 4″ (102 mm) wide. Thin, rubbery crust. Yellow, olive, greenish, tan, brown, purple, or black; *individuals outlined in yellow or white.* Otherwise similar to the Pacific Star Tunicate.

Habitat: On rocks, shells, pilings, bases of seaweeds, eelgrass, and other solid surfaces, on open shores and in brackish estuaries; near low-tide line and below in shallow water.

Range: Bay of Fundy to North Carolina.

Comments: This species can be cultured easily by tying a bit of a colony to a glass slide placed in a pint of sea water. It has been used in studies of reproduction, development, and the genetics of color patterns.

---

119 Orange Sheath Tunicates
(*Botrylloides* spp.)
Class Ascidiacea

Description: Colony ⅛″ (3 mm) high, more than 2″ (51 mm) wide. *Thin, rubbery crust.* Usually orange, but variable. *Individuals arranged in twisting rows,* more than 1/16″ (2 mm) long, 1/64″ (0.5 mm) wide.

Habitat: On rocks, pilings, floats, and other substrates in bays and harbors; near low-tide line and below in shallow water.

Range: Atlantic and Pacific coasts.

Comments: This is a broadly distributed group in which systematic work is needed to define the species and their distributions. A number of shelled gastropods and nudibranchs feed on *Botrylloides*, along with a flatworm that sucks individuals out of the tunic and swallows them whole.

94 **Taylor's Colonial Tunicate**
(*Metandrocarpa taylori*)
Class Ascidiacea

Description: Colony ¼″ (6 mm) high, 8″ (20 cm) wide. Individuals ¼″ (6 mm) high, ⅛″ (3 mm) wide. *Clusters of globes.* Bright red, orange, yellow, or greenish. Individuals separate, but joined by a thin basal tunic.

Habitat: On protected rock faces; from low-tide line to water 72′ (22 m) deep.

Range: British Columbia to s. California.

Comments: This species grows when individuals produce buds which separate from the parent but remain connected by a thin sheet of tunic that spreads on the substrate. The Fused Tunicate (*M. dura*) is a related, bright red species, occurring only below the low-tide line in the same range as *M. taylori.* It can be differentiated by its much thicker basal tunic and much larger colonies, which grow to ⅜″ (10 mm) high and to more than 24″ (61 cm) wide.

26 **Encrusted Tunicate**
(*Polycarpa obtecta*)
Class Ascidiacea

Description: 2″ (51 mm) long, 1¾″ (44 mm) wide. Tough-coated, sand-encrusted globe. Body yellowish, brownish-gray, or mud-colored; siphons red, purplish-brown, or brown. *Tunic wrinkled, usually encrusted* with sand and bits of shell; thick, rough, attached to substrate by *mossy fibers;* siphons prominent, 4-lobed.

Habitat: On rocks, grass stems, and other solid objects; near low-tide line and below in shallow water.

Range: Florida to Texas; West Indies to Brazil.

Comments: These solitary tunicates may be easily overlooked because of their encrusting habit. Those that settle on rocks well above the bottom, as on a seawall, may be relatively free of sand coating.

36, 99 **Green Encrusting Tunicate**
(*Symplegma viride*)
Class Ascidiacea

Description: Colony more than 3″ (76 mm) long and equally wide. Thin, gelatinous encrustation. Translucent, greenish; individuals purple, greenish, or black; area around siphons white, pale green, yellow, or orange. *Tunic soft, thin layer, individuals uniformly scattered* throughout, each with 2 siphons to surface.

Habitat: On rocky areas and in sea grass beds; near low-tide line and below in shallow water.

Range: Florida; West Indies.

Comments: This species encrusts rocks, shells, sea grass, and bryozoan stems, assuming the shape and contours of the substrate. Its individuals are uniformly scattered rather than arranged in flowerets.

41 **Stalked Tunicate**
(*Boltenia ovifera*)
Class Ascidiacea

Description: Body 3″ (76 mm) long, 2″ (51 mm) wide; stalk ¼″ (6 mm) wide, 12″ (30 cm) or more high. Oval body on long stalk. Tannish-yellow to orange or red; reddish around siphons. Body on long, sturdy, *upright stalk;* surface *lightly hairy,* sometimes wrinkled. *Siphons on one side, prominent, incurrent siphon aimed upward, excurrent siphon aimed downward.*

Habitat: On rock and gravel bottoms; from low-tide line to water 1640′ (500 m) deep.

Range: Arctic to Cape Cod.

Comments: The long stalk of older specimens of this tunicate is usually encrusted with bryozoans, hydroids, and algae. Arctic specimens blown ashore by storms are eaten by Eskimos.

111 **Cactus Tunicate**
(*Boltenia echinata*)
Class Ascidiacea

Description: 1″ (25 mm) high, 1⅜″ (35 mm) wide. Spiny globe. Body pink, salmon, or reddish; siphons rosy red; bristles pale buff. Tunic with low bumps, each bearing stiff spine with *4–8 radiating branches. Both siphons on top, near center, squarish.*

Habitat: On rocks, in protected places with freely circulating water; from low-tide line to water 976′ (297 m) deep.

Range: Arctic to Cape Cod.

Comments: Though this tunicate resembles a cactus, it is not prickly to handle. Its spines, while stiff, are somewhat flexible and not sharp-pointed.

110 **Bristly Tunicate**
(*Boltenia villosa*)
Class Ascidiacea

Description: Body 1⅝″ (41 mm) high, 1¼″ (32 mm) wide. Oval, sometimes with stalk 2⅜″ (60 mm) long. Reddish, orange, tan, or brown. Tunic leathery, *covered with spines.* May or may not have stalk. Otherwise similar to the Stalked Tunicate.
Habitat: On rocks and other hard substrates along open shores; from low-tide line to water 330′ (101 m) deep.
Range: British Columbia to s. California.
Comments: This tunicate has been used in studies of larval development and attachment.

---

92 **Sea Peach**
(*Halocynthia pyriformis*)
Class Ascidiacea

Description: 5″ (127 mm) high, 3″ (76 mm) wide. Barrel-shaped. Yellow to orange, strongly tinged with red at top and on one side. Tunic tough, *fuzzy;* siphons 4-lobed, both on top, *incurrent siphon larger and higher, excurrent siphon smaller, lower.* Attached to substrate by rootlike fibers.
Habitat: On rock and gravel bottoms; from low-tide line to water 637′ (194 m) deep.
Range: Arctic to Massachusetts Bay.
Comments: The shape, color, and fuzzy surface of this handsome solitary tunicate are reminiscent of a peach. The largest specimens are found near the Arctic.

---

97 **Orange Sea Grape**
(*Molgula citrina*)
Class Ascidiacea

Description: ⅝″ (16 mm) long, ½″ (13 mm) wide. Rounded, with slightly flattened sides.

Translucent; *orange sex organs visible. Siphons widely separated,* prominent. Tunic usually clean.

Habitat: Attached under rocks; near low-tide line and below in shallow water.

Range: Arctic to Rhode Island.

Comments: This species is a brooder, its tadpole larvae developing in its atrium. A close relative, the Common Sea Grape (*M. manhattensis*), is a spawner, shedding eggs and sperm into the water. It is larger, 2″ (51 mm) long and 1⅝″ (41 mm) wide, and gray to greenish, its siphons prominent but less far apart, and its tunic usually muddy. It lives on rocks, pilings, boat hulls, and various seaweeds in bays and estuaries, and is tolerant of a range of salinity, temperature, and pollution. It ranges from Maine to Texas, but does not occur in Florida, and has been introduced into several bays in central California, where it thrives.

489 Common Doliolid
(*Doliolum nationalis*)
Class Thaliacea

Description: ⅝″ (16 mm) long, ¼″ (6 mm) wide. Barrel-shaped. Transparent, colorless; muscle bands whitish; contents of digestive system visible, yellowish. Body an *open-ended barrel* with lobed margins and thin tunic; 8 prominent, circular *muscle bands,* end bands used to close openings. Front opening leads into pharynx with *numerous slits,* and a *long food groove* in bottom connected to rest of digestive tract.

Habitat: Floating in plankton; near ocean surface.

Range: Cape Cod to Florida and Texas; West Indies to tropical South America; Pacific coastal waters.

Comments: These transparent creatures are sometimes blown ashore by the thousands in summer and autumn. They reproduce both sexually and asexually, the latter by producing a long string of buds, reminiscent of the tail of a kite.

490 **Horned Salp**
(*Thalia democratica*)
Class Thaliacea

Description: 1″ (25 mm) long, ½″ (13 mm) wide. Barrel-shaped, with horns. Transparent, colorless; muscle bands whitish; contents of digestive system visible, yellowish. Body an open-ended barrel with thin tunic, pair of long horns on upper side of rear end; *5 prominent, circular muscle bands,* joined or near each other at the top; smaller ones at each end to close openings. Front opening leads into large pharynx with pair of single slanting gills bounded by slits on each side. Food groove long, rest of digestive system short, confined to region behind last body muscle band.

Habitat: Floating in plankton; near ocean surface.

Range: Cape Cod to Florida and Texas; West Indies to tropical South America; California.

Comments: The Horned Salp moves by jetting water out the open rear end through sequential contraction of the circular muscle bands. This creature occurs in such huge schools that scientists were able to collect over a bushel in Delaware Bay by towing a fine mesh net behind a boat for only 10 minutes.

488 **Common Salp**
(*Salpa fusiformis*)
Class Thaliacea

Description: 3¼″ (83 mm) long, 1⅝″ (41 mm) wide. Transparent cylinder. Colorless; muscle bands whitish; contents of digestive system visible, yellowish.

Body an open-ended cylinder, slightly wider in rear half; tunic variable, sometimes thick, sometimes thin. 9 prominent, *incomplete muscle bands,* not reaching midline underneath, *first 3 and last 2 muscle bands touching each other at top midline;* smaller muscles to close openings at ends. Front opening leads into large pharynx with a pair of single slanting gills bounded by slits on each side. Food groove long, joining rest of digestive system between last 2 muscle bands.

Habitat: Floating in plankton; near ocean surface.

Range: Alaska to California; entire Atlantic Coast.

Comments: The asexually reproducing form of the Common Salp, described above, develops a long chain of buds that trails from the lower rear end. The sexual form is smaller, 1″ (25 mm) long and ⅜″ (10 mm) wide, and its tunic tapers to a point at both ends.

Caribbean Lancelet
(*Branchiostoma caribaeum*)

Description: 2″ (51 mm) long, ¼″ (6 mm) high. Elongate. Translucent, whitish; colored digestive contents visible. *Fishlike body* without distinct head; flattened, vertical; rounded beak at front end, with chamber below leading into mouth; *row of tentacles on either side* curved toward midline. Ridgelike fin above from beak to tail; tail fin spearhead-shaped. *Paired fins,* from mouth chamber to excurrent opening of gill chamber, unite to form single fin in lower midline to tail. *V-shaped muscle segments* pointing forward on both sides of body. Large pharynx with many slits slanting downward and toward the rear; food groove in floor continuous with rest of digestive system. Row of rounded sex organs below pharynx.

Habitat: Fine sand and sandy-mud bottoms; near

low-tide line and below in shallow water.

Range: Chesapeake Bay to Florida and Texas; Mexico; West Indies.

Comments: The lancelet is a good swimmer, but spends most of its time buried in the sand with only its head end out. Cilia on the gills bring water, strained of coarse particles by the tentacles, into the pharynx through the mouth. The California Lancelet (*B. californiense*), which ranges south from central California, is distinguishable from *B. caribaeum* only in fine detail. It is usually found below the low-tide line in the sand bottoms of bays, but is sometimes encountered just at the low-tide line.

Part III
Appendices

The illustrations on the following pages show representative members of 18 marine invertebrate phyla, with labels identifying the body parts referred to in the text. These labelled drawings will help you to recognize the different seashore creatures and their kin.

Sponge

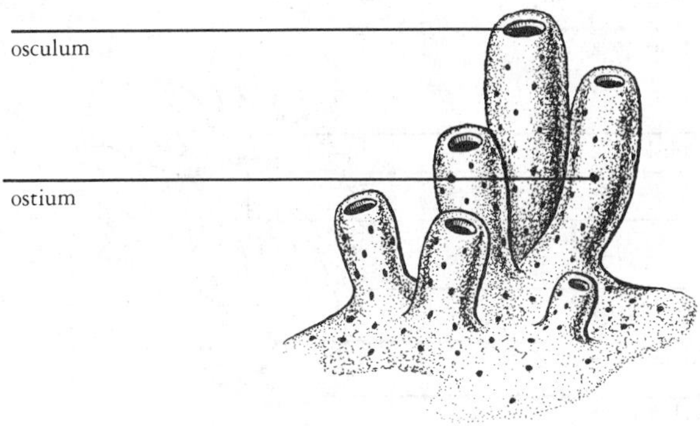

Hydrozoan Polyp

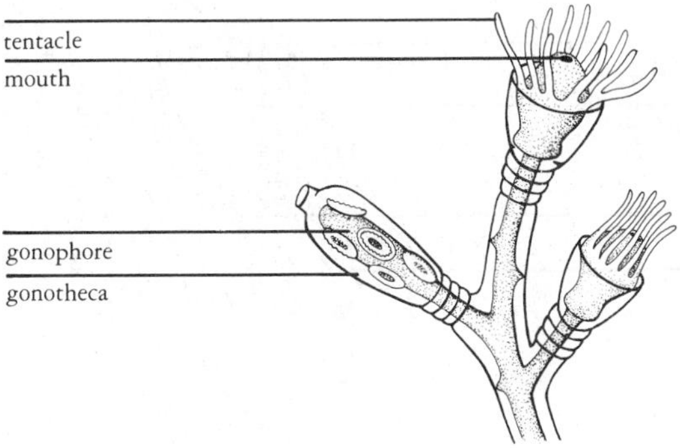

Scyphozoan Medusa

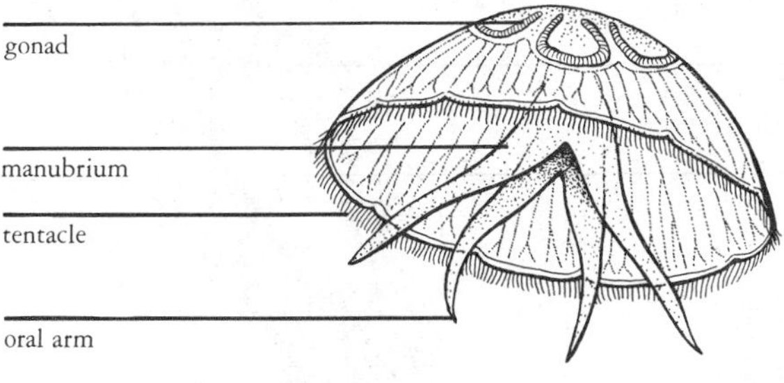

Sea Anemone

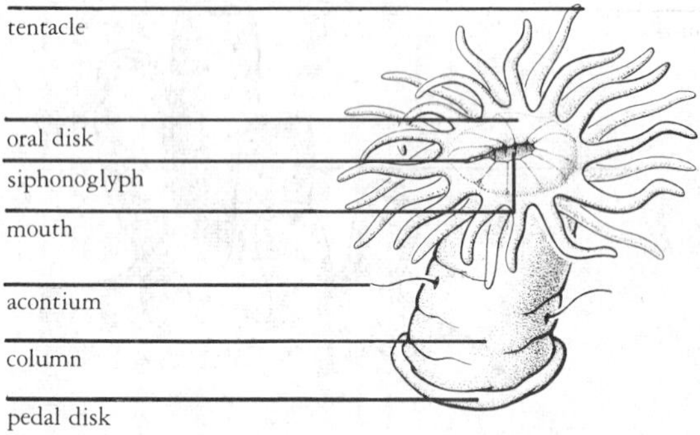

Comb Jelly

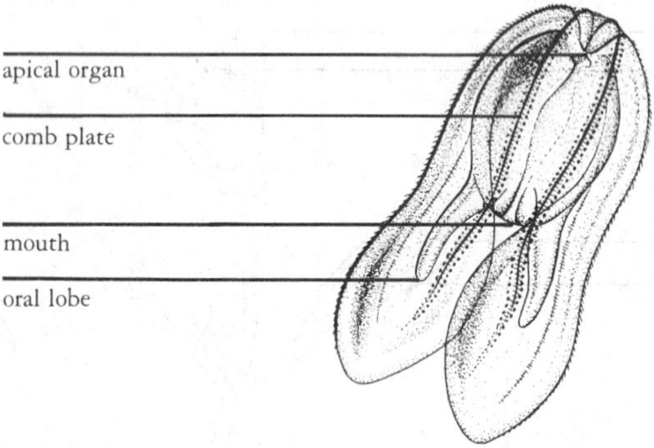

Flatworm

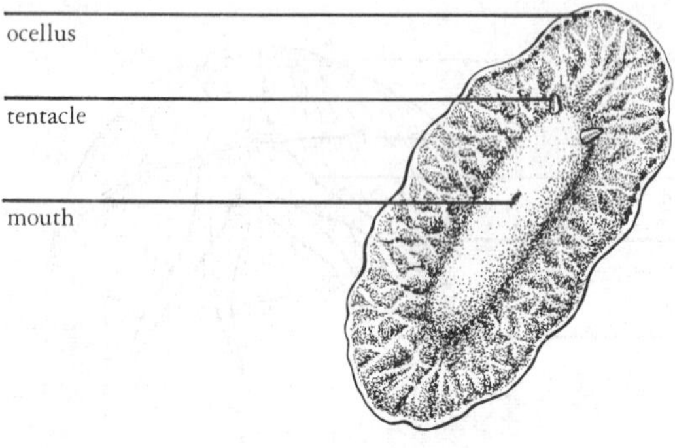

Ribbon Worm

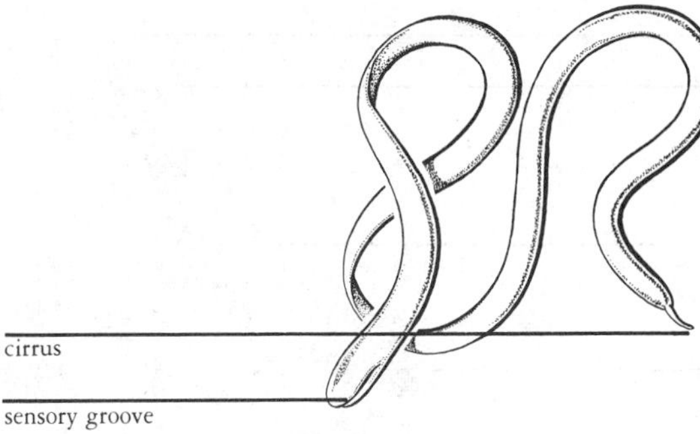

Errant Polychaete Worm (Clam Worm)

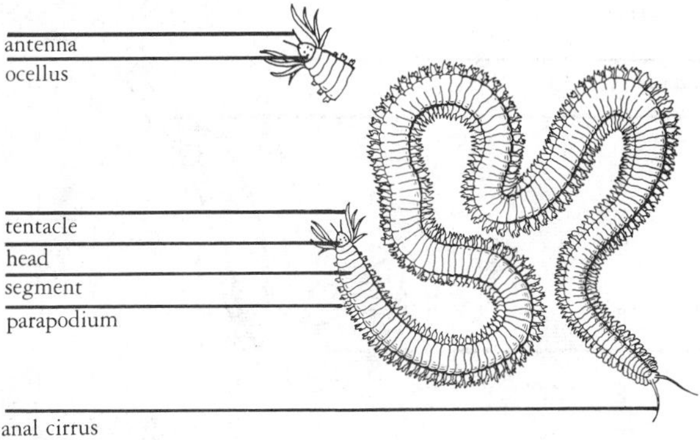

Sedentary Polychaete Worm (Parchment Worm)

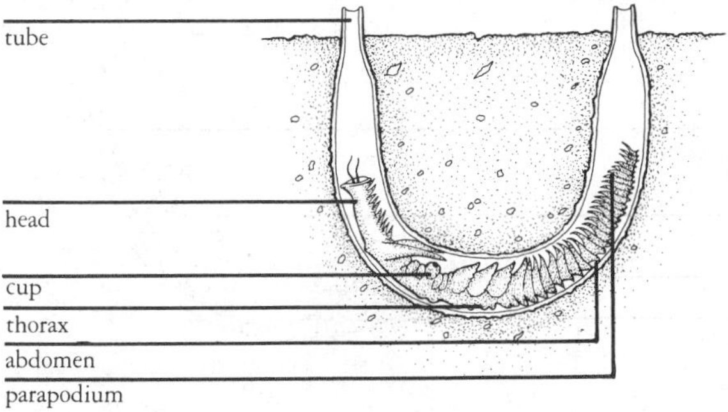

Priapulid Worm

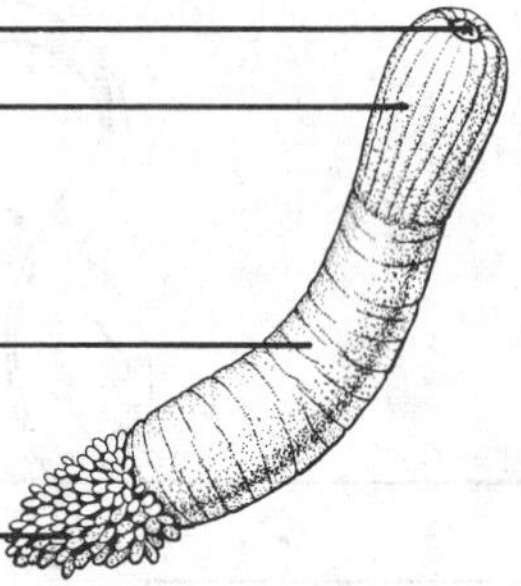

Peanut Worm

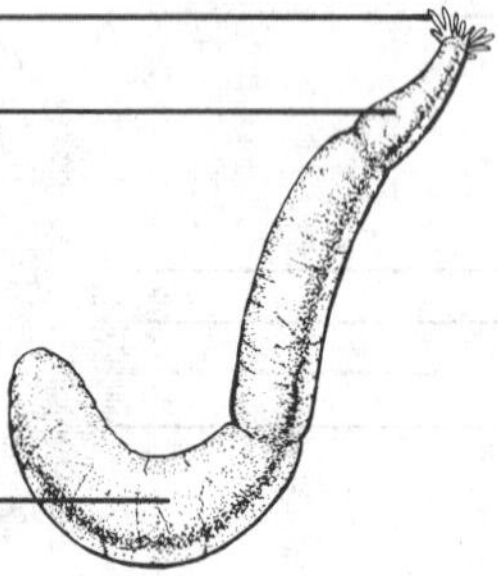

Echiurid Worm

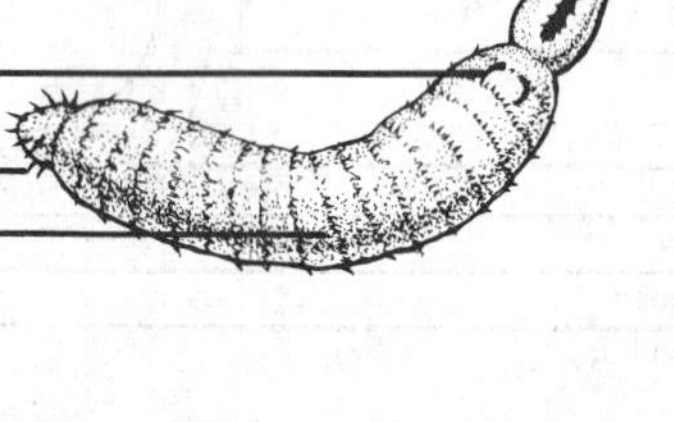

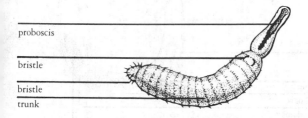

Chiton

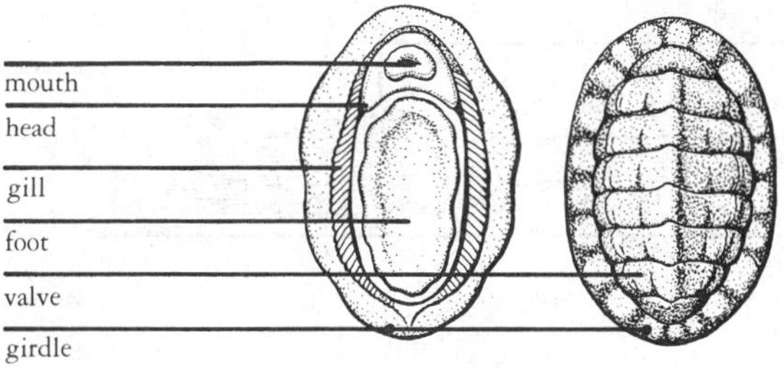

Snail

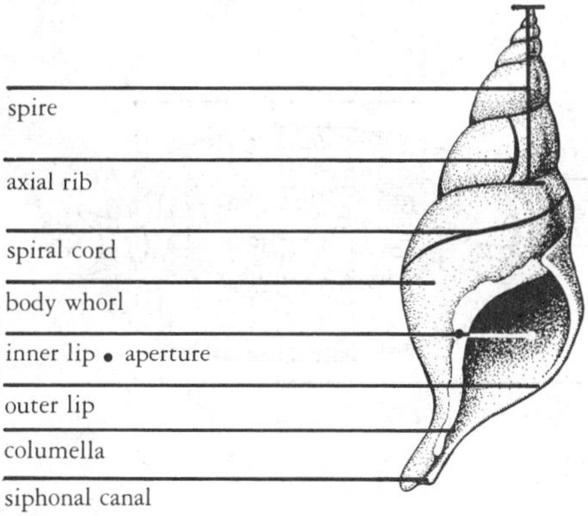

Snail

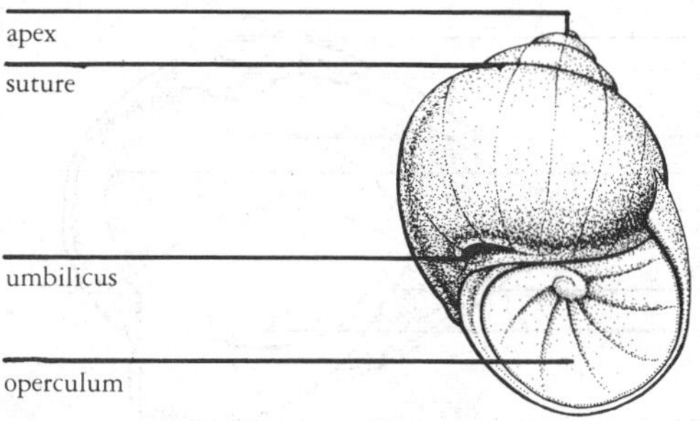

Nudibranch

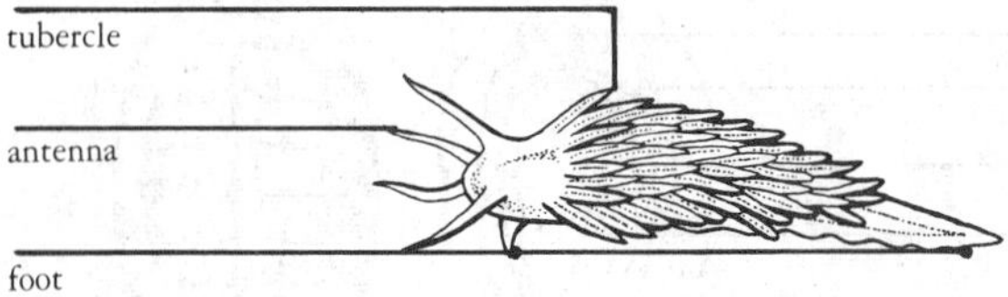

Nudibranch

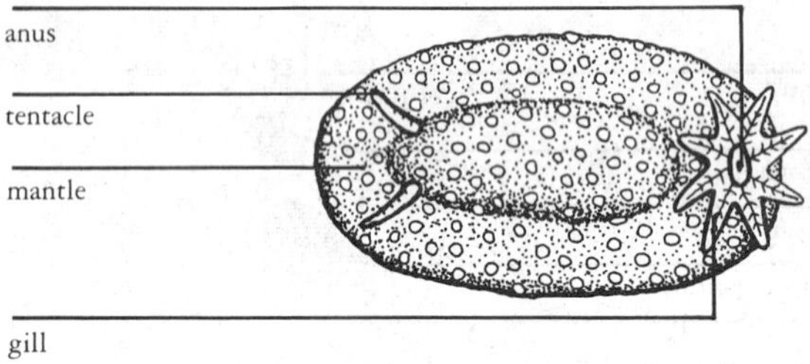

Clam

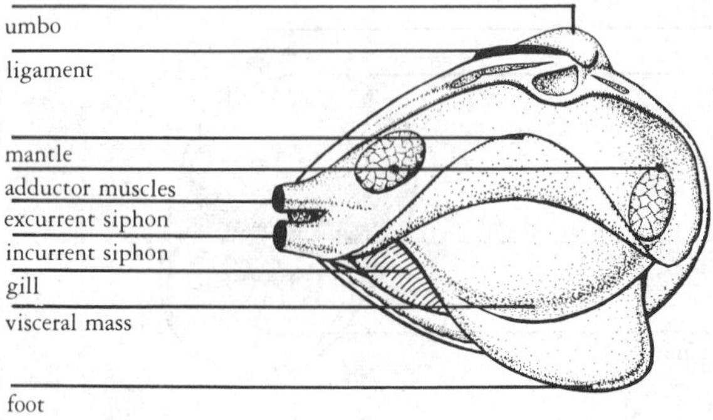

Clam

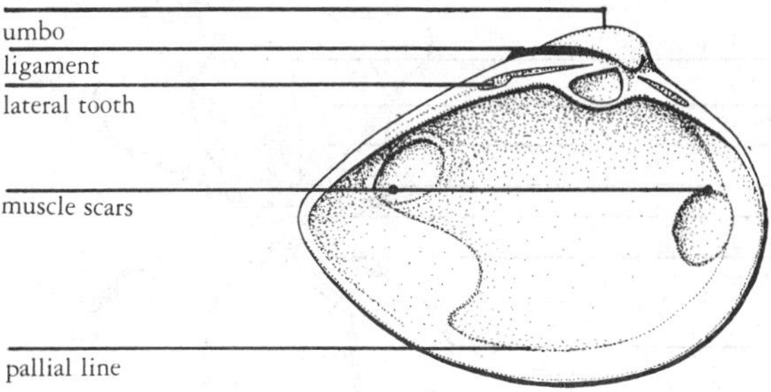

Squid

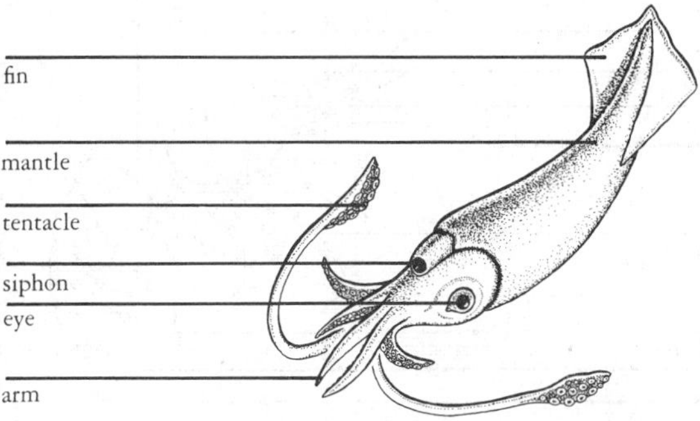

Horseshoe Crab

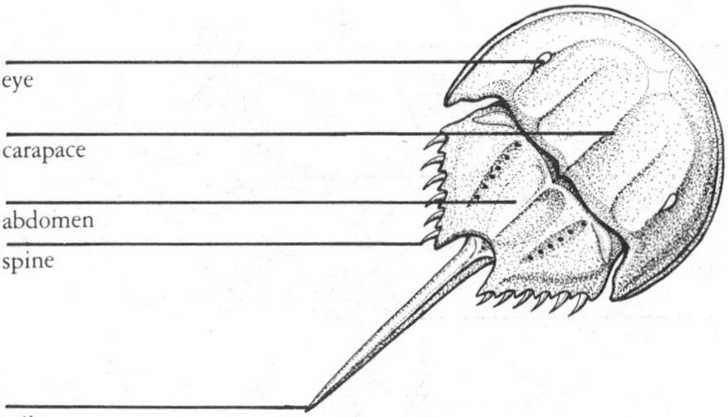

Sea Spider

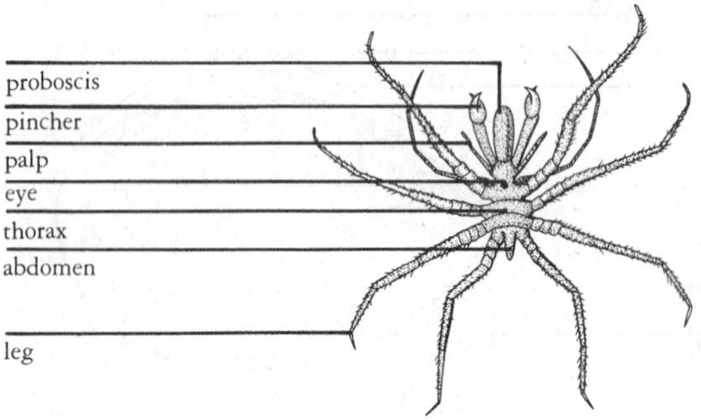

Shrimp

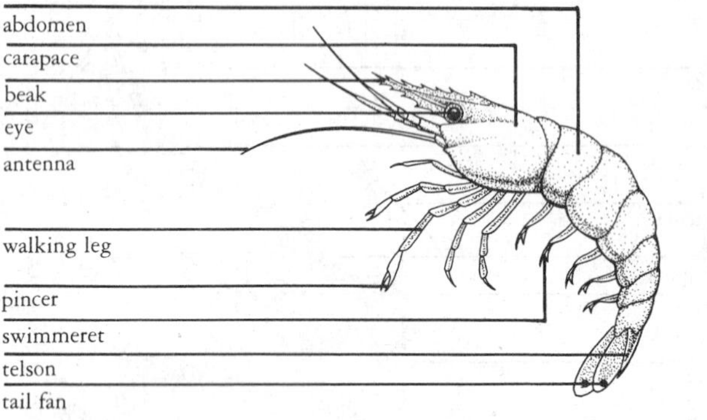

Crab

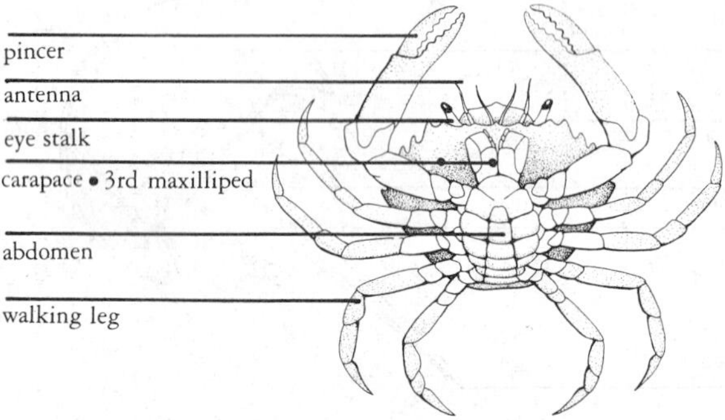

Sea Star

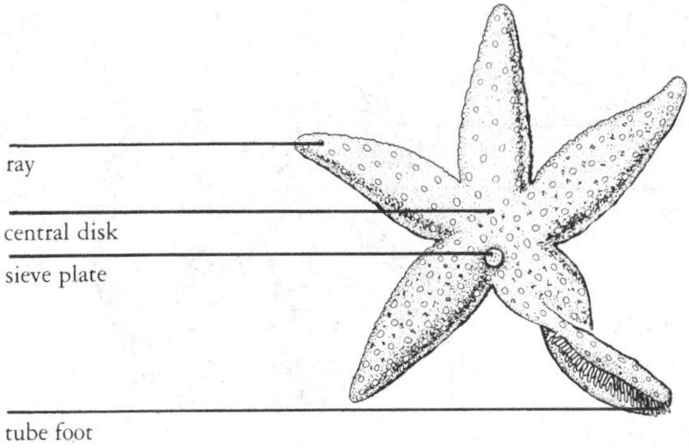

Brittle Star

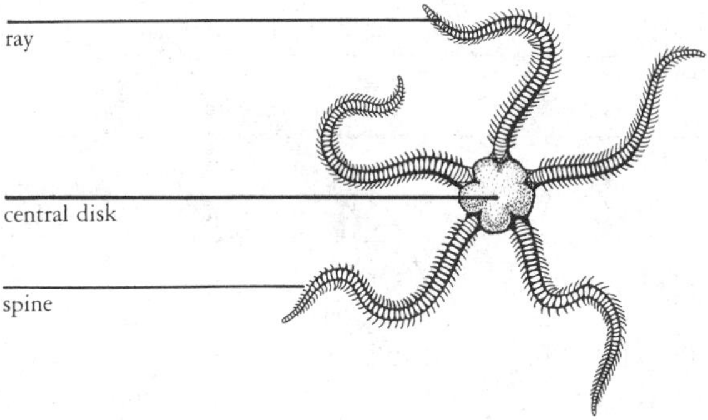

Sea Urchin

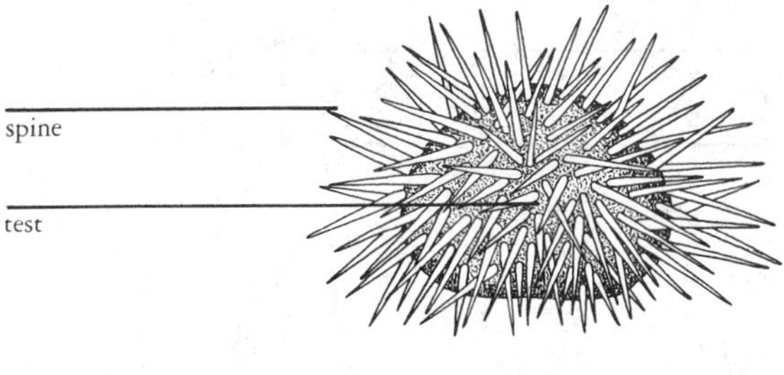

Sand Dollar

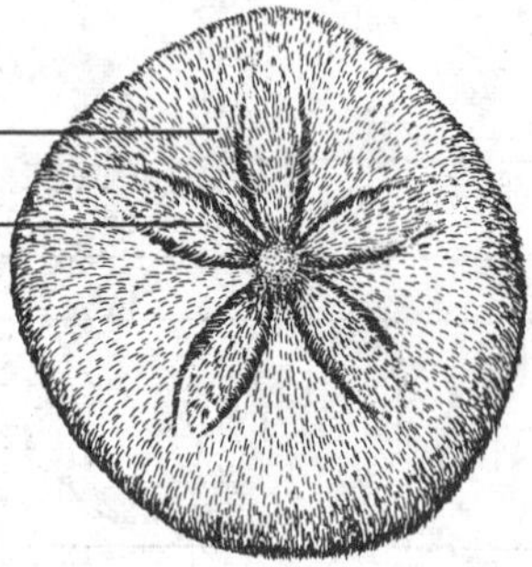

Sea Cucumber

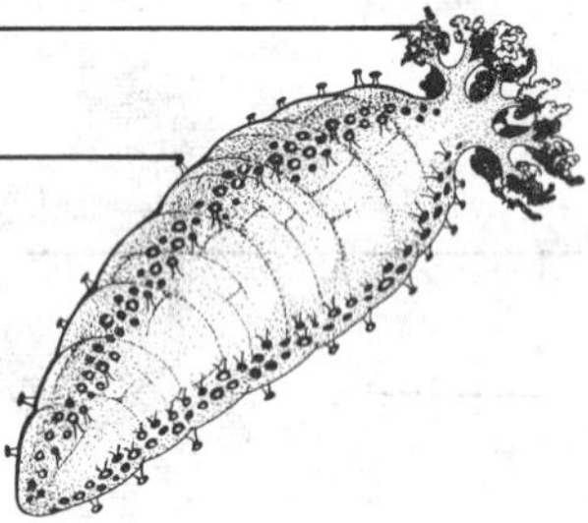

Bryozoan

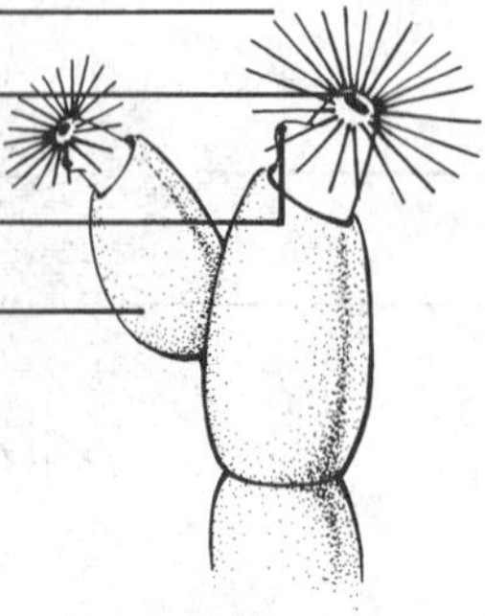

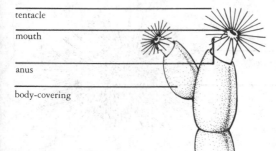

Entoproct

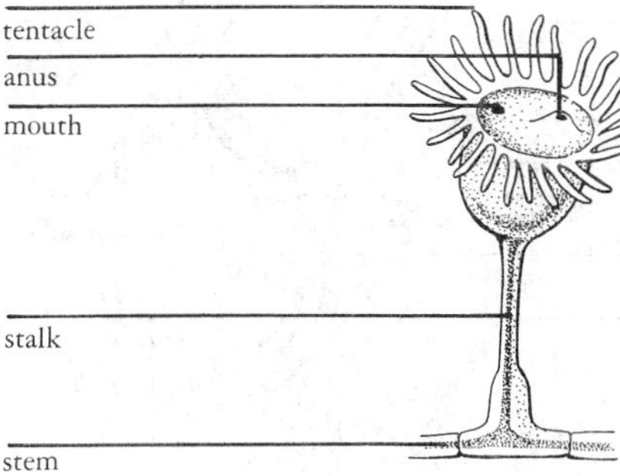

Phoronid Worm

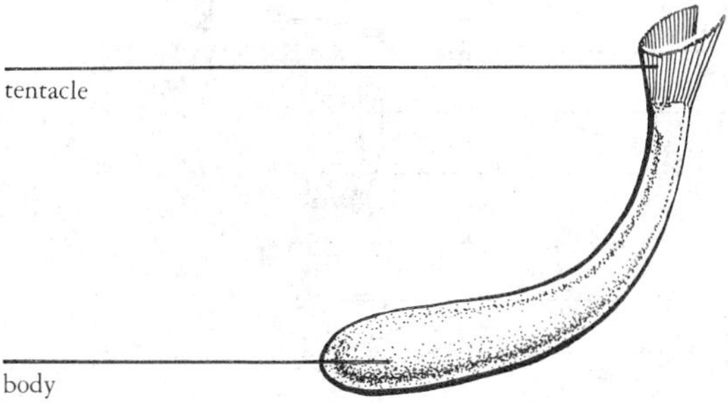

Lampshell

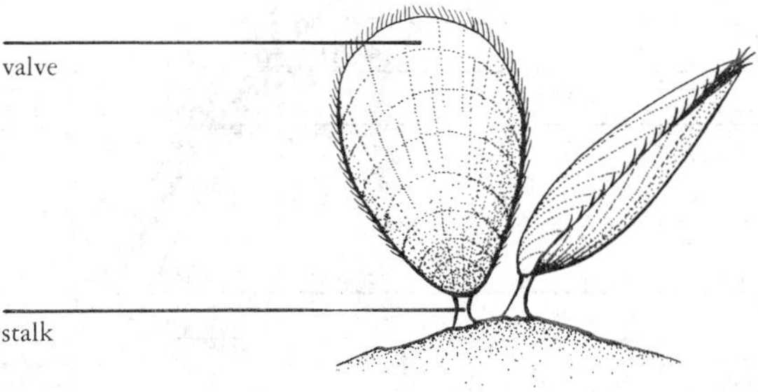

Acorn Worm

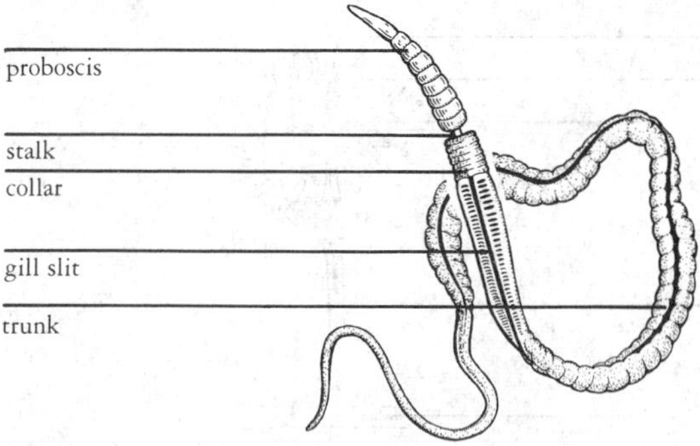

Tunicate

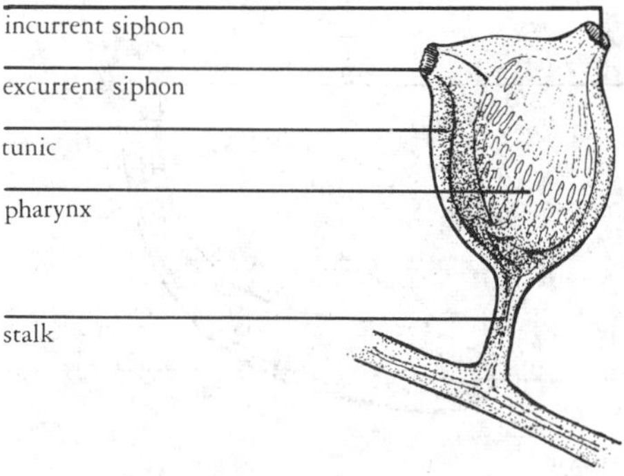

Lancelet

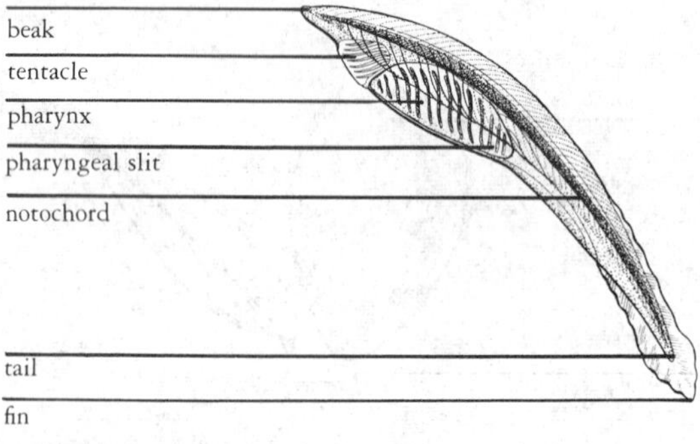

GLOSSARY

Abdomen The hindmost division of an animal's body.

Acontium (pl. acontia) In sea anemones, a thread armed with nematocysts and extruded from the mouth or through a pore in the body wall.

Adductor muscle In bivalve mollusks, a muscle that draws the two valves together, enclosing the soft parts.

Ambulacral groove A groove containing rows of tube feet on the lower surface of a sea star or brittle star.

Antenna (pl. antennae) A sensory appendage on the head of an arthropod or annelid.

Antennal scale The bladelike outer member of the 2nd antenna of a higher crustacean.

Aperture The opening of the body whorl of a snail or tusk shell through which the foot and head of the animal protrude.

Apical organ An organ sensitive to gravity, located at the end opposite the mouth in a ctenophore, and consisting of a capsule containing heavy granules that tumble in response to gravity as the animal's position changes, stimulating the animal to right itself.

Aristotle's lantern — The chewing mechanism of a sea urchin, consisting of 5 teeth operated by a complex system of levers and muscles.

Atrium (pl. atria) — In cephalochordates and tunicates, a chamber around the pharynx through which water flows from the pharyngeal gill slits to the outside, via the excurrent siphon.

Autotomy — The capacity of some animals to shed a part of the body or to break into several pieces, in the absence of external force.

Avicularium (pl. avicularia) — See Pincher.

Axis — The straight line with respect to which a body is symmetrical.

Bilateral symmetry — A body plan in which paired body parts lie on either side of a midline, so that each side is a mirror image of the other.

Binary fission — A mode of asexual reproduction in which the body of an organism divides into two nearly equal parts.

Bioluminescence — Light produced by a living organism through a biochemical reaction.

Body whorl — The last whorl of a snail's shell, ending at the aperture and containing the bulk of the animal.

Brackish — Containing some salt, but less than seawater.

Budding — A mode of asexual reproduction in which an outgrowth of an organism develops and forms a new individual.

Byssal threads, or Byssus — Tough, silken threads secreted by certain bivalve mollusks as a means of attachment to a substrate.

Carapace — That part of the exoskeleton of a horseshoe crab or higher crustacean extending over the head and thorax, but not the abdomen.

Cephalothorax The fused head and thorax of a crustacean.

Chela See Pincer.

Cilium (pl. cilia) A microscopic, hairlike projection on the free border of epithelial cells that beats with others in coordinated waves.

Cirrus (pl. cirri) A tactile projection that varies in shape according to the phylum; may be tentaclelike, fingerlike, or hairlike.

Cnidocyte In cnidarians, a cell containing a nematocyst, or stinging capsule.

Coelenteron The digestive cavity of a cnidarian.

Coelom A body cavity lying entirely within tissues of mesodermal origin.

Collar cell A cell in the lining of the cavity of a sponge, bearing a trumpetlike collar surrounding a flagellum.

Columella That part of the spiral shell of a snail surrounding the axis, about which the shell coils.

Comb plate A comblike membrane of fused cilia in a ctenophore; a ctena.

Commensalism An intimate relationship between animals of two species, in which each lives without obvious harm or benefit to the other. See Symbiosis.

Corallum The skeleton of calcium carbonate deposited by the base of a stony coral.

Ctena (pl. ctenae) One of the comblike flaps of fused cilia found in rows on the body of a ctenophore, and used in locomotion.

Detritus Particles worn off a solid body.

Dextral	In snails, having the aperture on the right-hand side of the columella.
Diatom	A microscopic, unicellular plant living either attached to a solid or as plankton, and capable of photosynthesis.
Dinoflagellate	A microscopic, unicellular, planktonic organism with two flagella, sometimes capable of photosynthesis.
Dioecious	Having separate sexes.
Ectoderm	The outermost layer in a developing embryo, lying over the mesoderm and the endoderm.
Endoderm	The innermost layer in a developing embryo, lying beneath the ectoderm and the mesoderm.
Ephyra (pl. ephyrae)	An 8-lobed, larval medusa, budded from a scyphistoma.
Epidermis	The outermost cellular layer of skin.
Epithelio-muscular cell	In cnidarians, an epidermal cell with long, contractile fibers on its inner surface, lying parallel to the surface of the body.
Epithelium (pl. epithelia)	A single layer of cells covering any exposed surface of an organism, or lining any cavity within it.
Exoskeleton	The rigid outer covering of an arthropod.
Fauna	The animal life of a region or locality.
Flagellum (pl. flagella)	A whiplike appendage of a cell, used in locomotion or in creating a current.
Gape	A region of incomplete closure between the two valves of a bivalve mollusk.
Gastrodermis	The cellular layer lining the digestive cavity.

Gemmule	An internal bud of certain sponges, consisting of a mass of cells surrounded by a tough sheath, which survives the winter or other rigors, and germinates to form a new individual under more favorable conditions.
Genus (pl. genera)	A group of very closely related species which share many structural features and ecological characteristics, and whose scientific names all begin with the same word, the generic name.
Girdle	A band of muscular tissue that surrounds and holds together a chiton's valves.
Gonad	Sex gland; ovary or testis.
Gonophore	A specialized part of a hydrozoan polyp, bearing sex organs.
Gonotheca (pl. gonothecae)	The case around the asexual reproductive bud of certain hydroids.
Growth line	In both gastropods and bivalves, the line that marks former resting stages in the growth of the shell.
Hermaphroditic	Having the organs of both sexes in one individual.
Holdfast	The rootlike part of an alga attaching to the substrate.
Hydranth	The feeding member of a hydroid colony, equipped with a mouth and tentacles.
Hypostome	The area surrounding the mouth of a cnidarian polyp.
Inner lip	The inner margin of the aperture of a snail bounded by the columella.
Ligament	In bivalves, a horny structure on the hinge area, either external or internal, connecting the valves, which acts as a spring to keep the valves open.

Lip, or Outer lip — The outer edge of the aperture of a coiled gastropod.

Madreporite — See Sieve plate.

Mandible — The jaw; the cutting or grinding mouthpart found in certain arthropods and annelids.

Mantle — A sheet of tissue that lines and secretes the shell of a mollusk, or covers the outside of a shell-less mollusk, and encloses the mantle cavity.

Mantle cavity — The space enclosed by the mantle of mollusks containing the gills and the visceral mass.

Manubrium — A fleshy stalk on the underside of a cnidarian medusa, bearing the mouth.

Margin — The edge of the shell of a limpet or the edges of a bivalve.

Maxilla — The mouthpart on a crustacean located directly behind the mandible.

Maxilliped — The thoracic appendage of a crustacean modified as an auxiliary mouthpart.

Medusa — One of the body forms of a cnidarian, cup-shaped or bowl-shaped with a mouth on a stalk on the underside, and capable of swimming by rhythmic contractions.

Mesentery — A membrane suspending an organ in the body cavity of an animal, its surface continuous with both the surface of the organ and the lining of the cavity.

Mesoderm — The middle layer in a developing embryo, lying between the ectoderm and the endoderm.

Mesoglea — The jellylike layer between the epidermis and gastrodermis in a cnidarian.

An extensible or permanently extendec structure on the head, commonly associated with the mouth of an anima used in feeding or sensing food or othe chemical substances.

The frontmost body division of certain arthropods.

A lobed segment anterior to the moutl in an annelid worm.

The part of a snail's shell formed in larval life, and later becoming the ape of the shell.

A one-celled animal belonging to the phylum Protozoa.

The terminal segment of an annelid's body, bearing the anus.

A canal branching from the central digestive cavity of a medusa, and extending to the margin of the bell.

A body plan in which repeated body parts are arranged around a central point, as in a wheel.

The flexible, rasplike "tongue" of a mollusk.

The arm, or radiating appendage, of echinoderm.

A ridgelike sculptural element that is usually axial in gastropods and radial bivalves.

The salt concentration of a solution; salinity of seawater is expressed in pa per thousand rather than in percent. Average salinity of seawater is 35 pa per thousand, or 35‰.

The ornamentation on the shells of mollusks.

Monoecious See Hermaphroditic.

Mucus A slippery protective substance secreted onto the exposed surfaces of many animals.

Muscle scar The site on the inner surface of a bivalve shell where the muscle was attached.

Mutualism An intimate relationship between organisms of two different species in which both gain from the association. See Symbiosis.

Nematocyst In cnidarians, an explosive capsule in a cell that erupts when stimulated, extending a long, stinging or entangling thread.

Nephridium (pl. nephridia) An excretory organ, or kidney, in certain invertebrates.

Ocellus (pl. ocelli) An eyespot; a simple photoreceptor, capable of distinguishing a source and change of light intensity, but not of forming an image.

Operculum A lid that closes an aperture; found in many snails and in certain tube-dwelling annelid worms.

Opisthosoma The body region to the rear of the prosoma in certain arthropods.

Oral arm One of the appendages surrounding the mouth of a scyphozoan medusa.

Oral disk The flattened area around the mouth of an anthozoan polyp.

Oral lobe A conical, rounded, or squarish structure lying above and in front of the mouth of an annelid worm; the prostomium.

Osculum (pl. oscula) The excurrent pore of a sponge.

Ostium (pl. ostia) One of the incurrent pores of a sponge through which water enters the animal.

Oviparous Egg-laying; a term applied to animals whose eggs are released from the body of the female before undergoing embryonic development.

Pallial line The line on the inner surface of a bivalve shell marking the site of attachment of the mantle.

Palp An appendage, usually sensitive to touch or taste, located near the mouth of an invertebrate.

Papilla (pl. papillae) A nipplelike or pimplelike projection or part.

Parapodium (pl. parapodia) One of the paired appendages on the segments of a polychaete annelid.

Parasite An animal living in or on the body of an organism of another species, to the detriment of the latter. See Symbiosis.

Parenchyma Undifferentiated tissue filling the space not occupied by organs in the body of a flatworm or nemertean worm.

Pedal disk The flat base of a sea anemone, by which it adheres to a solid surface.

Pedal laceration In sea anemones, a mode of asexual reproduction in which a bit of tissue separates from the base as the animal creeps along, and grows into a complete animal.

Pedicellaria (pl. pedicellariae) A small pincherlike or viselike structure on the surface of a sea star or sea urchin, used to keep the body surface clear of other organisms.

Periostracum The tough coat of organic material on the outside of a mollusk's shell.

Pharyngeal Pertaining to the pharynx.

Scyphistoma	In scyphozoans, a polyp stage that buds off medusae by dividing transversely.
Secondary consumer	An animal that feeds on primary consumers; *e.g.*, a carnivore or parasite.
Segment	One of the serially repeated divisions of the body of an annelid or arthropod.
Septum (pl. septa)	A sheet or wall of tissue separating two cavities in an animal's body.
Sessile	Attached to a substrate; sedentary.
Seta (pl. setae)	A bristle or hairlike structure.
Shoulder	In snails, the more or less flattened part of a whorl below the suture.
Sieve plate	A perforated plate on the body surface of an echinoderm through which it can take seawater into or expel it from the water vascular system.
Sinistral	In snails, having the aperture on the left-hand side of the columella.
Siphon	An opening, frequently tubular, through which an animal takes in or expels water; a siphon carrying water in is called an incurrent siphon and a siphon carrying water out is called an excurrent siphon. Found in mollusks and tunicates, among others.
Siphonal canal	In snails, a short channel, sometimes tubelike, at the lower end of the aperture, through which the siphon protrudes.
Siphonoglyph	A channel in the throat of an anthozoan, through which cilia carry food-laden mucus into the digestive cavity.
Species (pl. species)	A population of animals or plants whose members are at least potentially able to interbreed with each other, but are reproductively isolated from other populations.

Spicule A small structure, often needlelike or dartlike, supporting the tissues of various sponges, soft corals and compound tunicates.

Spiral cord Sculpture in spiral gastropods that follows the whorls, at right angles to the vertical axis of the shell.

Spire In gastropods, the whorls above the body whorl.

Spongin The tough, fibrous, skeletal meshwork of certain sponges.

Statolith A heavy granule in a gravity-sensing structure, such as the apical organ of a ctenophore.

Stylet In nemertean worms, a sharp, needlelike structure, used for puncturing.

Substrate The surface on which an organism lives.

Suture The seam between adjacent whorls of a snail's shell.

Swimmeret One of the small abdominal appendages of a crustacean.

Symbiosis An intimate biological relationship between two species; includes *parasitism,* where one lives at the expense of the other, *commensalism* where the presence of one neither helps nor damages the other, and *mutualism,* where both gain from the relationship.

Tail fan A fanlike structure at the tip of the tail of some crustaceans, consisting of a telson, or tailpiece, and a pair of flattened abdominal appendages.

Taxonomy The science of classifying organisms.

Telson The unpaired terminal structure attached to the last abdominal segment of a horseshoe crab or crustacean.

Tentacle A long, flexible structure, usually on the head or around the mouth of an invertebrate, used for grasping or feeding, or as a sense organ.

Test The skeleton of an echinoid echinoderm, consisting of rows of fused plates.

Thorax The division of an animal's body between the head and the abdomen.

Thread A fine, slender, sculptural element that may be axial or spiral in gastropods and radial or concentric in bivalves.

Trochophore The ciliated, swimming larva of certain annelids, echiurans, sipunculans, and mollusks.

Tube foot One of the numerous small appendages of an echinoderm, hydraulically operated and used in feeding or locomotion, or as a sense organ; often tipped with a suction disk.

Tubercle A bump, node, or low, rounded projection on the surface of an animal.

Tunic The covering of a tunicate's body; in compound tunicates, a thick mass in which many individuals are imbedded.

Umbilicus The hollow within the axis around which the whorls of a snail's shell coil; it may be closed by an overgrowth of shell material (the umbilical callus).

Umbo (pl. umbones) In bivalve mollusks, the oldest part of the shell, situated near the hinge.

Valve One of the separate parts of a mollusk or brachiopod shell.

Veliger A molluscan larva with a winglike swimming appendage on each side of its mouth.

Visceral mass That part of the molluscan body containing the visceral organs.

Viviparous Live-bearing; a term applied to animals whose eggs develop inside the body of the female, and in which larvae are born in an advanced stage.

Water vascular system The system of canals in an echinoderm that hydraulically operates the tube feet.

Whorl One of the turns of a snail's shell.

Zoochlorella (pl. zoochlorellae) A green, photosynthesizing, one-celled plant that lives symbiotically inside cells or tissues of various invertebrates.

Zooid An animal whose body is structurally continuous with that of others in a colonial species (e.g. bryozoans, compound tunicates, hydroids).

Zooxanthella (pl. zooxanthellae) A yellow, photosynthesizing, one-celled plant living symbiotically inside cells or tissues of various invertebrates.

PICTURE CREDITS

The numbers in parentheses are plate numbers. Some photographers have pictures under agency names as well as their own. Agency names appear in boldface. Photographers hold copyrights to their works.

William H. Amos (40, 85, 216, 278, 477, 587)

Animals Animals
E. R. Degginger (629) Zig Leszczynski (238, 549) Patti Murray (521) Anne Wertheim (283, 515) Jack Wilburn (663)

Ardea
P. Morris (513)

Charles Arneson (2, 8, 11, 15, 20, 61, 83, 95, 107, 108, 141, 174, 175, 188, 189, 191, 192, 211, 217, 219, 252, 261, 274, 279, 280, 292, 307, 309, 320, 372, 382, 386, 409, 410, 411, 419, 430, 440, 454, 482, 506, 508, 525, 526, 532, 533, 564, 565, 569, 597, 603, 617, 619, 620, 622, 627, 633, 648, 669, 680)

Peter Arnold, Inc.
Fred Bavendam (89, 203, 358, 501, 614) Bob Evans (284, 457, 479)

Robert G. Bachand (452, 481, 559)
David L. Ballantine (352, 658, 673)
Fred Bavendam (3, 52, 53, 70, 78, 90, 92, 105, 113, 138, 140, 149, 152, 170, 171, 182, 187, 223, 230, 235, 240, 249, 250, 256, 286, 355, 370, 408, 420, 424, 449, 461, 462, 480, 484, 496, 509, 511, 519, 540, 545, 547, 552, 558, 572, 574, 577, 585, 598, 602, 604, 612, 616, 618, 624, 625, 650, 653, 661, 664, 681, 686)
David W. Behrens (67, 106, 179, 213, 214, 220, 224, 234, 377, 436, 451, 459, 463, 464)
Michael Berrill (97)
Hans Bertsch (46, 227)
Gregory S. Boland (5, 25, 35, 130, 132, 173)
James H. Carmichael, Jr. (135, 198, 290, 296, 297, 298, 299, 300, 302, 303, 304, 306, 313, 317, 319, 321, 322, 323, 324, 325, 326, 327, 328, 329, 330, 331, 332, 333, 334, 335, 336, 337, 339, 340, 341, 344, 345, 346, 347, 348, 361, 362, 363, 365, 375, 378, 384, 385, 389, 393, 395, 396, 397, 398, 399, 401, 402, 403, 404, 405, 413, 416, 418, 423, 428, 431, 432, 433, 434, 438, 441, 443, 448, 450, 455, 465, 475, 611, 613, 638, 676, 682, 684)
Gary R. Carter (218)
Alfred D. Castro (147, 536)
Wen-Tien Chen (115)

Bruce Coleman, Incorporated
William H. Amos (79, 86, 369, 383, 510) Jen & Des Bartlett (427) James H. Carmichael, Jr. (193, 628) C. Braxton Dew (28) Jeff Foott (157, 184,

376) R. N. Mariscal (181, 262, 387, 600) Kjell B. Sandved (687) Nancy Simmerman (657) Kim Taylor (590) Ronald F. Thomas (439, 609)

Bruce Coleman, Limited
Jane Burton (123)

E. R. Degginger (144, 201, 231, 266, 293, 478, 527, 531, 557, 595, 634, 647, 666, 672, 685)
Douglas Denninger (76, 491)
Jack Dermid (63, 246, 272, 273, 289, 295, 338, 354, 426, 473, 514, 534, 665)

D.P.I.
Charles Seaborn (94, 122, 176, 207, 544, 586, 596, 646, 668, 674, 678, 679)
Nancy Sefton (19, 56, 60, 62, 91, 96, 236, 244, 357)

John DiMartini (37, 124, 582)
Jim Doran (507)
Jack Drafahl, Jr. (50, 102, 183, 364, 445, 522, 688)
Sue Drafahl (142, 143, 146, 675)
Douglas Faulkner (6, 33, 148, 285, 388, 524, 541, 542, 546, 550, 554, 556)
William E. Ferguson (390)
David H. Firmage (543)
Jeff Foott (104, 145, 156, 177, 186, 195, 197, 199, 233, 245, 351, 447, 460, 476, 553, 562, 571, 579, 608)
Julius Gordon and Townsend Weeks (367, 636)
Daniel W. Gotshall (27, 42, 51, 65, 109, 118, 209, 225, 400, 495, 561)
Al Grotell (21, 22, 30, 58, 129)
Arthur G. Guppy (667)
Paul H. Humann (1, 4, 7, 9, 13, 14, 16, 17, 18, 23, 24, 26, 31, 55, 57, 59, 64, 68, 69, 99, 100, 131, 133, 134, 136, 190, 210, 215, 232, 237, 242, 349, 359, 435, 517, 528, 529, 573, 615, 626, 640)

Jacana
Claude Carré (488, 489, 490, 503)
Hervé Chaumeton (34, 222) R. Konig (580, 632) Pierre Petit (512) Fred Winner (248) Yoff (243)

Paul D. Langer (159, 255, 267, 374, 380, 458, 575)
Robert Lipe (415, 429, 437)
Kenneth E. Lucas (126, 206, 221, 228, 265, 282, 287, 412, 444, 483, 520, 537, 584, 623, 644, 662, 683)
Kenneth E. Lucas/ Steinhart Aquarium (112)
Norman A. Meinkoth (43, 47, 54, 81, 93, 114, 119, 120, 127, 158, 164, 168, 172, 196, 205, 226, 253, 257, 260, 305, 308, 314, 342, 343, 356, 371, 381, 391, 392, 394, 425, 446, 466, 467, 468, 470, 487, 497, 498, 504, 516, 535, 538, 539, 560, 563, 566, 568, 578, 588, 589, 594, 605, 607, 641, 660)
Anthony Mercieca (690)

National Audubon Society Collection/ Photo Researchers, Inc.
Ken Brate (194) B. B. Cadbury (316) Richard Chesher (36, 421) Ron Church (422) Robert Dunne (570, 637) Gary G. Gibson (291, 310, 312, 368, 471, 583, 639, 651, 652, 655, 659) Walter E. Harvey (200) Robert C. Hermes (12, 635) George Holton (649) Russ Kinne (360) Stephen J. Krasemann (631) John Lidington (10, 32, 212, 241) Al Lowry (185) Tom McHugh (137, 315, 318, 581, 610, 671, 689) Tom McHugh/ Seattle Aquarium (151, 154, 165) Tom McHugh/ Steinhart Aquarium (485, 492) Noble Proctor (277) Harry Rogers (366) Daniel D. Sullivan (288) William E. Townsend (555) George Whiteley (251, 263, 271, 353)

Aaron Norman (350, 592)

Photo Classics
M. Woodbridge Williams (643)

Robert W. L. Potts (281)
Harold Wes Pratt (29, 38, 49, 72, 73, 74, 75, 77, 80, 84, 98, 117, 121, 139, 150, 161, 166, 169, 202, 208, 229, 239, 247, 258, 259, 264, 268, 269, 270, 276, 311, 406, 407, 414, 417, 453, 469, 472, 474, 486, 493, 494, 499, 502, 518, 576, 591, 593, 599, 601, 606, 621, 645, 677)
Betty Randall (128, 551, 567)
Jeffrey L. Rotman (39, 41, 88, 103, 275, 500, 530)
Kjell B. Sandved (442, 670)
Kenneth P. Sebens (44, 48, 66, 82, 110, 125, 153, 163, 167, 301, 654, 656)

Tom Stack and Associates
Kenneth R. H. Read (505)

John L. Tveten (548, 630, 642)

Valan Photos
L. Janosi (204)
Albert Kuhnigk (456)

Steven K. Webster (71, 160, 162, 180, 254, 373, 379)
Judith Winston (116)
Jon Witman (111, 155, 178, 294, 523)
Charles R. Wyttenbach (45, 87, 101)

INDEX

Numbers in boldface type refer to color plates. Numbers in italics refer to pages. Alternate common names appear in quotation marks.

A

Abalone(s), *455, 456*
Black, *469*
Japanese, *468*
Red, **392, 393,** *468*
Abietinaria, 352
spp., **71,** *351*
Acanthina
paucilirata, 495
punctulata, **409,** *495*
spirata, 495
spp., **451,** *494*
Acanthodoris pilosa, **222,** *521*
Acorn Worm(s), *725* (*see also,* Worms)
Golden, **272,** *726*
Kowalewsky's, *726*
Acropora
cervicornis, **55,** *385*
palmata, **32,** *385*
Actiniaria, *339, 372–383*
Actinopyga agassizii, **235,** *700*
Actinothoe modesta, **172,** *381*
Aeginella longicornis, **599,** *608*
Aequorea aequorea, **500,** *346*
Agaricia agaricites, **33,** *386*
Aglaophenia spp., **67, 69,** *354*
Aiptasia pallida, **167,** *380*
Alcyonidium
hirsutum, **49,** *708*
verrilli, 708
Alloioplana californica, **217,** *401*
Allopora porphyra, **124,** *358*
Alpheus armatus, **622,** *614*
Americardia media, **362,** *555*
Amphineura, *455*
Amphiodia occidentalis, **567,** *685*
Amphipod
Noble Sand, **588,** *606*
Red-eyed, **589,** *604*
Amphiporus,
Chevron, 255, 256, *409*
Amphiporus
angulatus, **255, 256,** *409*
cruentatus, 409
Amphitrite
johnstoni, **159,** *435*
ornata, **273,** *434*
Ampithoe rubricata, **589,** *604*
Anachis avara, **406,** *501*
Anadara ovalis, **369,** *535*
Ancula
Atlantic, **205,** *522*
Pacific, *522*
Ancula
gibbosa, **205,** *522*
pacifica, 522
Anemone(s), *339* (*see also,* Sea Anemones)
Aggregating, **185,** *376*
Club-tipped, **180,** *384*
Elegant Burrowing, **191,** *372*
Frilled, **171,** *382, 383*
Ghost, **169,** *383*
Giant Green, **186,** *376*
Leathery, **184, 195,** *374*
Lined, **166,** *372*
Mat, **12,** *371, 635*
Northern Red, **182,** *373, 379*
Pale, **167,** *380*
Pink-tipped, **187, 188,** *377*
Proliferating, **181, 197,** *377*
Red Stomphia, **178,** *378, 677*
Ringed, **190,** *381*
Silver-spotted, **196,** *375*
Smooth Burrowing, **172,** *381*
Speckled, **189, 192,** *379*
Strawberry, **183,** *374*

Striped, 168, *382*
Tricolor, 194, *379*
Warty Burrowing, *373*
Angel Wing 296, *573*
False, 297, *562*
Anisodoris nobilis, 228, *522*
Annelida, *411*
Annelids, *411*
Anodontia alba, 333, *552*
Anomia
aculeata, 547
simplex, 345, *546*
Anopla, *406–408*
Anoplodactylus
lentus, 576, *589*
oculospinatus, 589
Anthopleura
artemisia, 179, *375*
elegantissima, 185, *376*
xanthogrammica, 186, *376*
Anthozoa, *338, 364–394*
Aphrodita hastata, 414
Aplidium
constellatum, 98, *730*
solidum, 730
stellatum, 730
Aplysia
californica, 209, *519*
dactylomela, 210, *518*
vaccaria, 519
Arabella iricolor, 252, *423*
Arca zebra, 534
Archidoris
montereyensis, 231, *523*
odhneri, 224, *523*
Arcinella cornuta, 348, *554*
Arctica islandica, 339, *552*
Arctonoe
fragilis, 416
pulchra, 416
vittata, 416, 470
Arenicola
cristata, 246, *425*
marina, 426
Argopecten
gibbus, 543
irradians, 353, *542*
Ark
Blood, 369, *535*
Ponderous, 367, *536*
White-bearded, 323, *535*
Arthropoda, *583*
Arthropods, *583*
Articulata, *721, 723*
Aschelminthes, *445*
Ascidia
callosa, 735
ceratodes, 735
nigra, 735
Ascidiacea, *728, 729–742*
Astarte
Boreal, 341, *550*
Chestnut, *550*
Wavy, 340, *550*
Astarte
borealis, 341, *550*
castanea, 551
undata, 340, *550*
Asterias
forbesi, 557, 558, *679*
vulgaris, 547, 559, *678*
Asteroidea, *664*
Astichopus multifidus, 237, *699*
Astraea
gibberosa, 459, *476*
inaequalis, 476
Astrangia
danae, 391
solitaria, 391
Astrometis sertulifera, 563, *670*
Astropecten
armatus, 561, *669*
articulatus, 669
Astrophyton muricatum, 573, *683*
Atrina
rigida, 298, 354, *540*
serrata, 299, *541*
Auger
Common Atlantic, 398, *515*
Concave, 397, *515*
Aurelia aurita, 502, *363*
Axiognathus
pugetanus, 686
squamatus, 568, *686*
Axiothella rubrocincta, 426

B
Baby's Ear,
Common, 465, *492*
Balanoglossus
aurantiacus, 272, *726*
Balanophyllia elegans, 176, 177, *384*
Balanus
amphitrite, 279, *596*
aquila, 595
balanoides, 278, 286, *593*
balanus, 285, *594*
eburneus, 275, *594*
improvisus, 274, *596*
nubilis, 280, 283, *595*
Bankia setacea, 312, *575*
Barbatia candida, 323, *535*
Barentsia spp., 85, *718*
Barnacle(s), *586*
Bay, 274, *596*
Common Goose, 288, *591*
Dall's, *593*
Eagle, *595*
Float Goose, *592*
Giant Acorn, 280, 283, *595*
Ivory, 275, *594*
Leaf, 277, 287, *539, 592*
Little Gray, 276, *593*
Little Striped, 279, *596*
Northern Rock, 278, 286, *593*
Red-striped Acorn, 284, *597*
Rough, 285, *594*
Smooth Gray, *593*
Thatched, 281, *595*
Volcano, 282, *597*

West Indian Volcano, *598*
Bartholomea annulata, 190, *381*
Basket Star(s), *333, 665, 666* (*see also,* Star)
Caribbean, 573, *683*
Northern, 572, *683*
Bat Star, 537, *676, 678*
Batillaria
attramentaria, 484
minima, 401, *483*
Bdelloura candida, 258, *400*
Beach Fleas, *586*
Beroe
cucumis, 492, *398*
forskali, 398
ovata, 398
Bittersweet, Comb, 365, *536*
Bittium
Alternate, 405, *485*
Threaded, *485*
Variable, *485*
Bittium
alternatum, 405, *485*
eschrichtii, 485
varium, 485
Bivalves, *457, 533*
Bivalvia, *457, 533–576*
Bleeding Tooth, *477*
Blepharipoda
occidentalis, 688, *634*
Blood Star, 552, *675*
Blue Buttons, 504, *358*
Bolinopsis
infundibulum, 491, *397*
microptera, 397
Boltenia
echinata, 111, *741*
ovifera, 41, *741*
villosa, 110, *742*
Botrylloides spp., 119, *738*
Botryllus
schlosseri, 121, *738*
tuberatus, 120, *737*
Bougainvillia spp., 80, *344*
Bowerbankia
gracilis, 709
imbricata, 709
Brachiopoda, *721*
Brachiopod(s), *721*
Common Pacific, 360, *723*
Green-banded, *723*
Tongue-shell, 309, *722*
Branchiostoma, 729
californiense, 746
caribaeum, 745
Brittle Star(s), *333, 663, 664, 665, 666, 684*
Atlantic Long-spined, *687*
Burrowing, 567, *685*
Daisy, 570, *685*
Dwarf, 568, *686*
Esmark's, 571, *686*
Panama, 566, *684*
Puget Dwarf, *686*
Reticulate, 565, *687*
Ringed, *688*
Short-spined, *684*
Spiny, 562, *687*
Spiny, 569, *688*
Bryozoa, *707*
Bryozoan(s), *707*
Articulated, 65, *710*
Black-speckled, 116, *714*
Bowerbank's Graceful, *709*
Bowerbank's Imbricated, *709*
Branched-spine, *709*
Bushy Twinned, 76, *711*
California Spiral-tufted, 73, *713*
Common Red Crust, 114, *715*
Coralline, 51, *710*
Ellis', 47, *714*
Gulfweed, *712*
Hairy, 113, *712*
Lacy-crust, 112, *711*
Lattice-work, 50, *716*
Loose Sea Lichen, *713*
Porcupine, 34, *708*
Pussley, *708*
Rubbery, 49, *708*
Sea Lichen, 48, *712*
Single-horn, 117, *715*
Spiral-tufted, 73, *713*
Staghorn, *715*
Bubble
California, 464, *517*
Common West Indian, 443, *516*
Solitary Paper, *518*
White Paper, 463, *517*
Buccinum undatum, 408, *502*
Bugula
californica, 713
turrita, 73, *713*
Bulla
gouldiana, 464, *517*
occidentalis, 443, *516*
Bunodactis stella, 196, *375*
Bunodosoma cavernata, 193, *378*
Bursatella leachi, 148, *520*
Busycon
canaliculatum, 417, *503*
carica, 504
contrarium, 427, 428, *504*
By-the-wind Sailor, 515, 516, *357*

C
Caberea ellisii, 47, *714*
Cadlina
White Atlantic, 223, *524*
Yellow-edged, 225, *524*
Cadlina
laevis, 223, *524*
luteomarginata, 225, *524*
Cake Urchin(s), *666, 667*
Calappa
flammea, 671, *636*
gallus, 673, *637*

Calcispongiae, *322–324*
Calliactis tricolor, 194, *379*
Callianassa
affinis, 619, *622*
californiensis, 620, *623*
Callinectes sapidus, 657, *639*
Calliostoma
annulatum, 457, *474*
canaliculatum, 474
ligatum, 474
occidentale, 458, *475*
Callyspongia vaginalis, 23, *333*
Campanularia spp., 75, 77, *347*
Cancellaria reticulata, 416, *512*
Cancer
antennarius, 644, *641*
borealis, 653, *642*
irroratus, 650, 654, *643*
magister, 655, *643*
oregonensis, 652, *642*
productus, 651, *644*
Caprella
laeviuscula, 600, *608*
linearis, 601, *607*
Carcinus maenas, 664, *641*
Cardita, Broad-ribbed, 324, *551*
Carditamera floridana, 324, *551*
Carpilius corallinus, 648, *645*
Cardisoma guanhumi, 632, *652*
Cassiopeia xamachana, 509, *363*
Cassis
madagascariensis, 434, *492*
m. spinella, 492
Catablema vesicarium, 497, *345*
"Cats Eyes," *396*
Celleporaria brunnea, 116, *714*
Cephalochordata, *727, 728*
Cephalopoda, *576–581*
Ceratostoma foliatum, 436, *495*
Ceratozona squalida, 375, *463*
Cerebratulus
californiensis, 409
lacteus, 257, *408*
Ceriantharia, *339, 383*
Cerianthid, Northern, 170, *383*
Cerianthus borealis, 170, *383*
Cerith
California False, *484*
Dwarf, *485*
Florida, 402, *484*
Ivory, *484*
Lettered, *484*
Cerithidea
californica, 483
costata, 396, *483*
scalariformis, 483
Cerithium
eburneum, 484
floridanum, 402, *484*
literatum, 484
variabile, 485
Cestoidea, *399*
Chaetopleura apiculata, 467
Chaetopterus
variopedatus, 429
Chama
arcana, 554
macerophylla, 347, *554*
pellucida, 554
Chelyosoma productum, 122, *733*
"Cherrystones," *559*
Chicoreus florifer, 438, *496*
Chiridota laevis, 267, *705*
Chiton(s), *454*
"Bee," *467*
Black Katy, 381, *466*
California Nuttall's, 372, *464*
Common Eastern, *467*
Florida Slender, 378, *465*
Giant, *416*
Gum Boot, 379, *464*
Hartweg's, 377, *465*
Heath's, *465*
Lined, 371, *462*
Mertens', *464*
Mesh-pitted, *463*
Mossy, 376, *466*
Mottled Red, 370, 374, *461*
Northern Red, *462*
Rough-girdled, 375, *463*
Veiled, 373, *467*
White, 380, *463*
Chlamys islandicus, 355, *542*
Chloeia viridis, 243, *425*
Chondrilla nucula, 24, *334*
Chondrophorans, *337*
Chordata, *727*
Chordates, *727*
Chrysaora
melanaster, 362
quinquecirrha, 506, 510, *361*
Chthamalus
dalli, 593
fissus, 593
fragilis, 276, *593*
Ciona intestinalis, 105, *733*
Cirolana
harfordi, 594, *600*
polita, 601
Cirratulus cirratus, 161, *431*
Cirriformia
grandis, 158, *431*
luxuriosa, 160, *432*
Cistenides brevicoma, 434
Clam(s), *453, 457*
Angel Wing, 296, *573*
Atlantic Nut, 329, *533*
Atlantic Razor, 304, *568*
Baltic Macoma, 343, *564*

Bean, *567*
Bent-nosed Macoma, **320**, *565*
Black, **339**, *552*
Broad-ribbed Cardita, **324**, *551*
"Butternut," *561*
Butter, *561*
California Jackknife, **307**, *567*
Carolina Marsh, **337**, *551*
Carpenter's Tellin, **328**, *564*
Common Pacific Littleneck, **368**, *560*
Common Razor, **308**, *569*
Common Washington, **335**, *561*
Dwarf Tellin, **326**, *563*
False Angel Wing, **297**, *562*
File Yoldia, **300**, *534*
Flat-tipped Piddock, **310**, *574*
Gaper, **318**, *570*
Great Piddock, **314**, *574*
Jackknife, **305**, *567*
Long Neck, *572*
Modest Tellin, **327**, *563*
Pacific Razor, **303**, *568*
Pismo, **322**, *561*
Red Nose, **306**, *571*
Soft-shelled, **316**, *572*
Steamer, *572*
Striated Wood Piddock, **313**, *573*
Sunray Venus, **302**, *558*
Surf, **319**, *569*
Thin Nut, **301**, *533*
Veiled, *534*
White Sand Macoma, **317**, *566*
Clava leptostyla, **87**, *343*
Clavelina
huntsmani, **102**, **103**, **104**, *729*
picta, **100**, *729*
Cleaning Shrimp (*see also,* Shrimp)
Grabham's, **615**, *616*
Pederson's, **616**, *613*
Red-lined, **613**, *616*
Spotted, **617**, *614*
Clibanarius vittatus, **684**, *627*
Clinocardium
ciliatum, 556
nuttallii, **364**, *555*
Cliona celata, **127**, *330*
Clymenella torquata, **264**, **271**, *426*
Clypeaster rosaceus, 693
Cnidaria, *335*
Cnidarians, *335, 395, 399*
Cockle(s), *453, 457*
Atlantic Strawberry, **362**, *555*
Common Egg, **330**, *556*
Giant Atlantic, **366**, *556*
Iceland, *556*
Morton's Egg, **342**, *557*
Nuttall's, **364**, *555*
Yellow, **363**, *557*
Codakia orbicularis, **332**, *553*
Coenobita clypeatus, **685**, *626*
Collisella
digitalis, 473
pelta, **387**, *473*
scabra, 473
Columbella mercatoria, **455**, *500*
Colus stimpsoni, **424**, *502*
Comb Jelly(ies), *395*
Beroë's, **492**, *398*
Common Northern, **491**, *397*
Forskal's, *398*
Leidy's, **493**, *397*
McCrady's, *397*
Ovate, *398*
Short-lobed, *397*
Comet Star (*see also,* Stars)
Common, *677*
Pacific, **553**, *677*
Conch(s), *453*
Crown, **429**, *504*
Fighting, *488*
Florida Horse, **426**, *507, 509*
Queen, **435**, *488, 627*
Condylactis gigantea, **187**, **188**, *377*
Cone
Alphabet, **430**, *513*
California, **445**, *513*
Mouse, **431**, *514*
Stearns', **432**, *514*
Conus
californicus, **445**, *513*
mus, **431**, *514*
spurius, **430**, *513*
stearnsi, **432**, *514*
Copepod, Splash Pool, *590*
Coquina, **321**, *566*
Corallimorpharia, *340, 384*
Coral(s), *335, 336*
Brain, *340*
Clubbed Finger, **13**, *388*
Common Star, **10**, *391*
Dwarf Cup, *391*
Elkhorn, **32**, *385*
Fire, **25**, *358*
Flower, **173**, **198**, *394*
Ivory Bush, **35**, *392*
Knobbed Brain, **4**, *388*
Labyrinthine Brain, **3**, *389*
Large Flower, **1**, *394*
Large Star, **11**, **174**, *371, 390*
Lettuce, **33**, *386*
Meandrine Brain, **5**, *392*
Northern Stony, *391*
Orange Cup, **176**, **177**, *384*
Pillar, **6**, *393*
Porous, **9**, *387*

Red Soft, 37, 39, *364*
Reef Starlet, 14, *387*
Rose, 175, *390*
Smooth Brain, 2, *389*
Soft, *338*
Staghorn, 55, *385*
Starlet, 16, *386*
Stokes' Star, 7, *393*
Stony, *338, 339, 384, 385, 394*
Corymorpha palma, 341
Corynactis californica, 180, *384*
Coryphella
rufibranchialis, 202, *529*
salmonacea, 203, 204, *529*
Cowrie
Atlantic Deer, *489*
Atlantic Gray, *489*
Chestnut, 444, *489*
Crab(s), *586* (*see also,* Fiddler, Hermit Crab, Horseshoe Crab, Mole Crab, Porcelain Crab)
Acadian Hermit, 686, *629*
Arrow, 574, *661*
Atlantic Decorator, *660*
Atlantic Mole, 690, *633*
Atlantic Rock, 650, 654, *643*
Bar-eyed Hermit, 680, *629*
Black-clawed Mud, 646, *646*
Black Land, *653*
Blue, 657, *639*
Blue-handed Hermit, *632*
Brackish-water Fiddler, 629, *656*
Butterfly, 668, 674, *626*
Calico, *637*
California Fiddler, 630, *654*
Commensal, 634, *648*
Common Spider, 656, *657*
Coral, 648, *645*
Doubtful Spider, *657*
Dungeness, 655, *643*
Flame-streaked Box, 671, *636*
Flat-browed, 633, *640*
Flat-clawed Hermit, 676, *402, 415, 631*
Flat Mud, 645, *646*
Flat Porcelain, 641, *624*
Flattened, *651*
"Friendly," *652*
Fuzzy, 679, *625*
Ghost, 631, *653*
Giant Hermit, 682, *627*
Gibbes', *639*
Grainy Hermit, 683, *632*
Great Land, 632, *652*
Green, 664, *641*
Hairy Hermit, 681, *630*
Hermit, *343, 344, 494, 503, 624, 636, 637, 641*
Horseshoe, 666, *400, 583, 584, 586*
Jonah, 653, *642*
Lady, 637, *638*
Land Hermit, 685, *626*
Lesser Sponge, 669, *635*
Lesser Toad, *657*
Little Hairy Hermit, *632*
Long-clawed Hermit, 677, *631*
Marsh, *652, 655*
Masking, 675, *658*
Mottled Shore, *650*
Mountain, 638, *653*
Mud Fiddler, *652, 655*
Oregon Cancer, 652, *642*
Pacific Mole, 689, *633*
Pacific Rock, 644, *498, 641*
Pourtales' Long-armed, *661*
Purple Shore, 639, 663, *649*
Purse, 635, *636*
Red, 651, *644*
Sand Fiddler, 628, *654*
Sargassum, 658, *638*
Saw-toothed, *662*
Say's Mud, 636, *647*
Say's Porcelain, 647, *623*
Sharp-nosed, 678, *659*
Shield-backed Kelp, 659, *659*
Sitka, *626*
Spiny-handed, *639*
Spiny Mole, 688, *634*
Spiny Spider, 640, *658*
Sponge, 670, *634*
Star-eyed Hermit, 687, *628*
Stone, 642, *647*
Striped Hermit, 684, *493, 627*
Striped Shore, 662, *650*
Thick-clawed Porcelain, 672, *625*
Toad, 660, 661, *656*
Turtle, 667, *626*
Warty, *645*
Wharf, 665, *651*
Yellow Box, 673, *637*
Yellow Shore, 643, *650*
Crangon
franciscorum, 619
septemspinosa, 593, *618*
Crassispira ostrearum, 404, *516*
Crassostrea
gigas, 290, *548, 550*
virginica, 289, *547*
"Crawfish," *620*
Crepidula
adunca, 488

fornicata, 462, *487*
plana, 487
Crisia spp., 65, *710*
Crossaster papposus, 545, *673*
Crustacea, *583, 585, 590–661*
Cryptochiton stelleri, 379, *464*
Cryptolithodes
sitchensis, 667, *626*
typicus, 668, 674, *626*
Cryptosula pallasiana, 114, *715*
Ctenodiscus crispatus, 535, *670*
Ctenophora, *395*
Cucumaria
frondosa, 152, *701*
miniata, 157, *701*
piperata, 702
Cushion Star, 541, *671*
Cyanea capillata, 511, *362*
Cyanoplax hartwegii, 377, *465*
Cymatium
martinianum, 494
pileare, 419, *493*
Cypraea
cervus, 489
cinerea, 489
spadicea, 444, *489*
Cyrtopleura costata, 296, *573*

D
Dardanus
fucosus, 680, *629*
venosus, 687, *628*
Demospongiae, *322, 325–334*
Dendraster excentricus, 531, *694*
Dendrobeania
laxa, 713
murrayana, 48, *712*
Dendrodoa carnea, 93, *736*
Dendrogyra cylindrus, 6, *393*
Dendronotus frondosus, 208, *528*
Dendrostrea frons, 359, *549*
Dermasterias imbricata, 536, *677*
Diadema antillarum, 524, *689*
Diadumene leucolena, 169, *383*
Diaperoecia californica, 51, *710*
Diaulula sandiegensis, 221, *525*
Dichocoenia stokesii, 7, *393*
Didemnum
albidum, 125, *731*
candidum, 731
carnulentum, 118, *731*
Dinocardium robustum, 366, *556*
Diodora
aspera, 391, *469*
cayenensis, 384, *469*
Diopatra
cuprea, 270, *422*
ornata, 422
Diploria, 340, 392
clivosa, 4, *388*
labyrinthiformis, 3, *389*
strigosa, 2, *389*
Distaplia stylifera, 91, *732*
Divaricella
quadrisulcata, 331, *553*
Dodecaceria
corallii, 433
fewkesi, 433
Dogwinkle
Atlantic, 456, *499*
Channeled, *497*
Emarginate, 414, *497*
Dolabrifera dolabrifera, 211, *518*
Doliolid, Common, 489, *743*
Doliolum nationalis, 489, *743*
"Dolly Varden," *637*
Donax
gouldii, 567
variabilis, 321, *566*
Doriopsilla
albopunctata, 227, *525*
Doris
Crimson, 226, *327, 328, 527*
Hairy, 222, *521*
Monterey, 231, *523*
Ringed, 221, *326, 525*
Rough-mantled, 229, 230, *526*
Salted, 227, *525*
White Knight, 224, *523*
Dosinia, Disc, 334, *558*
Dosinia discus, 334, *558*
Dove Snail (*see also,* Snail)
Greedy, 406, *501*
Lunar, 453, *501*
Mottled, 455, *500*
Drill, Atlantic Oyster, 407, *499*
Dromia erythropus, 670, *634*
Dromidia antillensis, 669, *635*
Drupe, Spotted Thorn, 409, *495*

E
Earthworms, *411*
Echinarachnius parma, 530, *694*
Echinaster sentus, 549, *676*
Echinodermata, *663*
Echinoderms, *663*
Echinoidea, *664, 666–668, 688–698*
Echinometra lucunter, 519, *692*
Echiura, *451*
Ecteinascidia turbinata, 95, 96, *734*
Ectoprocta, *707*
Edwardsia elegans, 191, *372*
Electra pilosa, 113, *712*
Emerita, 634
analoga, 689, *633*
talpoida, 690, *633*

Emperor Helmet, 434, *492*
Encope michelini, 533, *695*
Enopla, *405*, *406*, *409–410*
Ensis directus, 308, *569*
Entoprocta, *717*
Entoproct(s), *717*
Bowing, *718*
Thick-based, 85, *718*
Eolid, Elegant, 200, *530*
Epiactis prolifera, 181, 197, *377*
Epitonium
angulatum, 395, *486*
greenlandicum, 394, *487*
tinctum, *486*
Eriphia gonagra, *645*
Errantia, *412–425*
Escargots, *453*
Eteone
heteropoda, *413*
lactea, *413*
Euapta lappa, 244, *706*
Eucidaris tribuloides, 517, *688*
Eucratea loricata, 76, *711*
Eudendrium
album, *346*
californicum, *346*
carneum, *345*
ramosum, 82, *345*
tenue, *346*
Eudistylia polymorpha, 143, 146, 147, *438*
Eulalia
aviculiseta, *414*
viridis, 248, *414*
Eunicea
calyculata, *369*
mammosa, *369*
palmeri, *369*
spp., 56, *368*
Eupentacta quinquesemita, 151, *702*
Eupolymnia crescentis, *435*
Eurylepta
aurantiaca, *403*
californica, 214, *403*
Eurypanopeus depressus, 645, *646*
Eurythoe complanata, 242, *424*
Eusmilia fastigiata, 173, 198, *394*
Evasterias troschelii, 554, *679*

F
Fagesia lineata, 166, *372*
Fasciolaria
hunteria, 422, *508*
tulipa, 421, *508*
Feather Duster
Banded, 144, *440*
Black-eyed, *440*
Giant, 143, 146, 147, *438*
Large-eyed, 138, 139, *439*
Magnificent, 141, *440*
Slime, 140, *439*
Fiddler
Brackish-water, 629, *656*
California, 630, *654*
Mud, *655*
Sand, 628, *654*
File Shell
Antillean, 325, *546*
Hemphill's, *545*
Rough, 350, *545*
Filograna implexa, 83, 149, *441*
Fire Worm
Green, 241, *424*
Orange, 242, *424*
Red-tipped, 243, *425*
Fissurella
barbadensis, *470*
volcano, 382, *470*
Flabellinopsis iodinea, 200, *530*
Flamingo Tongue, 449, *490*
Flatworm(s), *399*
Crozier's, 215, *404*
Horned, 214, *403*
Leopard, 232, *404*
Monterey, 213, *403*
Oval, 217, *401*
Speckled, *402*
Tapered, 219, *402*
Zebra, 216, *402*, *632*
Flea
Big-eyed Beach, 587, *606*
California Beach, 586, *607*
Long-horned Beach, *606*
Water, *586*
Flustrellidra
corniculata, *709*
hispida, 34, *708*
Fusitriton oregonensis, 418, *494*

G
Gammarus oceanicus, 591, 598, *604*
Gastropoda, *455–457*, *468–532*
Gecarcinus
lateralis, *653*
ruricola, 638, *653*
Geodia gibberosa, 15, *334*
Geoduck, 315, *571*
Gersemia rubiformis, 37, 39, *364*
Glottidea
albida, 309, *722*
pyramidata, *723*
Glycera
americana, *419*
dibranchiata, 251, *418*
Glycymeris pectinata, 365, *536*
Gonionemus vertens, 505, *355*
Gonodactylus oerstedii, 597, *598*
Gorgonia spp., 64, *366*
Gorgonocephalus arcticus, 572, *683*
Grapsus grapsus, 649, *648*
Gymnolaemata, *708*, *709*, *711–716*

H
Halecium
beani, *351*

gracile, *351*
halecinum, **72**, *351*
tenellum, *351*
Halichondria
bowerbanki, *329*
panicea, **123**, *328*
Haliclona
oculata, **52**, *325*
permollis, **126**, *326*
rubens, **31**, *326*
Haliclystus, *360*
auricula, *360*
salpinx, **40**, *359*
Haliotis
cracherodii, *469*
kamtschatkana, *468*
rufescens, **392**, **393**, *468*
Haliplanella luciae, **168**, *382*
Halisarca
dujardini, **115**, *325*
nahantensis, *325*
Halocynthia pyriformis, **92**, *742*
Halosydna brevisetosa, **245**, *417*
Haminoea
solitaria, *518*
vesicula, **463**, *517*
Hapalogaster mertensii, **679**, *625*
Harmothoe
extenuata, *418*
imbricata, **239**, *417*
Heart Urchins, *666*, *667*
Hemichordata, *725*
Hemigrapsus
nudus, **639**, **663**, *649*
oregonensis, **643**, *650*
Henricia, Pacific, **551**, *675*
Henricia
leviuscula, **551**, *675*
sanguinolenta, **552**, *675*
Hepatus ephеliticus, *637*
Hermit Crab(s), *343*, *344*, *494*, *624*, *636*, *637*, *641* (*see also*, Crabs)
Acadian, **686**, *629*
Bar-eyed, **680**, *629*
Blue-handed, *632*
Flat-clawed, **676**, *631*
Giant, **682**, *627*
Grainy, **683**, *632*
Hairy, **681**, *630*
Land, **685**, *626*
Little Hairy, *632*
Long-clawed, **677**, *631*
Star-eyed, **687**, 628
Striped, **684**, *493*, *627*
Hermodice carunculata, **241**, *424*
Heteromysis formosa, **603**, *610*
Hiatella arctica, **306**, *571*
Hinnites giganteus, **351**, *543*
Hippasteria phrygiana, **540**, *671*
Hippiospongia lachne, *330*
Hirudinea, *411*
Holothuria floridana, **236**, *700*
Holothuroidea, *664*, *668*, *698–706*
Homarus americanus, **624**, *619*
Hopkins' Rose, **199**, *526*
Hopkinsia rosacea, **199**, *526*
Hornmouth, Leafy, **436**, *495*
Horn Snail
Black, **401**, *483*
California, *483*
Costate, **396**, *483*
Ladder, *483*
Horseshoe Crab, **666**, *400*, *583*, *584*, *586*
Hyas
araneus, **660**, **661**, *656*
coarctatus, *657*
Hybocodon pendula, **89**, *340*
Hydractinia
echinata, **38**, *343*
milleri, *344*
Hydras, *335*, *337*
Hydrocorallines, *337*
Hydroides uncinata, *441*
Hydroid(s), *335*, *337*, *717*
Bean's Halecium, *351*
Bougainvillia, **80**, *344*
Bushy Wine-glass, **79**, *348*
California Stick, *346*
Club, **87**, *343*
Feathered, **86**, *342*
Feathery, **67**, **69**, *354*
Fern Garland, **71**, *351*
Flared Halecium, *351*
Forked Garland, *353*
Garland, **81**, *352*
Graceful Halecium, *351*
Halecium, **72**, *351*
Pacific Solitary, *341*
Red Stick, *345*
Ringed Tubularian, *341*
Silvery, **74**, *353*
Slender Stick, *346*
Solitary, **89**, *340*
Sparsely-branched Tubularian, *341*
Stick, **82**, *345*
Tall Tubularian, *341*
Tropical Garland, **70**, *354*
Tubularian, **88**, *341*
Turgid Garland, *354*
Two-branched Wine-glass, *348*, *349*
White Stick, *345*
Wine-glass, **75**, **77**, *347*
Zig-zag Wine-glass, **78**, *348*, *349*
Hydromedusa(e), *337*
Angled, **505**, *355*
Clapper, **84**, **494**, *342*
Eight-ribbed, **498**, *350*
Elegant, **499**, *350*
Many-ribbed, **500**, *346*

White-cross, **501**, *349*
Hydrozoa, *337, 340–358*
Hypselodoris californiensis, **233**, *526*

I
Idotea
baltica, **585**, *601*
kirchanskii, **584**, *602*
phosphorea, *601*
resecata, *601*
wosnesenskii, **583**, *602*
Illex illecebrosus, **484**, *578*
Inarticulata, *721, 722*
Ircinia
campana, **19**, **21**, *332*
fasciculata, **22**, *331*
strobilina, **18**, *330, 331*
Ischadium demissum, **295**, *538*
Ischnochiton
albus, **380**, *463*
papillosus, *463*
ruber, *462*
Isodictya
deichmani, *328*
palmata, **28**, *328*
Isognomon alatus, **357**, *539*
Isopod(s), *586*
Baltic, **585**, *601*
Bay Greedy, *601*
Cut-tailed, *601*
Harford's Greedy, **594**, *600*
Kirchansky's, **584**, *602*
Sharp-tailed, *601*
Vosnesensky's, **583**, *602*

J
Janthina janthina, **466**, *486*
Jassa falcata, **590**, *605*
Jellyfish(es), *335, 337* (*see also*, Stalked Jellyfish)
Cannonball, **507**, **514**, *364*
Constricted, **497**, *345*
Crown, **503**, *360*
Moon, **502**, *363*
"One-armed," *340*
Penicillate, **495**, *346*
Purple, **508**, *361*
Purple Banded, *361*
Upside-down, **509**, *363*
Jewel Box
Clear, *554*
Florida Spiny, **348**, *554*
Leafy, **347**, *554, 555*
Left-handed, *554*
Jingle Shell
Common, **345**, *546*
False Pacific, **344**, *547*
Prickly, *547*
Junonia, **423**, *511*

K
Katharina tunicata, **381**, *466*
Kelletia kelletii, *681*
Kitten's Paw, **361**, *541*
Krill, Horned, **605**, *611*

L
Lacuna vincta, **474**, *477*
Laevicardium
laevigatum, **330**, *556*
mortoni, **342**, *557*
Lampshell, Northern, **358**, *723*
Lancelet(s), *727, 728, 729*
California, *746*
Caribbean, *745*
Larvaceans, *727*
Leather Star, **536**, *374, 677*
Lebbeus groenlandicus, **612**, *617*
Leech(es), *411*
Gray Oyster, *401*
Limulus, **258**, *400*
Oyster, **218**, *401*
Red Oyster, *401*
Lepas
anatifera, **288**, *591*
fascicularis, *592*
Lepidametra commensalis, *434*
Lepidonotus
squamatus, **240**, *415*
sublevis, *415*
variabilis, *415*
Lepidozona mertensii, *464*
Leptasterias
hexactis, **555**, *681*
littoralis, *680*
pusilla, *681*
tenera, **560**, *680*
Leptogorgia
setacea, *365*
virgulata, **63**, *365*
Leptosynapta
albicans, *705*
inhaerens, **260**, *705*
roseola, *705*
Leucandra, *322*
Leucandra heathi, **109**, *324*
Leucilla, *322*
nuttingi, **46**, *324*
Leucosolenia, *322*
botryoides, **54**, *323*
eleanor, *323*
Libinia
dubia, *657*
emarginata, **656**, *657*
Ligia
occidentalis, **581**, **582**, *603*
oceanica, **580**, *603*
Lima
hemphilli, *545*
pellucida, **325**, *546*
scabra, **350**, *545*
Limpet(s), *455, 456*
Atlantic Barbados Keyhole, *470*
Cayenne Keyhole, **384**, *469*
Giant Keyhole, **388**, *471*
Keyhole, *416*
Owl, **390**, *473*
Plate, **385**, **386**, *472*
Ribbed, *473*
Rough, *473*

Rough Keyhole, **391**, *469*
Seaweed, **389**, *472*
Shield, **387**, *473*
Tortoise-shell, **383**, *471*
Volcano, **382**, *470*
Limulus polyphemus, **666**, *584, 586*
Linckia
columbiae, **553**, *677*
guildingii, 677
Lineus
Red, **259**, *407*
Social, *407*
Striped, *407*
Lineus
bicolor, 407
ruber, **259**, *407*
socialis, 407
Lingula, 721
Lion's Mane, 511, *362*
Lion's Paw, 352, *544*
Lissothuria nutriens, **156**, *704*
"Littlenecks," *559*
Littorina
angulifera, 479
irrorata, **473**, *479*
littorea, **472**, *478*
obtusata, **469**, *479*
planaxis, 478
saxatilis, **470**, *480*
scutulata, **471**, *480*
sitkana, 478
ziczac, 478
Lobster(s), *586*
California Rock, **623**, *620*
Northern, **624**, *619*
Ridged Slipper, *621*
Spanish, **626, 627**, *621*
West Indies Spiny, **625**, *620*
Loligo pealei, **486**, *576*
Lolliguncula brevis, 577
Lophopanopeus bellus, **646**, *646*
Lottia gigantea, **390**, *473*
Lovenia cordiformis, **527**, *697*
Loxorhynchus crispatus, **675**, *658*
Lucina tigrina, 553
Lucine
Buttercup, **333**, *552*
Cross-hatched, **331**, *553*
"Great White," *553*
Tiger, **332**, *553*
Luidia
Banded, **564**, *668*
Striped, *669*
Luidia
alternata, **564**, *668*
clathrata, 669
Lumbrineris fragilis, 423
Lunatia
heros, **461**, *491*
triseriata, 491
Lysiosquilla
scabricauda, **596**, *600*
Lysmata
californica, **608**, *615*
grabhami, **615**, *616*
wurdemanni, **613**, *616*
Lytechinus
anamesus, 690
variegatus, **521, 528**, *690*

M
Macoma
Baltic, **343**, *564*
Bent-nosed, **320**, *565*
White Sand, **317**, *566*
Macoma
balthica, **343**, *564*
nasuta, **320**, *565*
secta, **317**, *566*
Macrocallista nimbosa, **302**, *558*
Manicina areolata, **175**, *390*
Margarites
groenlandicus, **468**, *475*
helicinus, 475
Marginella, Common, 439, *512*
Martesia striata, **313**, *573*
Meandrina meandrites, **5**, *392*
Mediaster aequalis, **550**, *676*
Megabalanus californicus, **284**, *597*
Meganyctiphanes norvegica, **605**, *611*
Megathura crenulata, **388**, *471*
Melampus
bidentatus, **446, 448**, *532*
olivaceus, 532
Melibe leonina, **220**, *531*
Melicertum octocostatum, **498**, *350*
Mellita
quinquiesperforata, **534**, *695, 696*
sexiesperforata, **532**, *696*
Melongena corona, **429**, *504*
Membranipora
membranacea, **112**, *711*
tuberculata, 712
Menippe mercenaria, **642**, *647*
Meoma ventricosa, **529**, *698*
Mercenaria
campechiensis, **336**, *560*
mercenaria, **338**, *559*
"Mermaid's Gloves," *328*
Merostomata, 584, 586
Mertensia ovum, 396
Metandrocarpa
dura, 739
taylori, **94**, *739*
Metridium senile, **171**, *382*
Microciona prolifera, **29**, *327*
Micrura
leidyi, 408
verrilli, 408
Millepora alcicornis, **25**, *358*

Miter
Beaded, 403, *510*
Ida's 400, *511*
Mithrax spinosissimus, 640, *658*
Mitrella lunata, 453, *501*
Mnemiopsis, 395
leidyi, 493, *397*
mccradyi, 397
Modiolus
capax, 537
modiolus, 294, *537*
rectus, 537
Moira atropos, 696
Mole Crab
Atlantic, 690, *633*
Pacific, 689, *633*
Spiny, 688, *634*
Molgula
citrina, 97, *742*
manhattensis, 743
Mollusca, *453, 454*
Mollusks, *453*
Montastrea
annularis, 10, *391*
cavernosa, 11, 174, *390*
Mopalia
lignosa, 467
muscosa, 376, *466*
Mud Star, 535, *670*
Murex
Apple, 437, *496*
Lace, 438, *496*
Muricea
Drooping, *367*
Spiny, 61, *367*
Muricea
appressa, 368
muricata, 61, *367*
pendula, 367
Mussa angulosa, 1, *394*
Mussel(s), *453, 457, 459*
Blue, 293, *538*
California, 292, *539, 592*
Fat Horse, *537*
Horse, 294, *537*
Ribbed, 295, *538*
Straight Horse, *537*
Mya arenaria, 316, *572*
Mysis spp., 604, 606, *609*
Mytilus
californianus, 292, *539*
edulis, 293, *538*
Myxicola infundibulum, 140, *439*

N
Nassa, Giant Western, 411, *506*
Nassarius
fossatus, 411, *506*
obsoletus, 452, *505*
trivittatus, 412, *506*
vibex, 413, *507*
Nausithoe punctata, 503, *360*
Navanax, 234, *518, 520*
Nemertean
Blood, *409*
California
Four-eyed, *410*
Leidy's, *408*
Milky, 257, *408*
Six-lined, 254, *406*
Swimming, *409*
Tube, *406*
Verrill's, *408*
Wandering, *410*
Neofibularia nolitangere, 20, *332*
Neopanope texana, 636, *647*
Nephtys
bucera, 419
caeca, 419
californiensis, 420
Neptune, Corded, 420, 425, *503*
Neptunea
decemcostata, 503
lyrata, 420, 425, *503*
Nereis
brandti, 421
pelagica, 250, *421*
virens, 249, *420*
Nerita, Checkered, *477*
Nerita
peloronta, 477
tessellata, 477
Nodipecten nodosus, 352, *544*
Noetia ponderosa, 367, *536*
Nomeus gronovii, 357
Notoacmaea
incessa, 389, *472*
scutum, 385, 386, *472*
testudinalis, 383, *471*
Notoplana
acticola, 219, *402*
atomata, 402
Nucella
canaliculata, 497
emarginata, 414, *497*
Nucula proxima, 329, *533*
Nuculana tenuisulcata, 301, *533*
Nudibranch (*see also,* Doris)
Blue-and-gold, 233, *526*
Diamondback, *672*
Elegant Eolid, 200, *530*
Hermissenda, 201, *531*
Hopkins' Rose, 199, *526*
Lion, 220, *531*
Red-gilled, 202, *529*
Salmon-gilled, 203, 204, *529*
Sea Clown, 206, *527*
Sea Lemon, 228, *522*
Nutmeg, Common, 416, *512*
Nuttallina californica, 372, *464*

O
Obelia, 347, 349
bidentata, 349
dichotoma, 348, 349
geniculata, 78, *348, 349*
spp., 79, *348*
Octocorallia, *338, 364*
Octopods, *453, 460*
Octopus, *461*
Briar, *579*
Common Atlantic, 480, *579*

Giant Pacific, **483**, *580*
Joubin's, **481**, *580*
Long-armed, **482**, *579*
Mud Flat, *581*
Two-spotted, **478**, **479**, *581*
Octopus
bimaculatus, **478**, **479**, *581*
bimaculoides, *581*
briareus, *579*
dofleini, **483**, *580*
joubini, **481**, *580*
macropus, **482**, *579*
vulgaris, **480**, *579*
Oculina diffusa, **35**, *392*
Ocypode quadrata, **631**, *653*
Oligochaeta, *411*
Oliva
reticularis, **440**, *509*
sayana, **441**, **442**, *509*
Olive
Lettered, **441**, **442**, *509*
Netted, **440**, *509*
Purple Dwarf, **447**, *510*
Variable Dwarf, *510*
Olivella
biplicata, **447**, *510*
mutica, *510*
Onchidoris bilamellata, **229**, **230**, *526*
Ophelia
denticulata, *427*
limacina, *427*
Ophiocoma echinata, **569**, *688*
Ophioderma
brevispina, *684*
panamense, **566**, *684*
Ophiodromus pugettensis, **253**, *420*
Ophionereis
annulata, *688*
reticulata, **565**, *687*
Ophiopholis aculeata, **570**, *685*
Ophioplocus esmarki, **571**, *686*
Ophiothrix
angulata, *687*
spiculata, **562**, *687*
Ophiuroidea, *664*, *665*
Ophlitaspongia pennata, **128**, *327*
Opisthobranchia, *457*
Orchestoidea
californiana, **586**, *607*
Oreaster reticulatus, **541**, *671*
Ostrea lurida, **291**, *549*
Ovalipes ocellatus, **637**, *638*
Oyster Drill,
Atlantic, **407**, *499*
Oyster(s), *453*, *457*, *459*, *460*
Atlantic Pearl, **346**, *540*
Atlantic Thorny, **349**, *545*
Coon, **359**, *549*
Eastern, **289**, *547*, *597*
Flat Tree, **357**, *539*
Giant Pacific, **290**, *548*
"Japanese," *548*
Native Pacific, **291**, *549*

P
Pachycheles rudis, **672**, *625*
Pachygrapsus
crassipes, **662**, *650*
transversus, *650*
Paddle Worm
Black-striped, *414*
Green, **248**, *414*
Leafy, **247**, *413*
Milky, *413*
Varied-footed, *413*
Pagurus
acadianus, **686**, *629*
arcuatus, **681**, *630*
granosimanus, **683**, *632*
hirsutiusculus, *632*
longicarpus, **677**, *631*
pollicaris, **676**, *631*
samuelis, *632*
Palaemonetes vulgaris, **607**, *613*
Palythoa mammillosa, **8**, *371*
Pandalus
borealis, *618*
danae, **610**, *618*
montagui, **614**, *617*
Panopea generosa, **315**, *571*
Panulirus
argas, **625**, *620*
interruptus, **623**, *620*
Paranemertes peregrina, *410*
Parastichopus
californicus, **207**, *699*
parvimensis, *699*
Parthenope
pourtalesii, *661*
serrata, *662*
Patiria miniata, **537**, *678*
Peanut Worm(s), *447*
Agassiz's, **266**, *448*
Antillean, *448*
Bushy-headed, *448*
Eelgrass, *449*
Gould's, **263**, *447*
Pectinaria
californiensis, *434*
gouldii, **268**, *433*
Pedicellina cernua, *718*
Pelagia
colorata, *361*
noctiluca, **508**, *361*
Pelecypoda, *454*, *457*, *458*
Pen Shell
Saw-toothed, **299**, *541*
Stiff, **298**, **354**, *540*
Penaeus
aztecus, *612*
duorarum, **609**, **611**, *612*
setiferus, *612*
Pennaria tiarella, **86**, *342*
Periclimenes pedersoni, **616**, *613*
Periwinkle(s), *453*
Angulate, *479*
Checkered, **471**, *480*
Common, **472**, *478*

Eroded, *478*
Marsh, 473, *479*
Northern Yellow, 469, *479*
Rough, 470, *480*
Sitka, *478*
Zebra, *478*
Perophora
annectens, *734*
viridis, 101, *734*
Persephona punctata, 635, *636*
Petricola pholadiformis, 297, *562*
Petrochirus diogenes, 682, *627*
Petrolisthes cinctipes, 641, *624*
Phalium granulatum, 454, *493*
Phascolopsis gouldii, 263, *447*
Phascolosoma
agassizii, 266, *448*
antillarum, *448*
Phidolopora pacifica, 50, *716*
Phoronid Worm, Green, 43, *719*
Phoronida, *719*
Phoronids, *719*
Phoronopsis viridis, 43, *719*
Phoxichilidium femoratum, 575, *590*
Phragmatopoma
californica, *430*
lapidosa, *431*
Phyllodoce spp., 247, *413*
Phyllonotus pomum, 437, *496*
Phymanthus crucifer, 189, 192, *379*
Physalia physalis, 512, 513, *356*
Piddock
Flat-tipped, 310, *574*
Great, 314, *574*
Rough, *575*
Striated Wood, 313, *573*
Pinctada radiata, 346, *540*
Pinnotheres spp., 634, *648*
Pisaster
giganteus, 556, *681*
ochraceus, 548, *682*
Pista
cristata, *436*
elongata, *436*
pacifica, *436*
Placiphorella velata, 373, *467*
Placopecten magellanicus, 356, *544*
Plagiobrissus grandis, 526, *697*
Plagusia depressa, *651*
Platyhelminthes, *399*
Pleurobrachia
bachei, *396*
pileus, 496, *396*
Pleuroploca gigantea, 426, *507*
Plexaurella
Double-forked, 58, *369*
Gray, *369*
Plexaurella
dichotoma, 58, *369*
flexuosa, *368*
homomalla, 57, *368*
grisea, *369*
Plicatula gibbosa, 361, *541*
Plocamia karykina, 30, *328*
Pododesmus macrochisma, 344, *547*
Polinices
duplicatus, 467, *490*
lewisii, *491*, *566*
Pollicipes polymerus, 277, 287, *592*
Polycarpa obtecta, 26, *740*
Polychaeta, *411*, *412*
Polycirrus eximius, 164, *437*
Polycladida, *399*, *401–404*
Polyclinum planum, 27, *732*
Polydora ligni, 269, *428*
Polymastia
pachymastia, *329*
robusta, 53, *329*
Polymesoda caroliniana, 337, *551*
Polyorchis penicillatus, 495, *346*
Polyplacophora, *454–455*, *461–467*
Pomatostegus stellatus, 136, *442*
Porania insignis, 538, *674*
Porcelain Crab
Flat, 641, *624*
Say's, 647, *623*
Thick-clawed, 672, *625*
Porcellana sayana, 647, *623*
Porifera, *321*
Porites
astreoides, 9, *387*
porites, 13, *388*
Porpita linneana, 504, *358*
Portuguese man-of-war, 512, 513, *356*
Portunus
depressifrons, 633, *640*
gibbesii, *639*
sayi, 658, *638*
spinimanus, *639*
Potamilla reniformis, 138, 139, *439*
Praunus flexuosus, 602, *610*
Priapulus caudatus, 262, *445*
Prosobranchia, *457*
Protothaca staminea, 368, *560*
Prunum apicinum, 439, *512*
Psammonyx nobilis, 588, *606*
Pseudoceros
crozieri, 215, *404*
montereyensis, 213, *403*
pardalis, 232, *404*
Pseudochama exogyra, *554*
Pseudopterogorgia spp., 60, 66, *366*

Pseudosquilla ciliata, 592, 599
Pseudosquillopsis marmorata, 599
Psolus, Scarlet, 153, 155, *703*
Psolus
chitinoides, 154, *704*
fabricii, 153, 155, *703*
Pteraster militaris, 539, *673*
Pterogorgia
anceps, *367*
citrina, 59, 62, *367*
Ptilosarcus gurneyi, 44, *370*
Pugettia producta, 659, *659*
Pulmonata, *457*
Pycnogonida, *585, 587–590*
Pycnogonum
littorale, 577, *587*
stearnsi, 579, *588*
Pycnopodia helianthoides, 544, 682

Q
Quahog, 338, *559*
Northern, *560*
Southern, 336, *560*

R
Red Nose, 306, *571*
Renilla
koellikeri, *370*
muelleri, *370*
reniformis, *370*
Rhynchocoela, *405*
Roperia poulsoni, 410, *498*
Rostanga pulchra, 226, *527*

S
Sabella
crassicornis, 144, *440*
melanostigma, *440*
Sabellaria
cementarium, *430*
floridensis, *430*
Sabellastarte magnifica, 141, *440*
Saccoglossus kowalewskii, *726*
Sally Lightfoot, 649, *648*
Salpa fusiformis, 488, *744*
Salp(s), *727, 728*
Common, 488, *744*
Horned, 490, *744*
Sand Dollar(s), *663, 664, 666, 667*
Common, 530, *694*
Eccentric, 531, *694*
Michelin's, 533, *695*
Sarsia tubulosa, 84, 494, *342*
Saxidomus
giganteus, *561*
nuttalli, 335, *561*
Scallop(s), *457, 458*
Atlantic Bay, 353, *542*
Atlantic Deep-sea, 356, *544*
Calico, *543*
Giant Rock, 351, *543*
Iceland, 355, *542*
Lion's Paw, 352, *544*
Scaphella junonia, 423, *511*
Schizoporella
floridana, *715*
unicornis, 117, *715*
Scleractinia, *339, 384–394*
Sclerodactyla briareus, 150, *702*
Scotch Bonnet, 454, *493*
Scud, 591, 598, *604*
Scyllarides
aequinoctialis, 626, 627, *621*
nodifer, *621*
Scypha, 322, *323*
ciliata, 45, *323*
Scyphozoa, *337, 359–364*
Scyra acutifrons, 678, *659*
Sea Anemone(s) *335, 336, 338, 339, 613, 614* (*see also,* Anemone)
Buried, 179, *375*
Warty, 193, *378*
Sea Biscuit
Brown, *693*
Long-spined, 526, *697*
West Indian, 529, *698*
Sea Cat, Warty, 211, *518*
Sea Cucumber(s), *416, 663, 664, 668, 677*
Agassiz's, 235, *700*
Dwarf, 156, *704*
Fissured, 237, *699*
Florida, 236, *700*
Four-sided, 238, *698*
Hairy, 150, *702*
Orange-footed, 152, *701*
Peppered, *702*
Red, 157, *701*
Silky, 267, *705*
Slipper, 154, *704*
Sticky-skin, 244, *706*
Stiff-footed, 151, *702*
Sea Egg, 525, *691*
Sea Fan(s), 64, *338, 366*
Bushy, *368*
"Sea Fern," *353*
Sea Fingers, Corky, 68, *365*
Sea Gooseberry, 496, *396*
Arctic, *396*
Sea Grape
Common, *743*
Orange, 97, *742*
Sea Hare(s), *455, 457*
California, 209, *519*
California Black, *519*
Green, *518*
Ragged, 148, *520*
Spotted, 210, *518*
Sea Hearts, *664*
Sea Lemon, 228, *522*
Sea Lilies, *663, 664*
Sea Mouse, *414*
Sea Nettle, 506, 510, *361*
Lined, *362*

Sea Pansy(ies), *338*
Common, *370*
Müller's, *370*
Western, *370*
Sea Peach, **92,** *742*
Sea Pen(s), *338*
Gurney's, **44,** *370*
Sea Plumes, **60, 66,** *366*
Sea Pork
Common, *730*
Northern, **98,** *730*
Pacific, *730*
Sea Roach(es), *586*
Northern, **580,** *603*
Western, **581, 582,** *603*
Sea Rod
Bent, *368*
Black, **57,** *368*
Eunicea, **56,** *368*
Mammillated, *369*
Palmer's, *369*
Warty, *369*
Sea Slugs (*see* Slugs)
Sea Spider(s), *583, 585*
Anemone, **577,** *587*
California Ringed, *589*
Clawed, **575,** *590*
Lentil, **576,** *589, 590*
Ringed, **578,** *588*
Spiny, *589*
Stearns', **579,** *588*
Sea Squirts, *727, 728*
Sea Star(s), *416, 420, 663, 664* (*see also,* Basket Stars, Brittle Stars, Stars, Sun Stars)
Armored, **561,** *669*
Badge, **538,** *674*
Broad Six-rayed, **555,** *681*
Equal, **550,** *676*
Forbes' Common, **557, 558,** *679*
Giant, **556,** *681*
Green Slender, *680*
Northern, **547, 559,** *678*
Ochre, **548,** *679, 682*
Plated-margined, *669*
Slender, **560,** *680*
Small Slender, *681*
Spiny, **563,** *670*
Thorny, **549,** *676*
Troschel's, **554,** *679*
Winged, **539,** *673*
Sea Urchin(s), *494, 663, 664, 666* (*see also,* Urchin)
Atlantic Purple, **518,** *689*
Green, **523,** *691*
Little Gray, *690*
Purple, **522,** *677, 692*
Red, **520,** *692*
Sea Vase, **105,** *733*
Sea Whip(s), **63,** *338, 365, 366, 683*
Angular, *367*
Straight, 365
Yellow, **59, 62,** *367*
Sedentaria, *412, 425–443*
Semibalanus cariosus, **281,** *595*
Serpula vermicularis, **137, 142, 145,** *443*
Serpulorbis squamigerus, **476,** *481*
Sertularella, 352
speciosa, **70,** *354*
turgida, 354
Sertularia, 354
furcata, 353
pumila, **81,** *352*
Sesarma
cinereum, **665,** *651*
reticulatum, 652
Shark Eye, **467,** *490*
Shipworm
Common, **311,** *576*
Gould's, *576*
Pacific, **312,** *575*
Shrimp, *586* (*see also,* Skeleton Shrimp)
Banded Coral, **618,** *615*
Bay Ghost, **620,** *623*
Beach Ghost, **619,** *622*
Bent Opossum, **602,** *610*
Blue Mud, *622*
Brown, *612*
Brown Pistol, **622,** *614*
California Skeleton, *608*
Common Mantis, **595,** *599*
Common Shore, **607,** *613*
Coon-stripe, **610,** *618*
Flat-browed Mud, **621,** *622*
Franciscan Bay, *619*
Grabham's Cleaning, **615,** *616*
Greenland, **612,** *617*
Linear Skeleton, **601,** *607*
Long-horn Skeleton, **599,** *608*
Maine, *618*
Montague's, **614,** *617*
Opossum, **604, 606,** *609*
Pederson's Cleaning, **616,** *613*
Pink, **609, 611,** *612*
Red-lined Cleaning, **613,** *616*
Red Opossum, **603,** *610*
Red Rock, **608,** *615*
Sand, **593,** *618*
Scaly-tailed Mantis, **596,** *600*
Smooth Skeleton, **600,** *608*
White, *612*
"Shrimp snapper," *599*
Siderastrea
radians, **16,** *386*
siderea, **14,** *387*
Siliqua
costata, **304,** *568*
patula, **303,** *568*
Sinum perspectivum, **465,** *492*
Siphonophorans, *337*
Siphonophore, Chain, **487,** *356*
Sipuncula, *447*

Skeleton Shrimp
California, *608*
Linear, **601**, *607*
Long-horn, **599**, *608*
Smooth, **600**, *608*
Slipper Shells, *456*
Slug
Bushy-backed Sea, **208**, *528*
Common Lettuce, **212**, *521*
Garden, *455*
Sea, *455, 457*
Snail(s), *455, 456, 457*
Banded Tulip, **422**, *508*
Black Horn, **401**, *483*
Black Turban, **460**, *476, 632*
Blue Top, *474*
"Boat," *487*
Boring Turret, **399**, *481*
Brown Turban, *476*
California Horn, *483*
Channeled Top, *474*
Chink, **474**, *477*
Common Purple Sea, **466**, *486*
Common Slipper, **462**, *487*
Common Worm, **475**, *482*
Costate Horn, **396**, *483*
Eastern White Slipper, *487*
Florida Rock, *498*
Greedy Dove, **406**, *501*
Greenland Top, **468**, *475*
Hays' Rock, *499*
Hooked Slipper, *488*
Ladder Horn, *483*
Lewis' Moon, *491, 566*
Lunar Dove, **453**, *501*
Mottled Dove, **455**, *500*
Northern Moon, **461**, *491*
Olive Ear, *532*
Pearly Top, **458**, *475*
Poulson's Rock, **410**, *498*
Purple-ringed Top, **457**, *474*
Red Top, **459**, *476*
Rock, **415**, *498*
Salt-marsh, **446**, **448**, *532*
Scaled Worm, **476**, *481*
Smooth Top, *475*
Spotted Moon, *491*
True Tulip, **421**, *508*
Tulip, *629*
Variegated Turret, *481*
Snail Fur, **38**, *343*
Solaster
dawsoni, **546**, *672*
endeca, **542**, **543**, *672*
stimpsoni, *672*
Solemya velum, *534*
Spheciospongia vesparia, **17**, *329, 331*
Spio
filicornis, *428*
setosa, *428*
Spirobranchus
giganteus, **130**, **131**, **132**, **133**, **134**, **135**, *443*
spinosus, *444*
Spirorbis
borealis, **477**, *442*
spirillum, *442*
Spisula solidissima, **319**, *569*
Spondylus americanus, **349**, *545*
Sponge(s), *321*
Boring, **127**, *330*
Bowerbank's Crumb of Bread, *329*
"Cake," *331*
Chicken Liver, **24**, *334*
Common Palmate, **28**, *328*
Crumb of Bread, **123**, *328, 523*
Deichman's Palmate, *328*
Do-not-touch-me, **20**, *326, 332, 334*
Dujardin's Slime, **115**, *325*
Eleanor's Organ-pipe, *323*
"Eyed," *325*
Finger, **52**, *325*
Fire, **129**, *326, 333*
Heath's, **109**, *324*
Little Vase, **45**, *323*
Loggerhead, **17**, *329*
Loggerhead, **18**, *331*
Nahant Slime, *325*
Nipple, **53**, *329*
Nutting's, **46**, *324*
Organ-pipe, **54**, *323*
Purple, **126**, *326*
Red, **31**, *326*
Red Beard, **29**, *327*
Sheep's Wool, *330*
Smooth Red, **30**, *328*
Stinker, **22**, *331*
Tube, **23**, *333*
Vase, **19**, **21**, *332*
Velvety Red, **128**, *327, 328*
Western Nipple, *329*
White, **15**, *334*
Squid(s), *453, 460*
Atlantic Long-fin, **486**, *576*
Brief, *577*
Short-fin, **484**, *578*
Squilla
Ciliated False, **592**, *599*
Lesson's False, *599*
Swollen-clawed, **597**, *598*
Squilla empusa, **595**, *599*
Stalked Jellyfish
Eared, *360*
Trumpet, **40**, *359*
Star (*see also*, Basket Stars, Brittle Stars, Sea Stars, Sun Stars)
Bat, **537**, *676, 678*
Blood, **552**, *675*
Common Comet, *677*
Cushion, **541**, *671*
Horse, **540**, *671*
Leather, **536**, *677*
Mud, **535**, *670*

Pacific Comet, 553, *677*
Sunflower, 544, *682*
Starfish, *664*
Staurophora mertensi, 501, *349*
Stelleroidea, *664, 668–688*
Stenocianops furcata, 660
Stenorhynchus seticornis, 574, *661*
Stenolaemata, *708, 710*
Stenoplax
floridana, 378, *465*
heathiana, 465
Stenopus hispidus, 618, *615*
Stephanomia cara, 487, *356*
Sthenelais
boa, 418
fusca, 418
Stichopus
California, 207, *699*
Parvima, *699*
Stichopus badionotus, 238, *698*
Stomolophus meleagris, 507, 514, *364*
Stomphia, Red, 178, *378*
Stomphia coccinea, 178, *378*
Strombus
alatus, 488
gigas, 435, *488*
Strongylocentrotus
droebachiensis, 523, *691*
franciscanus, 520, *692*
purpuratus, 522, *692*
Styela
clava, 108, *736*
montereyensis, 42, *736*
plicata, 106, 107, *737*
Stylasterine, Purple, 124, *358*
Stylochus
ellipticus, 218, *401*
frontalis, 401
oculiferus, 401
zebra, 216, *402*
Sun Star (*see also,* Stars, Sea Stars)
Dawson's, 546, *672, 673*
Smooth, 542, 543, *672*
Spiny, 545, *673*
Stimpson's, *672*
Sundial, Common, 433, *482*
Sunray Venus, 302, *558*
Symplegma viride, 36, 99, *740*
Synapta
Common White, 260, *705*
Pink, *705*

T
Tagelus
californianus, 307, *567*
plebeius, 305, *567*
Talorchestia
longicornis, 606
megalophthalma, 587, *606*
Tanystylum
californicum, 589
orbiculare, 578, *588*
Tealia
coriacea, 184, 195, *374*
crassicornis, 182, *373*
felina, 182, *373*
lofotensis, 183, *374*
Tegula
brunnea, 476
funebralis, 460, *476*
Tellin
Alternate, *563*
Candy Stick, *563*
Carpenter's, 328, *564*
Dwarf, 326, *563*
Iris, *563*
Modest, 327, *564*
Tellina
agilis, 326, *563*
alternata, 563
carpenteri, 328, *564*
iris, 563
modesta, 327, *564*
similis, 563
Terebra
concava, 397, *515*
dislocata, 398, *515*
Terebratalia transversa, 360, *723*
Terebratulina
septentrionalis, 358, *723*
Teredo navalis, 311, *576*
Tetraclita
rubescens, 282, *597*
stalactifera, 598
Tetrastemma spp., *410*
Thais
haemastoma, 415, *498*
lapillus, 456, *499*
Thalassema mellita, 261, *451*
Thalia democratica, 490, *744*
Thaliacea, *728, 744*
Thelepus
crispus, 162, 165, *437*
setosus, 163, *438*
Themiste
pyroides, 448
zostericola, 449
Thorn Drupe, Spotted, 409, *495*
Thuiaria argentea, 74, *353*
Tigriopus californicus, 590
Tima formosa, 499, *350*
Tivela stultorum, 322, *561*
Tonicella
lineata, 371, *462*
marmorea, 370, 374, *461*
rubra, 462
Trachycardium
muricatum, 363, *557*
Trematoda, *399*
Tresus nuttallii, 318, *570*
Tricladida, *399, 400*
Tridachia crispata, 212, *521*
Triopha, Spotted, *528*